AF335277

Chemical Resistance of Polymers in Aggressive Media

Chemical Resistance of Polymers in Aggressive Media

Yu. V. Moiseev

and

G. E. Zaikov

Institute of Chemical Physics
Academy of Sciences of the USSR
Moscow, USSR

Translated from Russian by

R. J. Moseley

Consultants Bureau • New York and London

Library of Congress Cataloging in Publication Data

Moiseev, Yurii Vasil'evich.
  Chemical resistance of polymers in aggressive media.

  Includes bibliographies and index.
  1. Polymers and polymerization – Deterioration. I. Zaikov, Guennadii Efremovich.
II. Title.
TA455.P58M5813 1987   620.1′920422   87-9075
ISBN 0-306-10997-2

This volume is published under an agreement
with the Copyright Agency of the USSR (VAAP)

© 1987 Consultant Bureau, New York
A Division of Plenum Publishing Corporation
233 Spring Street, New York, N.Y. 10013

All rights reserved

No part of this book may be reproduced, stored in a retrieval system, or transmitted
in any form or by any means, electronic, mechanical, photocopying, microfilming,
recording, or otherwise, without written permission from the Publisher

Printed in the United States of America

FOREWORD
to the
ENGLISH EDITION

The increasing use of polymers in aggressive technological environments is a cause of concern to the polymer technologist and presents a challenge to the polymer scientist. A primary aim of the authors is to provide a scientific basis for understanding these phenomena and ultimately to controlling them to the advantage of the technologist. Although sensitivity of polymers to the hydrolytic environment may be a disadvantage to the materials technologist, it is increasingly being used advantageously by the biotechnologist in the design of biocompatible materials with controlled degradability for such applications as sutures, implants, etc., and a knowledge of the physical chemistry of the breakdown of synthetic polymers in this "aggressive" environment is as important as it is in preventing breakdown in engineering applications.

The authors recognize that, unlike the chemistry of oxidative degradation, where inhibitors can be highly effective because they are able to interfere with a kinetic chain reaction, no such conceptionally simple answer is available in the control of the ionic degradation reaction considered in this book. Total design of the component, from the chemical structure and morphology of the polymeric material to the final form of the polymer artifact, are the essential parameters upon which the durability of polymers in aggressive environments depend.

The authors approach this problem with elegant logic. In the early part of the book they examine the nature of acid—base catalysis and the essential physical chemistry of such reactions as amide and ester hyrolysis. They then address the more complex problem of how the nature of the polymeric medium imposes its own modifying control over the chemistry by considering such parameters as the rate and

v

mechanism of diffusion, the effect of modified neighboring groups, stereospecificity and conformation in the degradation process. The authors bring out clearly, with many relevant examples, how the behavior of even chemically homogeneous polymers is rarely simple, as in the case of partially crystalline polymers whose amorphous and crystalline regions may behave in entirely different ways leading to "accessible" and "inaccessible" domains in the matrix. In the final chapters fundamental questions are asked about the chemical nature of the mechanical behavior of polymers and how the attack of aggressive reagents modifies their technological performance.

This book breaks new ground in its integration of the science and technology of polymer behavior. It will be of interest to scientists working outside the immediate field suggested by the title. It gives me particular pleasure to recommend the book to Western readers since the translator has done an excellent job in making accessible in English what must already be a classic in Russian.

Gerald Scott
Department of Chemistry
University of Aston
Birmingham, England

FOREWORD

to the

RUSSIAN EDITION

About a decade and a half ago investigations into the chemical
resistance of polymers were commenced in the Department of Kinetics
of Chemical and Biological Processes of the Institute of Chemical
Physics of the Academy of Sciences of the USSR. These had in view
the need to form an up-to-date idea of the chemical resistance of
polymers, in the same way as for metals there had been created a
science of corrosion with a clear mathematical framework, a classification of the mechanisms, and establishment of the relationship
between the resistance of the various metals and their position in
the Periodic Table.

Strangely, there are as yet no published monographs on the fundamental basis of the degradation of polymers in aggressive media.
Accordingly, the publication of this monograph by Yu. V. Moiseev
and G. E. Zaikov is of particular importance.

The contact of polymers with liquid aggressive media is accompanied by a complex range of physical and chemical processes, such
as the adsorption and diffusion of the medium and also the chemical
and physical dissociation of the macromolecules. In most cases,
polymeric materials are heterogeneous systems. Accordingly, it was
necessary to study processes taking place in amorphous or ordered
regions in polymeric materials, in the same way as the reactions
with various chemical components forming part of the composition of
an actual polymeric material. The development of the idea of the
chemical resistance of polymers inevitably called for information
from such varied fields of knowledge as acid—base catalysis, the
theory of solutions, the diffusion of low-molecular substances, the
morphology of polymers, and the reaction kinetics of the condensed
phase.

The theoretical examination of a large amount of factual data has enabled the authors to formulate the basic principles of chemical degradation of polymers and to propose quantitative criteria for the assessment of their chemical resistance.

The authors emphasize not so much the description and explanation of phenomena as the quantitative procedure for studying the processes of chemical degradation. The book is written in a strictly logical style, which enables the reader to form a clear idea of the mechanisms of chemical degradation and of the possibilities of forecasting the chemical, physical, and mechanical properties of polymeric articles.

The monograph begins with a chapter stating the fundamentals of catalysis in liquid aggressive media, in solutions of acids, bases, or salts.

The second chapter considers the mechanisms of the breakdown in these media of model compounds containing amide, imide, and ester, acetal, and siloxane groups, and presents the kinetic parameters of the breakdown of these compounds in acidic or basic media. These two chapters are, as it were, the foundations of the monograph as a whole.

The third and fourth chapters analyze in detail the processes of chemical degradation of polymers and the main types of breakdown of macromolecules, depolymerization at the terminal bonds and random breakdown. The anomalies of degradation of polymer chains in the solid state are considered.

Polymeric materials are frequently used as protective coatings, and in this case it is necessary to know at what rate the components of these media diffuse into the polymers. In addition, to understand the mechanism of degradation, it is necessary to have information on the state of the medium in the polymer, i.e., to what degree the electrolytes dissociate into ions while the latter are solvated by the solvent. All these questions are clarified in the fifth chapter.

The sixth chapter studies the quantitative foundations of the degradation of polymers in these media, and kinetic equations are presented for the main types of degradation.

The seventh chapter is one of the longest, and contains a variety of kinetic data on the degradation of polyamides, polyesters and polyethers, cellulose and its derivatives, polycarbonates, polysiloxanes, and poly(vinyl chloride).

The next two chapters describe the influence of the medium on the mechanical and other service properties of polymer materials, and also the utilization of such a medium for the chemical and physical modification of polymers.

The tenth, final, chapter deals with the prediction of the service properties of polymer materials, something certainly of particular importance for their practical employment.

The monograph is of indubitable importance both for specialists studying the degradation of polymers and for scientists in the general field of the chemistry and physics of polymers. This book comprises information needed by chemists and laymen.

Academician N. M. Émanuél'

PREFACE

The problem of the chemical resistance of polymers confronts
research workers, engineers, and designers in various fields of industry. Quite apart from the large number of chemical reagents, a
degradative action on polymers is effected by, for instance, deter-
gents, seawater, and exhaust gases containing dioxides of nitrogen,
of sulfur, etc. The concern for this problem is shown by the large
number of papers published in the past decade — there are around
5000. However, there are no more than two monographs in the world
literature dealing with the chemical resistance of polymers: B.
Doležěl's "Corrosion of Plastics and Rubbers," published in 1964 and
containing material published up to 1962; and Yu. S. Zuev's "Breakdown of Polymers under the Action of Aggressive Media," published
in 1972 and dealing mainly with rubbers.

The inadequacy of theoretical work in the field of chemical degradation of polymers, and the abundance of experimental data,
prompted us to attempt to work out the general principles and quantitative pattern of degradation. The outcome is this monograph,
which, naturally, is not free from shortcomings, so that we are
grateful to receive any critical observations.

We take this opportunity to express our profound gratitude to
Prof. K. S. Minsker, Dr. A. A. Berlin, M. I. Vinnik, N. B. Librovich,
and V. S. Pudov for valuable advice in the manuscript stage, and also to M. I. Artsis, K. Z. Gumargaliev, and T. E. Rudakov for help
in its production.

Yu. V. Moiseev
G. E. Zaikov

xi

TRANSLATOR'S NOTE

Close collaboration at all stages of this often demanding work,
continuing an association with Dr. S. H. Morrell lasting over many
years, has been of great interest and inestimable value. I would
like also to express my gratitude to J. MacLachlan, Manager of the
Information Centre, Rubber and Plastics Research Association of
Great Britain, for the opportunity to examine a considerable amount
of the extensive literature cited by the Russian authors, and in
particular to R. H. Norman and Dr. B. G. Willoughby, who have helped Dr. Morrell and myself in this and many previous undertakings.
Discussions between subject specialists and translator, each understanding the other's viewpoint, have helped, it is believed, to give
a respectful translation of a work of dedication and enthusiasm in
which we have, as in other Russian works, noted a warm humanity
which recognizes that "man does not live by bread alone."

R. J. Moseley

xii

CONTENTS

Chapter 1
CATALYSIS IN AGGRESSIVE LIQUID MEDIA. THE BASIC CONCEPTS

Chapter 2
MECHANISMS OF DISSOCIATION IN AGGRESSIVE MEDIA AND THE REACTIVITY OF CHEMICALLY UNSTABLE BONDS OF VARIOUS POLYMERS

Chapter 3
SPECIAL FEATURES OF THE CHEMICAL DEGRADATION OF POLYMERS

Chapter 4
PRINCIPAL TYPES OF DECOMPOSITION OF POLYMER MOLECULES

Chapter 5
DIFFUSION OF AGGRESSIVE MEDIA IN POLYMERS

Chapter 6
DEGRADATION OF POLYMERS IN AGGRESSIVE MEDIA

Chapter 7
DEGRADATION OF HETERO-CHAIN POLYMERS

Chapter 8
INFLUENCE OF AGGRESSIVE MEDIA ON THE MECHANICAL
PROPERTIES OF POLYMERS

Chapter 9
INFLUENCE OF CHEMICAL DEGRADATION PROCESSES
ON THE SERVICE PROPERTIES OF POLYMER ARTICLES

Chapter 10
PREDICTION OF THE SERVICE PROPERTIES OF POLYMER ARTICLES

Chapter 1

CATALYSIS IN AGGRESSIVE
LIQUID MEDIA.
THE BASIC CONCEPTS

The aggressive media to which reference is made are solutions of acids, bases, or salts.

Acids and bases are the most common of these media. Salts are less common and less active than acids or bases. The action of acidic and basic salts on polymers may, as a first approximation, be regarded as the action of weak acids or bases. The mechanism of action of neutral salts on polymers has at present hardly been studied, and it is therefore not clear how to characterize quantitatively the catalytic action of solutions of such salts.

Certain acids and salts have an oxidative effect on polymers. The oxidative electrochemical potential may be taken as a measure of the oxidative activity of such media.

There are at present two well-known approaches to the classification of substances in relation to acids or bases.

According to Brønsted, an acid is a proton donor and a base a proton acceptor. According to Lewis, an acid is a reagent accepting an electron pair and a base a reagent donating an electron pair.

The Lewis approach is the more general since the proton, having high affinity for the electron pair, may be regarded as indeed a Lewis acid. However, such a general definition is not always convenient for practical purposes, particularly when considering chemical degradation since all electrophilic compounds rate as acids and all nucleophilic compounds as bases.

Henceforth, mainly the Brønsted definition will be used, but, in certain cases, where necessary, it is stated what sort of acid is meant, Brønsted or Lewis.

1.1. QUANTITATIVE ASSESSMENT OF THE STRENGTH OF ACIDS AND BASES

The strength of an acid or base is determined by the dissociation constant, i.e., the ratio of the ionized and un-ionized forms of the substance. The general form of acid—base equilibrium may be expressed as follows:

$$A_1 + B_2 \rightleftharpoons A_2 + B_1$$

where A is the acid, B the base.

If the reactions are accompanied by proton transfer (protolytic reactions), the equilibrium constant of this process

$$K = \frac{a_{B_1} a_{A_2}}{a_{B_2} a_{A_1}} \tag{1.1}$$

is equal to the ratio of two hypothetical constants, known as the acidity constants

$$K_{A_1} = \frac{a_{B_1} a_{H^+}}{a_{A_1}}, \qquad K_{A_2} = \frac{a_{B_2} a_{H^+}}{a_{A_2}}$$

where a represents the thermodynamic activities.

The constant K is the ratio of the strength of the acids A_1 and A_2 or the strength of the bases B_1 and B_2. In principle, one does not determine the "absolute strength" of an acid or base; it is possible only to measure the strength of an acid or base relative to a particular standard pair A_0—B_0. Ordinarily we use a solvent as this standard. In aqueous solutions there can exist two pairs: $H_2O \cdot H^+$—H_2O or H_2O—OH^-. Then the relative strength of an acid HA in aqueous solution is equal to the thermodynamic dissociation constant:

$$K_{\text{diss}} = \frac{a_A - a_{H_2O \cdot H^+}}{a_{HA} a_{H_2O}} \tag{1.2}$$

In sufficiently dilute solutions it is possible to replace activities by concentrations and take the concentration of water as constant. In this case

$$K_{\text{diss,c}} = \frac{c_A - c_{H^+}}{c_{HA}} \tag{1.3}$$

where $K_{\text{diss,c}}$ is the concentration dissociation constant of the acid.

The strength of acids and bases is most frequently indicated by the values pK = $-$log K_{diss} and pK_c = $-$log $K_{diss,c}$. Moreover, according to the conventional division, strong acids in aqueous solutions have values of pK less than zero, while weak acids have values higher than zero. Frequently the dissociation constants of acids are referred to as acidity constants, symbol K (and, correspondingly, K_b for bases, basicity constants).

1.1.1. Influence of the Medium on the Strength of Acids and Bases

The influence of the medium on protolytic equilibrium was first studied by Brønsted [1, p. 585]. Ismailov [2] studied the process of ionization not of acids and bases themselves, but of the products of their prior association with solvent molecules. For instance, the dissociation of an electrolyte KA in the presence of a protolytic solvent M appears as follows:

$$KA + M \underset{\longleftarrow}{\overset{K_{inst}}{\longrightarrow}} KA_M \underset{\longleftarrow}{\overset{K_d}{\longrightarrow}} K_M^+ + A_M^- \underset{\longleftarrow}{\overset{K_{ass}}{\longrightarrow}} K_M^+ A_M^-$$

where K_{inst}, K_d, and K_{ass} are the constants of instability of the solvated electrolyte, of its dissociation, and of the association of the resulting ions, respectively.

The dissociation constant of the electrolyte K_{diss} (this is equal to the ratio of the activities of the ions to the activity of the other particles of the electrolyte) is equal to

$$K_{diss}^{-1} = K_d^{-1}(1 + K_{inst}) + K_{ass} \tag{1.4}$$

Depending on the properties of the electrolyte and of the solvent, Eq. (1.4) has partial solutions. In solvents with high dielectric permittivity and high affinity for the ions, K_{diss} = K_d, and in solvents with low permittivity but having high affinity for the electrolyte, K_{diss} = K_{ass}^{-1}.

The complete equation linking K_{diss} with the permittivity of the medium and the physical characteristics of the ions which are formed on dissociation is as follows:

$$\log K_{diss} = \log K_{diss}^0 + \log \gamma_{KA,M}^0 - \log(1 + K_{inst} + K_d K_{ass}^{-1}$$

$$+ \frac{e^2 N_A}{4{,}58RT} \sum \frac{Z_i^2}{r_i}\left(1 - \frac{1}{\varepsilon}\right) + \sum n_i \frac{Z_i e \mu N_A}{2.3 r_i^2} \tag{1.5}$$

where K_{diss}^0 is the dissociation constant of KA into ions in a vacuum; $\gamma_{KA,M}^0$ is the activity coefficient at infinite dilution, with

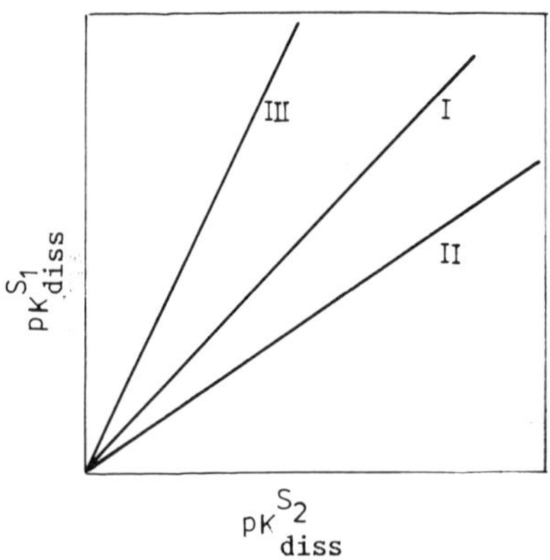

Fig. 1.1. Types of relationship between pK$^{S_1}_{diss}$ and pK$^{S_2}_{diss}$ (see text for explanation).

respect to vacuum or to the standard state; Z_i and r_i, respectively, the charge and radius of the i-th ion; n_i the number of solvent molecules which solvate the ion; μ the dipole moment of the solvent molecules; and ε the permittivity of the medium.

Analysis of Eq. (1.5) shows that log K_{diss} is a linear function of $1/\varepsilon$ in solvents which have closely similar solvation energies of the ions.

From Eq. (1.5) it is possible to obtain the relationship between the electrolytic dissociation constants of one and the same electrolyte in two different solvents S_1 and S_2, which is expressed graphically as follows (Fig. 1.1). The relationship between pK$^{S_1}_{diss}$ and pK$^{S_2}_{diss}$ has a slope equal to unity if the difference between the solvation energies of the electrolyte in solvents S_1 and S_2 is constant (line I). In the opposite case there is either differentiation of the strength of the electrolyte (line II) or else it is minimized in solvent S_2 as compared with S_1 (line III). In the above scheme, solvents having opposite chemical function with respect to the dissolved substance must generally bring about a rise in the strength of the electrolyte. For instance, inorganic acids which are strong in water are weak in acetic acid, since, as a consequence of the low value of ε in acetic acid, differences in the strengths of the acids are enhanced (the main role being played by the combination of the solvent with the acids).

Thus, generally, acidic solvents are bound to differentiate the strengths of acids and minimize the differences between bases. On the other hand, basic solvents are bound to minimize the differences between acids and enhance the differences between bases.

1.1.2. Methods of Determining the Strength of Acids and Bases

Methods of determining the dissociation constants of weak acids and bases in aqueous and nonaqueous solutions are described in detail in a monograph by Albert and Serjeant [3].

The determination of the dissociation constants of strong acids and bases in aqueous solutions is difficult since in dilute solutions these electrolytes dissociate practically completely into ions.

To determine dissociation constants in concentrated solutions of acids and bases, either the methods of vibrational spectroscopy (Raman and IR spectra) or NMR may be used. In both methods, the calculation of the thermodynamic dissociation constant from the determined values of the degree of dissociation involves the use of measured or calculated activity coefficients. Accordingly, with strong electrolytes we get a large scatter in the dissociation constants.

In Tables 1.1 and 1.2 we give the dissociation constants of the most important acids and bases in water. The dissociation constants of other acids can be assessed as follows. For instance, the first dissociation constant of oxygen-containing inorganic acids of the general formula $XO_m(OH)_n$ is determined by the value of m. If $m = 0$, then the acid is very weak and $K_1 < 10^{-7}$; with $m = 1$, the acid is weak and $K_1 \approx 10^{-2}$; with $m = 2$, $K_1 \approx 10^2$, and with $m = 3$ the acid is strong ($K_1 \approx 10^8$). The subsequent dissociation constants of the polybasic acids K_1, K_2, K_3 are in the ratio $1:10^{-5}:10^{-10}$ [4, p. 368 of Russian translation]

In the general form, the dependence of the dissociation constant of an organic acid on its structure is examined in [5, p. 237]. The acids and the corresponding conjugate base, an anion, may be represented by the formulas

$$X—E—H \text{ and } X—E^-$$

where E is the reactive center of the acid and X a substituent attached to this center.

Then the strength of the acid will be determined by the following factors:

the effective electronegativity of the reactive center in the initial ($—E—X$) and final ($—E^-$) states;

the resonance characteristics of the reactive center in the initial and final states;

the bond energy of an unshared electron pair in the anion center $—E$;

TABLE 1.1. Dissociation Constants of the Most Important Acids in Water

Acids		Temperature, °C	K_{diss}
Oxygen-containing:			
Nitrous (HNO_2)		12.5	$4.6 \cdot 10^{-4}$
Nitric (HNO_3)		25	$4.4 \cdot 10$
Hypobromous ($HOBr$)		25	$2.06 \cdot 10^{-9}$
Periodic (HIO_4)		25	$2.3 \cdot 10^{-2}$
Iodic (HIO_3)		25	$1 \ 69 \cdot 10^{-1}$
Hypoiodous (HOI)		25	$1.0 \cdot 10^{-10}$
Arsenic (H_3AsO_4)	(1)	18	$5.62 \cdot 10^{-3}$
	(2)	18	$1.70 \cdot 10^{-7}$
	(3)	18	$3.95 \cdot 10^{-12}$
Arsenous ($HAsO_3$)		25	$6 \cdot 10^{-10}$
Orthoboric (H_3BO_3)	(1)	20	$7.3 \cdot 10^{-10}$
	(2)	20	$1.8 \cdot 10^{-13}$
	(3)	20	$1.6 \cdot 10^{-14}$
Orthophosphorous (H_3PO_4)	(1)	25	$1.52 \cdot 10^{-3}$
	(2)	25	$6.23 \cdot 10^{-8}$
	(3)	18	$2.2 \cdot 10^{-13}$
Pyrophosphoric ($H_4P_2O_7$)	(1)	25	$1.0 \cdot 10^{-1}$
	(2)	25	$1.0 \cdot 10^{-2}$
	(3)	25	$1.0 \cdot 10^{-7}$
	(4)	25	$1.0 \cdot 10^{-9}$
Sulfuric (H_2SO_4)	(1)	25	$1.0 \cdot 10^{3}$
	(2)	25	$1.2 \cdot 10^{-3}$
Sulfurous (H_2SO_3)	(1)	18	$1.54 \cdot 10^{-2}$
	(2)	18	$1.02 \cdot 10^{-7}$
Carbonic (H_2CO_3)	(1)	25	$4.3 \cdot 10^{-7}$
	(2)	25	$5.61 \cdot 10^{-11}$
Perchloric ($HClO_4$)		25	$1.0 \cdot 10^{8}$
Hypochlorous ($HOCl$)		18	$2.9 \cdot 10^{-8}$
Chromic (H_2CRO_4)	(1)	25	$1.8 \cdot 10^{-1}$
	(2)	25	$3.2 \cdot 10^{-7}$
Hydrohalic:			
Hydroiodic (HI)		25	$1.0 \cdot 10^{10}$
Hydrochloric (HCl)		25	$1.0 \cdot 10^{7}$
Hydrofluoric (HF)		25	$6.76 \cdot 10^{-4}$
Carboxylic:			
Benzoic (C_6H_5COOH)		25	$6.3 \cdot 10^{-5}$
Formic ($HCOOH$)		25	$1.77 \cdot 10^{-4}$
n-Butyric (C_3H_7COOH)		25	$1.51 \cdot 10^{-5}$
Trichloroacetic (Cl_3CCOOH)		25	2.51
Trifluoroacetic (F_3CCOOH)		25	$5.9 \cdot 10^{-1}$
Acetic (CH_3COOH)		25	$1.75 \cdot 10^{-5}$
Chloroacetic (ClH_2CCOOH)		25	$1.36 \cdot 10^{-3}$
Alcohols:			
Methanol (CH_3OH)		25	$3.16 \cdot 10^{-16}$
Ethanol (C_2H_5OH)		25	$1.0 \cdot 10^{-18}$
Trichloroethanol (Cl_3CCH_2OH)		25	$5.3 \cdot 10^{-13}$
Trifluoroethanol (F_3CCH_2OH)		25	$6.3 \cdot 10^{-13}$
Trinitrophenol [$(NO_2)_3C_6H_2OH$]		25	$1.9 \cdot 10^{-1}$
Phenol (C_6H_5OH)		20	$1.28 \cdot 10^{-10}$

Note. 1) The degrees of dissociation of the monobasic acids are in parentheses. 2) The values of K_{diss} are taken from [3, 21, 22].

the effective electronegativity substituent X;

the resonance characteristics of the substituent X.

TABLE 1.2. Dissociation Constants of the Conjugate Acids Corresponding to the Most Important Bases in Water

Base	Temperature, °C	K_{diss}
Ammonia (NH_3)	25	$5.66 \cdot 10^{-10}$
Aniline ($C_6H_5NH_2$)	25	$2.34 \cdot 10^{-5}$
n-Butylamine ($C_4H_9NH_2$)	20	$1.67 \cdot 10^{-11}$
Hexamethylenediamine (HMD) ($NH_2C_6H_{12}NH_2$) (1)	20	$7.85 \cdot 10^{-12}$
(2)	20	$9.68 \cdot 10^{-11}$
Dimethylamine [$(CH_3)_2NH$]	25	$1.85 \cdot 10^{-11}$
Diphenylamine [$(C_6H_5)_2NH$]	25	$1.62 \cdot 10^{-1}$
Methylamine (CH_3NH_2)	25	$2.10 \cdot 10^{-11}$
Urea (NH_2CONH_2)	21	$7.94 \cdot 10^{-1}$
Pyridine (C_5H_5N)	25	$6.17 \cdot 10^{-6}$
Trimethylamine [$(CH_3)_3N$]	25	$1.55 \cdot 10^{-10}$
Ethylamine ($C_2H_5NH_2$)	20	$1.56 \cdot 10^{-11}$
Diethylenediamine ($NH_2C_2H_4NH_2$) (1)	20	$8.41 \cdot 10^{-11}$
(2)	20	$1.04 \cdot 10^{-7}$

<u>Note</u>. 1) The degrees of dissociation of the bases are in parentheses. 2) The values of K_{diss} are taken from [23].

In acids with constant substituents X the dissociation constant of the acids rises as follows: C–H acids < Si–H acids < N–H acids < P–H acids < O–H acids < F–H acids < S–H acids.

The dissociation constant of acids of the same type (E remaining constant) depends on the structure of the substituent as follows:

$$\log K_{diss} = \log K^0_{diss} + \rho^* \sum \sigma^* + \rho_R^- \sum \sigma_R^- + \rho_R^+ \sum \sigma_R^+ \qquad (1.6)$$

where σ^* and ρ^* are the induction parameters and constants, and σ_R and ρ_R the resonance parameters and constants.

The positive signs of the constants of the substituents correspond to a rise in the acidity.

1.1.3. The Ionic Composition of Solutions of Acids and Bases

In an examination of the mechanisms of the chemical degradation of polymers it is necessary to know not only how the acids and bases dissociate into ions but also how the latter are solvated by the solvent.

It has now been shown by careful investigation using IR and Raman spectroscopy [6, 7, p. 404 of Russian translation; 8, 9], x-ray and neutron analysis [10, 11], and NMR [12, p. 531 of Russian translation] that the main hydrated form of the proton is not H_3O^+ but $H_5O_2^+$. The two molecules of water in this structure maintain their individuality as vibrational systems, the "tunnel" transfer of the proton causing continuous absorption in the IR spectra of the acid solutions.

TABLE 1.3.   Concentration of $H_5O_2^+$ Ions and Anions in Aqueous Solutions of Sulfuric Acid at 25°C [24]

Concentration of H_2SO_4, mass %	$c_{H_5O_2^+}$, g-ion/ liter	c_{A^-}, g-ion/ liter	Concentration of H_2SO_4, mass %	$c_{H_5O_2^+}$, g-ion/ liter	c_{A^-}, g-ion/ liter
2.34	0.29	0.245	41,1	5,69	5,49
3.71	0,46	0,385	46,2	6,41	6,39
4.01	0,49	0.417	50,4	6,89	7,18
5.89	0,73	0,425	51,3	7,02	7,35
9,74	1,21	1,06	56,2	7,77	8,34
13,0	1,67	1.44	59,8	8,38	9,11
15,8	2,06	1,78	63.0	8,87	9,81
19,4	2,57	2,24	66,0	9,15	10,5
21,6	2,92	2,53	67,5	9,24	10,9
22,6	3,08	2,67	68,9	9,30	11,2
25,8	3,58	3,11	70,7	9,37	11,7
27,4	3,77	3,34	72,4	9,40	12,1
30,9	4,20	3,85	74,0	9,44	12,5
32,8	4,47	4,14	75,4	9,46	12,8
36,0	4,97	4,65	76,9	8,50	13,2
40,2	5,55	5,34	77,9	7,60	13,5
			80,6	5,30	14,2

In dilute and moderately concentrated acid solutions the $H_5O_2^+$ ion can be hydrated by one or two molecules of water, forming, for instance, the following structure [9]:

$$
\begin{array}{ccccc}
\text{H} & & \text{H} & \text{H} & \text{H} \\
\diagdown & & \diagdown & \diagup & \diagup \\
\text{O} & & \text{O}\cdots\text{H}^+\cdots\text{O} & & \text{O} \\
\diagup & & \diagup & \diagdown & \diagdown \\
\text{H} & & \text{H} & \text{H} & \text{H}
\end{array}
$$

The composition of aqueous solutions of various acids is given in [13, p. 427 of Russian translation].   Table 1.3 gives the ionic composition of solutions of sulfuric acid.

In aqueous basic solutions the ions are the basic hydrates [6]:

$$[HO\cdots H\cdots OH]^-$$

In kinetic equations the solvated proton will be designated as H_S^+.

1.1.4. Aprotic Acids and Bases

An acid—base reaction between an aprotic acid and a base consists of the transfer of an electron pair from the base to the acid with the formation of a chemical bond between the adducts, i.e., a collective electron pair:

$$A + :B \overset{K_{calc}}{\rightleftharpoons} A:B$$

In this the acid—base reaction differs from a redox reaction, in which one or more electrons pass completely from the reducing agent to the oxidizing agent. Aprotic acids fall into three groups:

acids with free orbitals (e.g., M^{n+}, MX_n);

π-acids (e.g., trinitrobenzene);

σ-acids (e.g., I_2).

Aprotic bases are subdivided similarly. The most common are π-bases, including alkenes, polyenes, and alkynes.

Aprotic acids and bases are characterized by a particular interaction, which makes it impossible to arrange them in a series according to strength since this would depend on which substance is taken as the standard for comparison. For many aprotic acids and bases there are known to be correlations of pK_{calc} with pK_a or pK_b for acids or bases of the same type of structure [14].

1.2. MEASURE OF CATALYTIC ACTION OF ACIDS AND BASES

The catalytic action of acids and bases is assessed by the capacity of the medium for releasing protons and hydroxide ions.

In dilute aqueous solutions of acids and bases hydronium and hydroxide ions are effective catalysts. In this case, the dissociation constants of the acids and bases reflect the measure of their catalytic action. In concentrated solutions we find that, along with these ions, undissociated molecules, whether of strong or weak acids and bases, are donors of protons and hydroxide ions.

In addition, catalytic action may be possessed by aprotic particles (cations and anions or, for instance, SO_3 in oleum).

Thus, generally, the catalytic action of acids and bases cannot be quantitatively characterized by their dissociation constants.

The catalytic action of acids and bases is determined by the activities (or concentrations) of ionized and un-ionized particles of the acids and bases in these media.

At the present time it is impossible to determine the activities of individual ions, in particular hydronium and hydroxide ions. This is due, firstly, to the problem of establishing the absolute zero of potential and, secondly, to the dependence of the activity of the ion on the nature of the counterion.

Let us now consider methods of assessing the activity of the proton and hydroxide ion.

1.2.1. Electrometric Determination of pH
in Aqueous Solutions

The pH of aqueous solutions of acids and bases is determined
by the expression

$$\text{pH} = -\log m_{H^+}\gamma_{H^+}\gamma_{Cl^-} = \frac{(E - E_0)\,F}{2.3RT} + \log m_{Cl^-} \tag{1.7}$$

where E is the emf of the galvanic cell: reference electrode—reference solution—saturated solution—solution under investigation—
working electrode.

The value E_0 is determined from a similarly constructed cell,
except that in place of the solution under investigation we use a
solution with a standard pH value. Such a method contains three
sources of error, connected with a) the determination of the pH of
standard solutions, b) the assessment of the value of log γ_{Cl^-}, and
c) the presence of a diffusion potential at the boundary of a saturated solution of KCl with the investigated (standard) solution.

The pH of standard solutions is established using cells without
transfer, so as to exclude errors connected with the presence of a
diffusion potential.

There are published descriptions of attempts to determine a_{H^+}
in concentrated acid solutions by an electrometric method using
different electrodes [15, 16]. However, they all have insufficient
theoretical foundation and accordingly have not become very popular.

1.2.2. Indicator Method of Determining Acidity
in Aqueous Solutions

The determination of the acidity of aqueous acid solutions
with the aid of indicators was first described by Hammett [17, p.
345 of Russian translation]. As indicators there were used a series
of substituted primary nitroanilines, for which the protolytic equilibrium may be expressed by the general scheme:

$$\text{In} + \text{H}^+ \rightleftharpoons \text{InH}^+$$

This mechanism of ionization of nitroanilines was confirmed indirect-
ly by the value of the i-factor (the number of types of particles
in the solution), which in 100% H_2SO_4 is 2. It should, however,
be borne in mind that either the hydrated protons or the undissociated molecules of the acid may be the proton donor.

In moderately concentrated solutions of acids the dissociation
constant of the conjugate acid of the indicator InH^+ may be expressed as follows:

$$K_{InH^+} = \frac{a_{InH^+}a_{H_2O}}{a_{In}a_{H_2O \cdot H^+}} \tag{1.8}$$

It is usually assumed that $a_{H_2O \cdot H^+} = a_{H^+}a_{H_2O}$ (the thermodynamic equilibrium constant is unity). By definition, the acidity function H_0 is

$$H_0 = -\log h_0 = -\log a_{H^+} + \log \frac{f_{InH^+}}{f_{In}} = pK_{InH^+} - \log \frac{c_{InH^+}}{c_{In}} \tag{1.9}$$

According to Hammett, H_0 characterizes the potential of the solution for the transfer of a proton to the base B for which, under the influence of the medium, f_{BH^+}/f_B changes in the same way as the analogous ratio for the indicators [17, p. 345 of Russian translation]. The Hammett function is designated as H_0 since basic indicators have no charge. There are also functions H-, for whose determination we use as an indicator an uncharged acid, while the corresponding base has a negative charge. The selection of the concentration scale in Eq. (1.9) is very important. Usually h_0 is expressed on a scale of molar concentrations.

<u>Experimental Determination of H_0</u>. Using spectrophotometry, we determine the concentration of the colored nonionized form of the indicator, while the concentration of the protonated form is calculated by difference.

The chosen concentration of indicator was the minimum, so that it had practically no influence on the properties of the medium. There are a number of restrictions on the use of the indicator method, connected with the influence of oxidizing or reducing agents, present in the solution, on the color of the indicator, with the influence of the ionic strength on the color of the indicator, or with the possibility of chemical interaction with reagents in the solution, e.g., with the albumens.

To determine H_0 it is necessary to know the value of pK_{InH^+} in Eq. (1.9). Since this is a thermodynamic constant, its determination must include extrapolation to infinitely dilute solutions. For indicators with $pK_{InH^+} > 1$ the measurements are effected in very dilute solutions, in which

$$\lim_{c \to 0} \frac{f_{InH^+}}{f_{In}} = 1 \quad \text{and} \quad H_0 = pH$$

For indicators where pK_{BH^+} lies between 0 and 1, we use the following approach. It has been shown empirically that $\log \frac{c_{InH^+}}{c_{In}}$ vs. $\log c_{H^+}$ is a linear function of the acid concentration for solutions of strong acids up to 2 M, so that pK_{InH^+} can be found by extrapolation of the above relationship to $c \to 0$.

TABLE 1.4. Values of the Acidity Function H_0 for Aqueous Solutions of the Most Important Inorganic Acids at 25°C [25, p. 79]

Concentration, mass %	HCl	H_2SO_4	$HClO_4$	H_3PO_4	Concentration, mass %	HCl	H_2SO_4	$HClO_4$	H_3PO_4
5	0.45	0,02	—0,02	—0,96	55		3,91	4,22	1,42
10	1.00	0,31	0,30	—0,64	60		4,46	5,25	1.66
15	1.54	0,66	0,63	—0,39	65		5,10	6,43	1.97
20	2.10	1,01	0.95	—0,15	70		5,80	7,72	2.29
25	2.70	1,37	1,25	—0,06	75		6.56	9,15	2.64
30	3.39	1,72	1.60	—0,25	80		7,34	10,31 (78,6)	3,05
35	4.08	2,06	1,97	0,48	85		8,14		3.48
40		2,41	2,40	0,72	90		8.92		
45		2,85	2,92	0,94	95		9.85		
50		3,38	3,52	1,17	100		11.94		

For indicators with $pK_{InH^+} < 0$, these methods become inapplicable, and so we use the following. Each indicator is compared in one and the same acid solution with an indicator having a higher pK_{InH^+}. For two indicators In_1 and In_2 we get the relationship

$$\log \frac{c_{InH^+}}{c_{In_1}} - \log \frac{c_{In_2H^+}}{c_{In_2}} = pK_1 - pK_2 \tag{1.10}$$

If pK_1 is known, from (1.10) it is possible to calculate the value of pK_2. In this way a number of indicators have been investigated, and a H_0 scale established for many acids (Table 1.4).

In concentrated solutions of sulfuric acid (>90%) undissociated molecules of sulfuric acid act, together with hydrated protons, as proton donors. This, on the one hand, explains the differing values of pK_{InH^+} for one and the same indicator in concentrated H_2SO_4 and, on the other hand, makes it possible to obtain simple relationships linking H_0 with the molar fractions (N) of the components of concentrated solutions of sulfuric acid. For instance, Brand's ratio [18] for solutions of 90-99.5% H_2SO_4 takes the form

$$H_0 = -8.36 + \log \frac{N_{H_2SO_4^-}}{N_{H_2SO_4}} \tag{1.11}$$

<u>Other Acidity Functions</u>. Knowledge of the acidity function, as determined from Eq. (1.9), depends, for a given acid solution, on the type of indicator when given an identical scheme of protonation.

A check on this suggestion led to other acidity functions for different classes of indicators, the most used in practice being H_A, which is obtained when using as indicators a series of substituted aromatic amides and N-oxides of the pyridine series [19, 20]. The

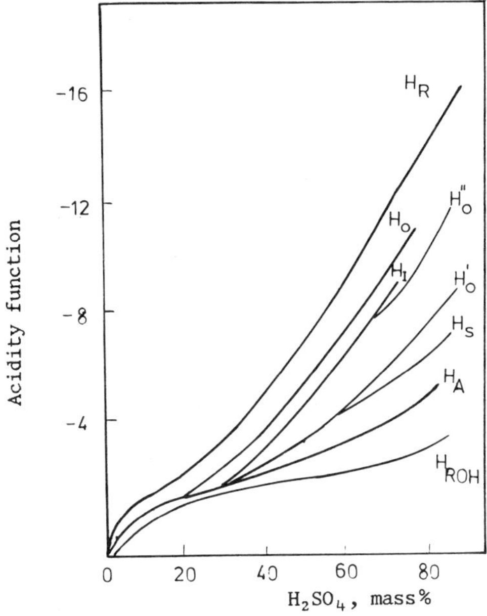

Fig. 1.2.   Changes in the various acidity functions in aqueous solutions of H_2SO_4.

changes in the various acidity functions in aqueous solutions of
H_2SO_4 are shown in Fig. 1.2.

1.2.3. Acidity of Nonaqueous Solutions

For comparison of the acidity of nonaqueous solutions we compare the acidity of solutions in a single solvent and in differing
solvents.

The acidity of any given solution is characterized by the concentration of the "lyonium" ions S^+ and by the index of acidity

$$pS^+ = -\log a_{S^+} = -\log c_{S^+}\gamma_{S^+} \qquad (1.12)$$

In the partial case with establishment of protolytic equilibrium

$$pS^+ = pH_S = -\log a_{H^+} \qquad (1.13)$$

each solvent is characterized by its own pH_S scale, the point of
neutrality being equal to the square root of the autoprotolysis
constant of the solvent.

For a comparison of the acidity of solutions in various solvents a number of criteria have been proposed:

1. Brønsted [1] proposed a comparison of the chemical potentials of the proton in a given medium; these are linked with the dissociation constant by a simple relationship:

$$\mu_{H^+} = RT \ln K_{diss} \tag{1.14}$$

However, as K_{diss} cannot be determined by a simple experiment, and has to be found by indirect methods, the accuracy of the results proves inadequate.

2. The method of the acidity function makes it possible to compare acidities if it is assumed that the basicity constants do not depend on the solvent. Nevertheless, such a suggestion is insufficiently founded.

3. Izmailov [2, p. 792] compared the acidity of nonaqueous solutions, taking as a standard the state of the protons in an infinitely dilute aqueous solution. The index of unit acidity pA is determined as

$$pA = -\log a_{H^+} = -\log a_{SH^+} - \log \gamma_{0\,H^+} = pH_p - \log \gamma_{0\,H^+} \tag{1.15}$$

1.3. MECHANISMS OF ACID–BASE CATALYSIS

The classification of the mechanisms of dissociation of chemically unstable bonds of various polymers in aggressive media is based on three fundamental characteristics: the form of the catalyst (A acid, B base), its specificity, and the molecularity of the limiting stage of the process.

Classification by the first and third of these was introduced first by Ingold [26].

Generally, the effective rate constant of the process is made up of the products of the catalytic activity of all the acids and bases present in the reaction mixture, including that of the solvent, and the concentration of these reagents:

$$k_{eff} = \sum_i k_{eff_i} c_{cat_i} \tag{1.16}$$

where cat_i represents a solvated proton (H_S), hydroxide ion (OH_S^-), undissociated forms of the acids (HA) and bases (B), the solvent (S), and other substances catalyzing this reaction.

The catalytic action of any of the above substances is linked with the fact that the presence of the catalyst within the activated complex for the limiting stage of the process stabilizes the activated state (or lowers the activation barrier). Identification of the intermediate compounds is of fundamental importance for determining the form of the dependence of the rate of the reaction on the catalyst concentration and other reagents.

1.3.1. Specific and General Catalysis

Distinction is made between specific and general acid and specific and general basic catalysis.

Specific acid (or base) catalysis take place if the catalytic effect is brought about by the presence of a proton (or of a base conjugated with a hydrogen-containing acid) in an activated complex.

General acid (or base) catalysis occurs if the catalysts forming part of the activated complex are other acids and bases along with the proton and hydroxide ion. The manifestation of specific and general catalysis may be illustrated by the example of the dissociation of a substrate R in an acid

$$R + HA \underset{k_{-1}}{\overset{k_1}{\rightleftarrows}} RH^+ + A^- \tag{a}$$

$$RH^+ + S \xrightarrow{k_2} P + H_S^+ \tag{b}$$

where HA represents the acidic substances in the solution and P the reaction products.

In the first stage of the process there is protonation of the reagent by the transfer of a proton from any acid substance in the solution; in the second stage the protonated form of the reagent is converted into reaction products with transfer of the proton to the solvent molecule.

Assuming that the process takes place in dilute solutions, it is possible to disregard the activity coefficients. The use of the principle of steady-state concentrations for a compound RH^+ gives

$$k_1 c_R c_{HA} - k_{-1} c_{RH^+} c_{A^-} - k_2 c_{RH^+} = 0$$

The rate of formation of the reaction products is

$$W = k_2 c_{RH^+} = \frac{k_1 k_2 c_R c_{HA}}{k_{-1} c_{A^-} + k_2} \tag{1.17}$$

(here with the constant k_2 we introduce the concentration of the solvent). Let us consider two cases.

1. With $k_{-1}c_{A^-} \gg k_2$ the equilibrium concentration of RH^+ is established only when reaction (a) takes place. In this case, the intermediate compound RH^+ is often called an Arrhenius complex. The rate of the process is

$$W = \frac{k_1 k_2}{k_{-1}} \frac{c_R c_{HA}}{c_{A^-}} \tag{1.18}$$

On incorporating into Eq. (1.18) the dissociation constant of the acid

$$K_{diss} = \frac{c_{A^-} c_{H_S^+}}{c_{HA}} \tag{1.19}$$

we obtain a final expression

$$W = \frac{k_1 k_2}{k_{-1} K_{diss}} c_R c_{H_S^+} \tag{1.20}$$

This mechanism leads to specific acid catalysis, in spite of the fact that the protonation of the reagent is effected by means of the transfer of a proton from any acid substance there may be in the solution.

2. With $k_2 \gg k_{-1}c_A^-$ the concentration of RH^+ may not be determined only when reaction (a) takes place; nevertheless, during the course of the reaction this concentration is maintained practically constant, which makes it possible to use the principle of steady-state concentrations. In this case, the intermediate compound is called a van't Hoff complex. The expression for the rate of the process takes the form

$$W = k_1 c_R c_{HA} \tag{1.21}$$

The process takes place according to the general mechanism of acid catalysis since the catalysis may be effected by differing acidic substances. According to the general mechanism of acid catalysis, there takes place reactions which are catalyzed by aprotic acids. In aggressive media there are various substances possessing acid or base characteristics and, therefore, general acid and base catalysis is a very common phenomenon. Nevertheless, it is not always easy to observe the catalytic action of substances other than hydrogen or hydroxide ions. The most reliable method consists of using as catalysts a series of buffer solutions with identical ratios of components but with differing concentrations: the solutions have to have identical ionic strengths so as to exclude salt effects. If in such solutions the reaction rate increases with the buffer concentration, then this is evidence that one or both components of the buffer are exhibiting a catalytic effect.

For many reactions there has been established a relationship known in the literature as the Brønsted relationship [27, p. 278]:

$$\log k_{eff_i} = \alpha \log K_{diss_i} + \beta \qquad (1.22)$$

where α and β are constants of the reaction.

When α is close to zero the main contribution to the catalysis is that of the solvent; with α close to unity the catalysis is specific; with intermediate values of α the catalysis is general.

Thus, specific catalysis and catalysis by the solvent are the two limiting cases of general catalysis.

1.3.2. Monomolecular and Bimolecular Mechanisms

A distinction is drawn between monomolecular and bimolecular reactions depending on the molecularity of the limiting stage. Trimolecular reactions are of low probability, and their occurrence in a condensed phase has not been finally proved.

<u>The Monomolecular Mechanism</u>. Let us consider a reaction taking place as follows:

$$R + cat \underset{}{\overset{K_X}{\rightleftharpoons}} X$$

$$X \rightleftharpoons M^{\neq} \xrightarrow{k_{act}} P + cat$$

where X is an intermediate compound, $M^{\neq}$ an activated complex.

The pseudo first-order reaction rate is expressed as follows:

$$W = k_{eff} = k_{act} \frac{a^{\neq}}{f^{\neq}} = k_{act}\, c_X \frac{f_X}{f^{\neq}} \qquad (1.23)$$

Taking into account the equilibrium constant for the formation of X and that X is an Arrhenius complex, we get

$$c_R = \frac{K_X c_X}{a_{cat}} \frac{f_X}{f_R} \qquad (1.24)$$

Taking into account the equation

$$c_0 = c_R + c_X \qquad (1.25)$$

we find the general equation for the monomolecular process:

$$k_{eff} = k_{act}\frac{f_X}{f^{\neq}}\bigg/ 1 + \frac{K_X}{a_{cat}}\frac{f_X}{f_R} \tag{1.26}$$

If the mechanism of ionization of the substrate is known, we can, using Hammett's postulate,

$$\frac{f_X}{f_R} = \frac{f_{InH^+}}{f_{In}} = \text{const} \tag{1.27}$$

introduce into Eq. (1.26) the corresponding acidity or alkalinity. In addition, usually the complex X and the activated complex are identical in composition, charge, and structure, i.e., $f_X/f^{\neq}$ = const, which considerably simplifies Eq. (1.26):

$$k_{eff} = \frac{k_{act}}{1 + \dfrac{K_X}{h_X}} \tag{1.28}$$

At the present time, independent methods have been used in a number of investigations [27] to determine the values of c_X and to prove the correctness of Eq. (1.26).

The Bimolecular Mechanism. In reactions proceeding by a bimolecular mechanism there are added to the activated complex the reagent, catalyst, and, in the case of hydrolysis, water. Sometimes the activated complex is formed by the interaction of X with a water molecule, sometimes by the interaction of the reagent with a complex of the catalyst and a water molecule.

The rate of the hydrolysis reaction is described by two equations:

$$W = k_{eff}c_0 = k_{act}\frac{a^{\neq}}{f^{\neq}} \quad \overrightarrow{\underset{}{}} \left|\begin{array}{l} \rightleftarrows\ k_{act}\dfrac{a_R a_{(cat\ H_2O)}}{f^{\neq}} \\[2em] \rightleftarrows\ k_{act}\dfrac{a_X a_{H_2O}}{f^{\neq}} \end{array}\right. \tag{1.29}$$

It is not possible to solve these equations in the general form since at the present time there are no reasonably quantitative methods for calculating the change in $f^{\neq}$. There are, however, a number of semiempirical approaches. Let us consider some of these, assuming for simplicity that the concentration of R is considerably higher than that of complex X.

The approach of Bunnett [28] is as follows. To all the particles in the reaction medium there are allocated definite hydration numbers. On writing out the equation

$$R(H_2O)_s + cat(H_2O)_n \overset{K_X}{\rightleftharpoons} X(H_2O)_p + (s+n-p)H_2O$$

$$X(H_2O)_p + r(H_2O) \overset{k_{act}}{\rightleftharpoons} M^{\neq}(H_2O)_t \longrightarrow \qquad (1.30)$$

and taking into account the indicator equilibrium

$$In(H_2O)_b + cat(H_2O)_n \rightleftharpoons In\,cat(H_2O)_c + (b+n-c)H_2O \qquad (1.31)$$

we get

$$k_{eff} = \frac{k_{act}}{K_X} h_X \left(\frac{f_{R(H_2O)_s}}{f_{M^{\neq}(H_2O)_t}} \frac{f_{In\,cat(H_2O)_c}}{f_{In(H_2O)_b}} \right) a_{H_2O}^{r+(b-c)-(s-p)} \qquad (1.32)$$

Using Hammett's postulate and assuming that $f_{In\,cat(H_2O)_c}/f_{M\neq(H_2O)_t}$ does not depend on the composition of the medium since the complex of reagent with catalyst (or of indicator with catalyst) and the activated complex are identical in charge and closely similar in structure, we get

$$k_{eff} = \frac{k_{act}}{K_X} h_X a_{H_2O}^r \qquad (1.33)$$

Using this equation, Bunnett described numerous reactions of the hydrolysis of various compounds [28]. However, this equation is frequently inapplicable for data obtained over a wide range of catalyst concentrations. Strictly speaking, in Eq. (1.32) we should not regard the ratio $f_{In_{cat}(H_2O)_c}/f_{M^+H_2O)_t}$ as constant since both complexes are bound to differ in composition (differing hydration numbers); however, lack of knowledge of this ratio is compensated for by the empirical parameter r, which formally expresses the number of water molecules contained in the activated complex.

The approach of Vinnik [29] involves assuming the formation, in aqueous solutions of strong acids, of an intermediate complex M (R with H_S^+) which decomposes into the reaction products according to a monomolecular process. Assuming that the ratio of the activity coefficients $f_M/f^{\neq}$ is constant, and that the activity coefficients of the activated complex and of complex M are expressed by the equation

$$f_M = f^{\neq} = f_R^{\beta} f_{H_S^+}^{\alpha} + const$$

where α and β are parameters not depending on the medium,

$$k_{eff} = \frac{k_{act}}{K} c_{H_S^+} f_R^{(1-\beta)} f_{H_S^+}^{(1-\alpha)} \qquad (1.34)$$

the value of f_R may be determined from the solubility data and it
is necessary only to find the parameters α and β. A similar approach
is proposed by Kresge [30] for acid—base exchange reactions.

More recently, Vinnik [31] has formulated a general law for
establishing the mechanism of the limiting stage of the reactions
of catalytic hydrolysis in aqueous solutions, namely: the activity
of the activated complex is to be expressed by way of the concentrations of the various forms of the reagents in such a way that the
multiplier obtained in the kinetic equation, with the activity coefficient, does not change with a change in the acid concentration. On
the basis of much experimental data Vinnik divided all hydrolysis
reactions by aqueous solutions of strong acids into four groups.
In groups I and II the activated complex is formed from an un-ionized
form of the reagent, a hydrated proton, and a nucleophile, namely
a water molecule (group I) or an anion (group II). In reactions
in the other groups an activated complex is formed from a protonated
form of the reagent with or without the participation of a water
molecule (group III) or with that of an anion (group IV).

Bimolecular processes include processes whose limiting stage
is the interaction of the reagent with the catalyst

$$R + cat \rightleftharpoons M^{\neq} \xrightarrow{k_{act}} X$$
$$X \longrightarrow P + cat$$

The pseudo first-order constant k_{eff} in this case equals

$$k_{eff} = k_{act} \frac{f_R a_{cat}}{f^{\neq}}$$

i.e., we get an equation which formally is no different from Eq.
(1.26), where $c_R \gg c_X$.

Thus, whereas for monomolecular processes there is a simple
general equation derived within a framework of reasonable assumptions, there is no such equation for bimolecular processes and for
the selection of the mechanism we need additional information in
each specific case.

Mono- and bimolecular processes are characterized by the
features shown in Table 1.5.

Table 1.6 gives a general classification of simple mechanisms
for the degradation of polymers in aggressive media. In the literature it is usual to designate monomolecular reactions involving
acid or base catalysis as A-1 or B-1, respectively; bimolecular reactions as A-2 or B-2.

TABLE 1.5. Characteristics of Mono- and Bimolecular Processes with
the Same Relative Concentration of the Reactive Form

Characteristic	Process	
	Monomolecular	Bimolecular
Change in k_{eff} with increase in concentration of a solvent which participates in the reaction	Changes little or falls slightly	Rises
Kinetic isotope effect (k_{D_2O}/k_{H_2O})	>1 in aqueous solutions of acids 2.5-3.0, inclusive	Varies only slightly from 1
Entropy of activation $\Delta S^{\neq}$	Positive	Negative
Volume of activation $\Delta V^{\neq}$	Most frequently positive	Negative
Change in k_{eff} with increase in volume of substituents	Does not change	Falls
Slope of the relationship between log k_{eff} and HX when $c_{ion} \gg c_B$	Most frequently 1	<1

1.4. THE MECHANISM OF CATALYSIS BY SALTS

The degradation of polymers frequently occurs in aggressive media
containing salts. Salts can considerably change the rate of degradation, either by being catalysts themselves or else by altering
the activity of acids and bases present in the solution.

1.4.1. Bifunctional Catalysis by Anions of Polybasic Acids

There are published descriptions of examples of catalysis by
the anions of polybasic acids of compounds containing amide or ester
groups which have a neutral pH. Table 1.7 gives some examples.

This type of catalysis is shown experimentally by a directly
proportional rise in k_{eff} with increase in the concentration of the
components of a buffer mixture with constant pH [34].

At the present time it is not clear what the nature of the motive forces in catalysis of this type is. The best known is the
mechanism of bifunctional or coordinate catalysis [35, p. 175 of
Russian translation], which includes the simultaneous attachment

TABLE 1.6. General Classification of Degradation Reactions in Acidic and Basic Media

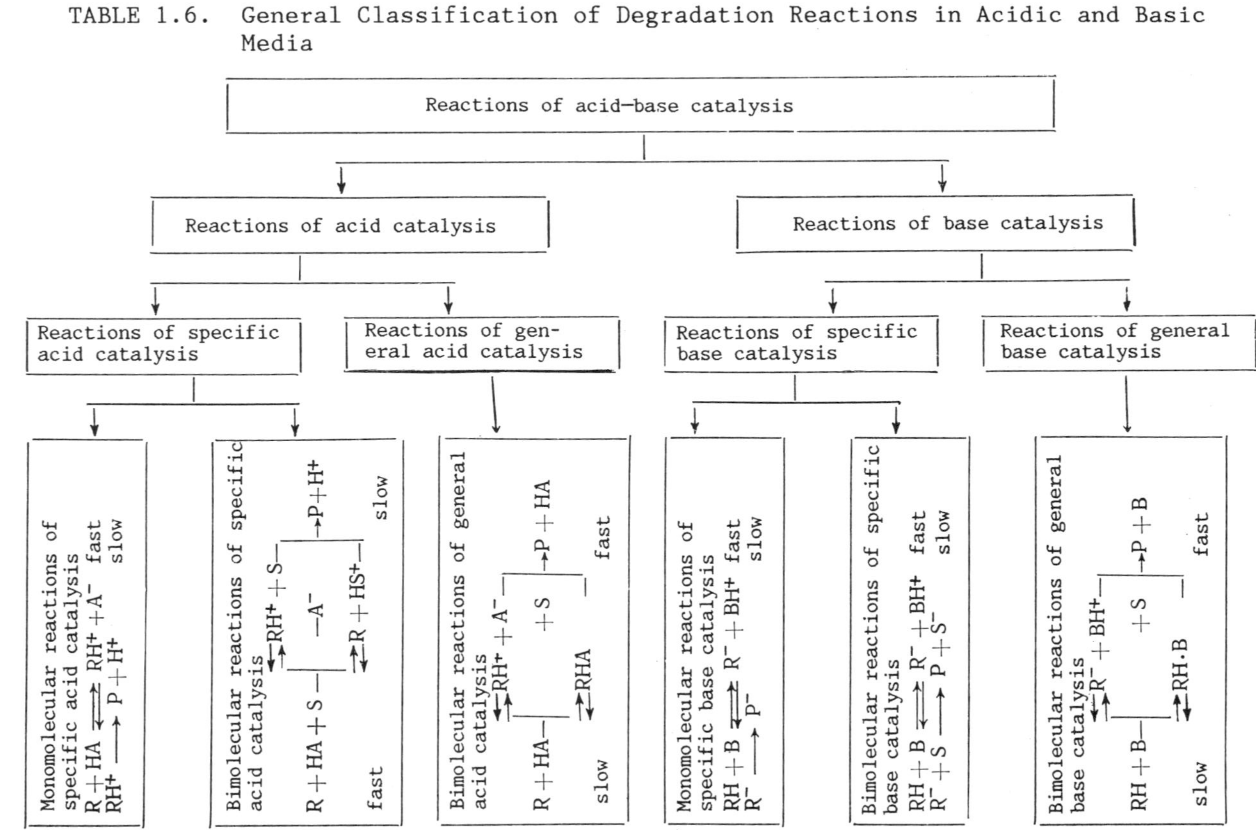

TABLE 1.7.   Catalysis by Anions of Polybasic Acids with Neutral pH

Catalyst	Substrate	References
Citrate ion	Chloramphenicol	[32]
Phosphate ion		
Bicarbonate ion	N-Phenylimino-lactone	[33]
Phosphate ion	Amido acids	[34]

and release of a proton by a bifunctional catalyst containing both acidic and basic groups. For instance, in phosphate, carbonate, or sulfate monoanions the oxygen atom is a base and the hydroxyl group an acid.

The general scheme for the catalytic decomposition of carbonyl-containing substrates may be represented as follows:

$$\sim Y - \overset{\overset{O}{\|}}{C} \sim \; \underset{\longleftarrow}{\overset{+H_2O}{\longrightarrow}} \; \sim Y - \overset{\overset{OH}{|}}{\underset{\underset{OH}{|}}{C}} \sim \; + \; \underset{HO}{\overset{^-O}{\diagdown}} X \diagup \; \longrightarrow$$

$$\longrightarrow \; \sim C \overset{\diagup O}{\diagdown OH} \; + \; HY \sim \; + \; \underset{^-O}{\overset{HO}{\diagdown}} X \diagup$$

1.4.2.   Influence of Salts on the Activity of Acidic and Basic Catalysts

The addition of salts to solutions containing acidic and basic catalysts may alter their catalytic activity.

If the functional group R of the substrate has a positive or negative charge, then the addition of neutral salts will alter the activity coefficients of the ions (a primary kinetic effect of the salt). This may be expressed quantitatively working from the Debye–Hückel equation, for instance for low ionic strengths (not above 0.01) by the limiting law

$$-\log f_i = z_i^2 A \sqrt{\mu} \qquad (1.35)$$

By using (1.35) it is possible to obtain an equation which describes the change in the rate constant of the reaction between ions X and Y

$$\log k = \log k_0 + 2Az_X z_y \sqrt{\mu} \tag{1.36}$$

where $\log k_0$ is the logarithm of the rate constant of the reaction in pure water.

From Eq. (1.36) it may be seen that in dilute salt solutions where the rate constant increases with increase in the ionic strength of the solution for all reactions of ions of one sign and decreases in reactions between ions of differing sign, the slope of the dependence of $\log k$ on $\sqrt{\mu}$, which is designated as φ, must be an integer.

In recent years, Pal'm et al. [36] have proposed a different interpretation of the primary effect of a salt. Studying the acidic and alkaline hydrolysis of esters of carboxylic, sulpho- and thiophosphoric acids with charged substituents, they observed three effects:

the values of φ are frequently not integers;

the value of φ depends to a considerable extent on the temperature (according to the theory of the primary effect of a salt there should be only a slight temperature dependence of φ);

the value of k reaches a particular limiting value k_∞, which is practically independent of the electrolyte concentration.

The experimental data, as Pal'm [36] postulates, may be explained by assuming that the salt effect is linked with the equilibrium formation of ion pairs. Then the difference between the rate constants k_0 and k_∞ at high electrolyte concentrations suggests that the former relates to a reaction between free ions and the latter to a reaction with the participation of ion pairs. In the general form this may be represented as follows:

the reaction between ions

$$X^{\pm}\!-\!G\!-\!Y + cat^{\pm} \longrightarrow [X^{\pm}\!-\!G\!-\!Z^{\pm}] \xrightarrow{k_0} P$$

the reaction with an ion pair

$$A^{\mp}X^{\pm}\!-\!G\!-\!Y + cat^{\pm} \longrightarrow [A^{\mp}X^{\pm}\!-\!G\!-\!Z^{\pm}] \xrightarrow{k_\infty} P$$

or

$$X^{\pm}-G-Y + cat^{\pm}A^{\mp} \longrightarrow [X^{\pm}-G-Z^{\pm}A^{\mp}] \xrightarrow{k_{\infty}} P$$

where Y and $Z^{\pm}$ are the reaction sites in, respectively, the initial and activated states; $X^{\pm}$ a positively or negatively charged substituent; G an isolating bivalent structural unit; and $A^{\neq}$ a counterion.

The presence of counterions in the activated complex compensates for a definite part of the electrostatic component in the free activation energy. The free activation energies which correspond to the values of k_0 and k_{∞} should differ by the aforementioned electrostatic component, which, in the case of practically full compensation, should differ by the amount

$$E_{el} = \frac{331 z_X z_Z e^2}{\varepsilon l_{XZ}} \tag{1.37}$$

where l_{XZ} is the distance between the ion charges of the substituent X and the reaction site Z, in Å.

The magnitude of the maximum salt effect is

$$\Delta \log k_{\infty} = -\frac{E_{el}}{2.3RT} \tag{1.38}$$

<u>The Influence of Salts on the Conversion of Functional Groups in the Presence of Weak Electrolytes</u>. The addition of salts to solutions containing weak acids or bases leads, as a rule, to a considerable rise in the rate of catalytic conversions. This, known in the literature as the secondary salt effect, is not linked with the kinetics of the process, but is brought about by an increase in the degree of dissociation of weak electrolytes in the presence of salts. In the general form the influence of the ionic strength on the dissociation constant of a weak electrolyte is determined by the expression

$$\log K_{diss} = \log K^0_{diss} - (z-1)\mu^{1/2}\left(1 + \mu^{1/2}\right) \tag{1.39}$$

where z is the charge of the dissociating ion.

It must be noted that if z = ±1 (e.g., in a buffer solution of ammonia with ammonium chloride), the secondary salt effect is, to a first approximation, zero.

<u>Influence of Salts on the Dissociation of Functional Groups with Alteration in the Solvation of the Reagents</u>. With a concentration of >0.1 M, the addition of a salt alters the acidity function H_X [30]. Salts bring about desolvation of the ions and form ion pairs. For the same reason there is in the presence of salts a

change in the reactivity of the water, which is bound to affect the rate of the reactions taking place according to mechanisms A-2 and B-2. The salts can directly combine with the reagents, changing their activity. For instance,

$$CH_3C \begin{matrix} \nearrow O \\ \searrow OH \end{matrix} + K^+A^- \rightleftharpoons CH_3C \begin{matrix} \nearrow O \cdots K^+A^- \\ \searrow OH \end{matrix}$$

The resulting compound is active as an aprotic acid.

Influence of Salts on the Stabilization of an Activated Complex. Vinnik [31] presented numerous examples of the accelerating action of a number of salts in hydrolysis reactions. In his opinion, anions of salts may enter an activated complex, exerting a nucleophilic effect in hydrolysis reactions. The influence of ions of differing charge density on the stabilization of an activated complex is described in a review [30].

REFERENCES

1. J. H. Brønsted, Z. Phys. Chem., A, <u>169</u>, 52 (1934).
2. N. A. Izmailov, Electrochemistry of Solutions, Khar'kov State University, Khar'kov (1959).
3. A. Albert and E. P. Serjeant, Ionization Constants of Acids and Bases, Chapman & Hall, London (1971).
4. L. Pauling, General Chemistry, W. H. Freeman (1970).
5. V. A. Pal'm, Introduction to Theoretical Organic Chemistry, Vyssh. Shkola, Moscow (1974).
6. T. Ackermann, Z. Phys. Chem. (Frankfurt), <u>27</u>, 253 (1961).
7. G. Zundell, Hydration and Intermolecular Interaction, Academic Press, New York (1969).
8. J. B. Bates and L. M. Toth, J. Chem. Phys., <u>61</u>, 129 (1974).
9. N. B. Librovich, V. D. Maiorov, and V. A. Savel'ev, Dokl. Akad. Nauk SSSR, <u>225</u>, 1358 (1975).
10. A. F. Beecham and A. C. Hurley, J. Chem. Phys., <u>49</u>, 3312 (1968).
11. S. C. Lee and R. Kaplow, Science, <u>169</u>, 477 (1970).
12. J. A. Pople et al., High-Resolution Nuclear Magnetic Resonance, McGraw-Hill, New York (1959).
13. R. A. Robinson and R. H. Stokes, Electrolyte Solutions, Butterworth (1959).
14. D. P. N. Satchell and R. S. Satchell, Quart. Rev., <u>25</u>, 171 (1971).
15. Yu. F. Rybkin and N. F. Shevchenko, Elektrokhimiya, <u>1</u>, No. 1, 45 (1965).
16. K. Yates et al., J. Am. Chem. Soc., <u>95</u>, No. 2, 418 (1973).
17. L. P. Hammett, Physical Organic Chemistry, McGraw-Hill, New York (1970).

18. J. C. D. Brand, J. Chem. Soc., No. 3, 997 (1950).
19. K. Yates, J. B. Stevens, and A. R. Katritzky, Can. J. Chem., 42, No. 8, 1957 (1964).
20. C. D. Johnson, A. R. Katritzky, and N. Shakir, J. Chem. Soc., B, No. 11, 1235 (1967).
21. G. Kortum, W. Vogel, and K. Andrussow, Dissociation Constants of Organic Acids in Aqueous Solution, Plenum, New York (1961).
22. J. A. Flaschka, A. J. Barhard, and P. E. Sturock, Quantitative Analytical Chemistry, Barnes & Noble, New York (1969).
23. D. D. Perrin, Dissociation Constants of Organic Bases, Plenum, New York (1965).
24. N. B. Librovich and M. I. Vinnik, Zh. Fiz. Khim., 50, No. 1, 150 (1976).
25. A. Gordon and R. Ford, Chemist's Companion: A Handbook of Practical Data, Techniques, and References, Wiley (1972).
26. C. K. Ingold, Structure and Mechanism in Organic Chemistry, Cornell University Press (1969).
27. N. M. Émanuél' and D. G. Knorre, Course in Chemical Kinetics, Vyssh. Shkola, Moscow (1969).
28. J. E. Bunnett, J. Am. Chem. Soc., 82, No. 2, 499 (1960); 83, No. 24, 4978 (1961).
29. M. I. Vinnik et al., Zh. Fiz. Khim., 41, No. 1, 252 (1967).
30. K. Yates and R. A. McClelland, in: Progress in Physical Organic Chemistry, Vol. II, Interscience, New York (1974), p. 323.
31. M. I. Vinnik, Izv. Akad. Nauk SSSR, Ser. Khim., No. 5, 998 (1973).
32. I. Higuchi, A. D. Marcus, and C. D. Bias, J. Am. Pharm. Assoc., 43, 129 (1954).
33. B. A. Cunninghaur and G. L. Schmir, J. Am. Chem. Soc., 88, No. 2, 551 (1966).
34. M. F. Aldersley et al., J. Chem. Soc., B, 12, 1487 (1974).
35. W. P. Jencks, Catalysis in Chemistry and Enzymology, McGraw-Hill (1969).
36. V. A. Pal'm et al., Reakts. Sposobn. Org. Soedin., 10, No. 1, 223 (1973).

Chapter 2

MECHANISMS OF DISSOCIATION
IN AGGRESSIVE MEDIA
AND THE REACTIVITY
OF CHEMICALLY UNSTABLE BONDS
OF VARIOUS POLYMERS

In investigating the degradation of polymers in aggressive media, it is very important to know the mechanism of dissociation of chemically unstable bonds and the values of the parameters in the kinetic equations.

We shall now consider the dissociation, in such media, of model compounds with various chemically unstable bonds.

2.1. COMPOUNDS CONTAINING AN AMIDE BOND

Compounds containing an amide bond decompose under the action of acids and bases. Let us consider the mechanism of decomposition of amides in these media.

2.1.1. Decomposition of Amides
by Acid Catalysis

The mechanism of the decomposition of amides by acid catalysis has been the subject of numerous investigations. With the majority of amides the amide bond decomposes by acid hydrolysis or solvolysis, but some amides may decompose at an alkyl—nitrogen bond [1]. However, these instances are rare and are not considered in particular.

Protonation of Amides. The electronic structure of an amide bond predetermines two possible sites for protonation: on the carbonyl oxygen and on the amide nitrogen (see p. 30). The presence of the O-protonated form is usually shown by the NMR spectra of the amides in strongly acidic media (100% H_2SO_4 [2] or FSO_3H [3]) and by some less direct evidence, which is assessed in a review by O'Connor [4].

29

$$\sim\overset{\text{O}}{\overset{\|}{\text{C}}}-\ddot{\text{N}}\text{H}\sim \;-\;\left[\begin{array}{c}\rightleftharpoons \\ +\,\text{H}^+ \\ \rightleftharpoons\end{array}\right.\quad\begin{array}{c}\sim\overset{\text{O}}{\overset{\|}{\text{C}}}-\overset{+}{\text{N}}\text{H}_2\sim \\ \overset{+}{\text{O}}\text{H} \\ \sim\text{C}\!\cdots\!\text{NH}\sim\end{array}$$

The concentration of the O-protonated form of the amides may be
determined by NMR spectroscopy [5], IR spectroscopy [6], or UV spectroscopy
[7, 8]. The data on the kinetics of protonation of N,N-dimethylform-
amide [9] and on catalytic H—D exchange in amides indicate the presence of the N-protonated form, which is mainly formed in moderately concentrated acid solutions. Kresge et al. [10], investigating the influence of a strained ring on the basicity of the amide
bond

$$(\overset{\text{\textbar}}{\text{C}}\text{H}_2)_n \quad \overset{\text{\textbar}}{\text{N}}\text{—CO—R}$$

showed that the size of the ring has no influence on the basicity
constant of the amides and that, consequently, the O-protonated form
is also the main form in moderately concentrated acid solutions.
It has not been possible to determine the N-protonated form of amides
quantitatively by present methods, but this does not exclude the
possibility of its formation in small amounts. According to Bruylants et al. [11], the O- and N-protonated forms are in equilibrium,
the concentration of the N-protonated form increasing with the addition of electronegative substituents to the nitrogen atom and also
being dependent on the nature of the solvent. In [7, 12] it is
shown that the protonation of amides at the CO group is not expressed by the acidity function H_0, and accordingly a new function H_A has
been proposed [8, 13], which is determined by means of amide indicators. Thus, the concentration of the O-protonated form of the amide
can be calculated from the formula

$$c_{\text{BH}_0^+} = c_\text{B}\,\frac{h_\text{A}}{K_{\text{BH}_0^+}} \tag{2.1}$$

where $K_{\text{BH}_0^+}$ is the basicity constant of the amide when protonated
on the oxygen.

The protonation of a number of amides is not expressed by either
the acidity function H_0 or by H_A; this may be linked with the possibility of the formation of ion pairs in aqueous solutions of strong
acids [14].

If it is assumed that protonation of the nitrogen, by analogy
with secondary amines, is expressed by H_0, then the concentration
of the N-protonated form is

$$c_{BH_N^+} = \left(c_0 - c_{BH_0^+}\right) \frac{h_0}{K_{BH_N^+}} \qquad (2.2)$$

where $K_{BH_N^+}$ is the basicity constant of the amide in protonation on the nitrogen.

<u>Molecularity of the Limiting Stage</u>. It is now thought that for the majority of amides the limiting stage of the reaction is bimolecular. This is generally confirmed by:

the negative values of the entropy of activation [15];

the negative values of the volume of activation [16];

the influence of the steric and polar effects of the substituents, which is characteristic of bimolecular processes [17];

the proportionality between k_{eff} and the concentration of hydrated protons in dilute acid solutions [18, 19];

the reduction in k_{eff} with the replacement of water by non-aqueous solvents [20-23].

Within the framework of a bimolecular mechanism the possible courses of hydrolysis of amides may be represented as follows:

$$(I)$$

$$+H^+ \parallel K_{BH_0^+}$$

$$(II)$$

$$+H^+ \parallel K_{BH_N^+}$$

$$(III)$$

When reaction I takes place, the BH_0^+ form is in equilibrium with the tetrahedral intermediate compounds which have high reactivities [24]. The proof of the existence of these compounds is usually oxygen exchange with the water in the solution. In the acid hydrolysis of amides such exchange does not occur, inasmuch as [24] $K\sim NH^+ \gg K\sim OH$ and the tetrahedral intermediate compound decomposes only one way, with the formation of a protonated acid molecule and an amine. In the case of thioamides the compound may decompose by two routes ($K\sim NH_2^+ \sim K\sim SH$) with the formation of a thio acid and an oxygen-containing amide [25]

With accuracy up to the equilibrium constants of formation of the tetrahedral intermediate compound, the equation for k_{eff} will have the form

$$k_{eff} = \frac{k_{act\,1} a_{H_2O}}{1 + K_{BH_0^+}/h_A} \; \frac{f_{BH_0^+}}{f^{\neq}} \tag{2.3}$$

With practically complete protonation of the reagent ($1 \gg K_{BH_0^+}/h_A$)

$$k_{eff} = k_{act\,1} a_{H_2O} \; \frac{f_{BH_0^+}}{f^{\neq}}$$

For route II the unprotonated form

$$k'_{eff} = \frac{k_{act\,2} c_{H_S^+}}{1 + h_A/K_{BH_0^+}} \; \frac{f_B f_{H_S^+}}{f^{\neq}} \tag{2.4}$$

is the reactive species.

The value of $k_{act\,2}$ may either include the equilibrium constant of formation of a complex of the reagent with a hydrated proton, or not include it, if the reaction takes place with collision. According to Eq. (2.4), there is with protonation of the amide group on the oxygen a reduction in k_{eff}, and when $1 \ll k_A/K_{BH^+}$,

$$k_{eff} = k_{act\,2} K_{BH_0^+} \; \frac{c_{H_S^+}}{h_A} \; \frac{f_B f_{H_S^+}}{f^{\neq}} \tag{2.5}$$

In reaction III the reactive form is BH_N^+. The high reactivity of the BH_N^+ ions has been confirmed in numerous model compounds [26-28]. If it is assumed that the protonation of the amide group on the nitrogen is expressed by H_0, then

$$k_{eff} = \frac{k_{act\,3} \dfrac{h_0}{K_{BH_N^+}} a_{H_2O}}{1 + h_A/K_{BH_N^+} + h_0/K_{BH_N^+}} \; \frac{f_{BH_N^+}}{f^{\neq}} \tag{2.6}$$

Taking into account that the concentration of the N-protonated form is low, Eq. (2.6) may be simplified as follows:

$$k_{\text{eff}} = \frac{k_{\text{act3}}\,\dfrac{h_0}{K_{\text{BH}_N^+}}\,a_{\text{H}_2\text{O}}}{1 + h_A/K_{\text{BH}_0^+}}\;\frac{f_{\text{BH}_N^+}}{f^{\neq}} \qquad\qquad (2.7)$$

In concentrated acid solutions, in which the amides are practically in the 0-protonated form ($h_A/K_{\text{BH}_0^+} \gg 1$), Eq. (2.7) takes the form

$$k_{\text{eff}} = k_{\text{act3}}\,\frac{K_{\text{BH}_0^+}}{K_{\text{BH}_N^+}}\,\frac{h_0 a_{\text{H}_2\text{O}}}{h_A}\,\frac{f_{\text{BH}_N^+}}{f^{\neq}} \qquad\qquad (2.8)$$

The ratio $K_{\text{BH}_0^+}/K_{\text{BH}_N^+}$ is given in [29-31] for various amides: it lies in the range 10^3-10^7.

In [32] an equation is proposed for the possibility of the occurrence of the hydrolysis reaction of the amides by routes I and III simultaneously:

$$k_{\text{eff}} = k_{\text{act1}}\alpha a_{\text{H}_2\text{O}} + k_{\text{act3}}(1 - \alpha)\,c_{\text{H}_S^+}a_{\text{H}_2\text{O}} \qquad\qquad (2.9)$$

where α is the proportion of the 0-protonated form of the amide.

Nevertheless, in deriving this equation it is not indicated by what mechanism the N-protonated form is formed, and it is assumed that the ratio of the activity coefficients does not depend on the medium.

There are instances of the use of the Bunnett approach for describing experimental data on the hydrolysis of amides. For instance, Yates [13] suggests that the reaction is by route I, while O'Connor [33] assumes that the N-protonated form is reactive (route III).

Let us now consider the use of the above equations to describe the experimental data on the acid hydrolysis of amides. The majority of amides are characterized by k_{eff} passing through a maximum relative to the concentration of acids.

Figure 2.1 shows the changes in k_{eff} for the hydrolysis of polycaproamide (PCA) and of its cyclic trimer in relation to the H_A of aqueous solutions of sulfuric acid [34]. The limited solubility of PCA did not allow us to carry out the investigation at H_2SO_4 concentrations lower than 40%.

Let us assume that the ratio of the activity coefficients in all the equations does not change greatly with the acid concentration. It is easy to see that Eq. (2.3) does not describe the sharp reduction in k_{eff} of the hydrolysis of the cyclic trimer of PCA in concentrated acid solutions ($a_{\text{H}_2\text{O}}$ decreases 5.5-fold, but k_{eff}

decreases 100-fold in the range from 20 to 60% H_2SO_4). This may be either because Eq. (2.3) is inapplicable or because, with a rise in the acid concentrations, there is a change in the ratio $f_{BH^+}/f^{\neq}$.

Equation (2.8) does not predict the reduction in k_{eff} for the hydrolysis of the amides since the increase in h_A does not outstrip the increase in the product $h_0 a_{H_2O}$ with the rise in acid concentration. But, in this case, again there is an absence of data denying the possibility of reduction in the ratio of the activity coefficients with rise in the acid concentration. Equation (2.4) satisfactorily describes the experimental data without taking into account the ratio of the activity coefficients over the whole range of acid concentrations. To determine k_{act} and $K_{BH_0^+}$ it is convenient to represent Eq. (2.4) in the form

$$\frac{c_{H_S^+}}{k_{eff}} = \frac{1}{k_{act}\,\mathrm{const}} + \frac{h_A}{k_{act}K_{BH_0^+}\,\mathrm{const}} \qquad (2.10)$$

The value of k_{act} is found from this equation with accuracy up to a constant magnitude (const $= f_B \cdot f_{H_S^+}/f^{\neq}$); the value of $K_{BH_0^+}$ is a true one.

Thus in the acid hydrolysis of amides, route II is more favored than routes I and III. This assumption is supported by two factors: (a) by means of Eq. (2.4) it is possible to describe the experimental results in the acid hydrolysis of the majority of amides; the values of $K_{BH_0^+}$, as determined from the kinetic data and independently from the spectral data, practically coincide [35]; (b) for amides whose ionization takes place in highly concentrated acid solutions, for instance in the case of nitroacetanilides [36, 37], the increase in k_{eff} over a wide range of concentrations is described by the equation

$$k_{eff} = k_{act} \cdot c_{H_S^+}\,\mathrm{const} \qquad (2.11)$$

which is a special case of Eq. (2.4) under conditions of low ionization of amides $(1 \gg h_A/K_{BH_0^+})$.

Figure 2.2 shows the kinetic data on the hydrolysis of diglycine in the concentration range 0.5-65.5% H_2SO_4 on the coordinates of (2.11) [38].

For some amides decomposition in acids may take place by a monomolecular mechanism, for instance that of anilides in concentrated solutions of sulfuric acid (>70 mass %) [39] or the hydrolysis of dinitro-2-pyridylalanylglycine in moderately concentrated solutions of sulfuric acid [40].

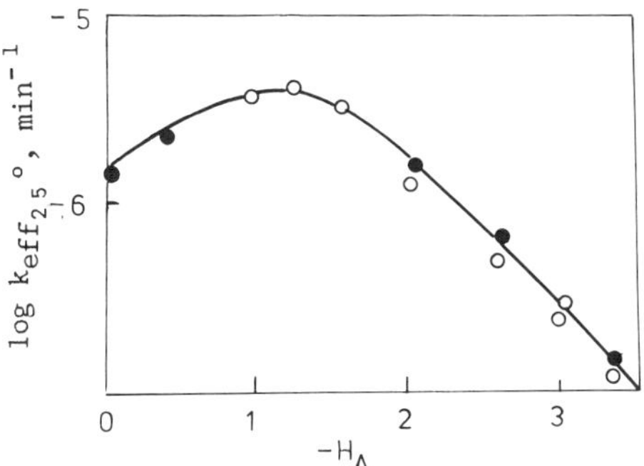

Fig. 2.1. Dependence of log $k_{eff\,25^\circ}$ on the amide
acidity function of aqueous solutions
of sulfuric acid for the hydrolysis of
a cyclic trimer of polycaproamide ($\bullet$)
and of polycaproamide of M_V 20,000 (o).

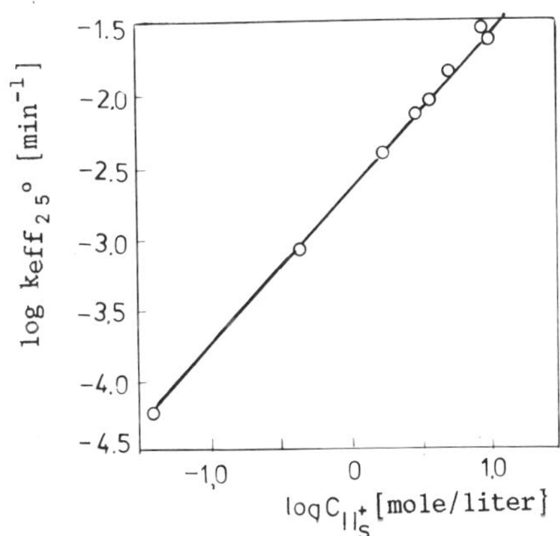

Fig. 2.2. Relationship between log $k_{eff\,25^\circ}$ and
log C_{HS}^+ for the hydrolysis of di-
glycine in aqueous solutions of sulfuric acid [38].

The hydrolysis of the latter compound, pparently takes place
as follows:

$$\sim\!\underset{\underset{O}{\overset{|}{\|}}}{\overset{\overset{H}{|}}{N}}\!\!-\!C\sim \;\underset{}{\overset{+H^+}{\rightleftarrows}}\; \sim\!\underset{\underset{H\ \ O}{|\ \ \|}}{\overset{\overset{H}{|}}{N^+}}\!\!-\!C\sim \;\rightleftarrows\; M^{\neq} \;\xrightarrow{k\ act}\; \sim\!NH_2 + \overset{+}{\underset{\underset{O}{\|}}{C}}\!\sim \;\xrightarrow[-H^+]{+H_2O}\; \dashrightarrow\; \sim\!NH_2 + HOOC\!\sim$$

and the changes in k_{eff} are described by the equation (Fig. 2.3)

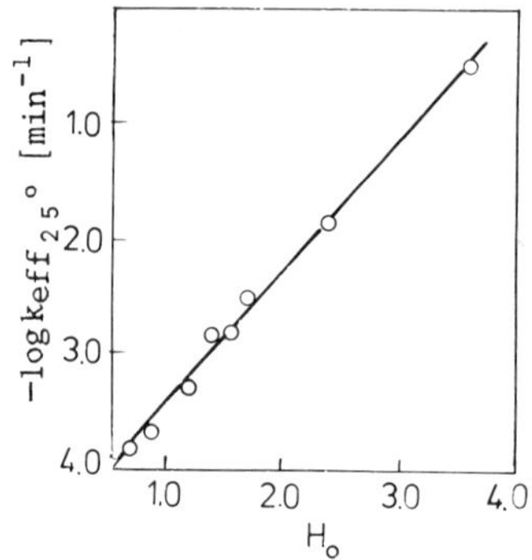

Fig. 2.3. Relationship between log $k_{eff_{25°}}$ and
H_0 for the hydrolysis of dinitro-2-
pyridylalanylglycine in aqueous solutions of sulfuric acid [40].

$$k_{eff} = \frac{k_{act}}{K_{BH_N^+}} h_0 \qquad\qquad (2.12)$$

Thus, the decomposition of the majority of amides in acidic media
is described by mechanism A-2 and is satisfactorily described by
Eq. (2.4). Attention must be drawn to the following, practically
important, factor: For the majority of compounds containing an
amide bond k_{eff} passes through a maximum in relation to the acid
concentration, and the hydrolytic stability of these compounds rises
in concentrated acid solutions.

2.1.2. Catalytic Decomposition of Amides
by Bases

As in acid decomposition, so also in basic media the majority
of amides decompose at the amide bond, although cases are known
where amides decompose at an alkyl–nitrogen bond, e.g., when the
alkyl substituent on the nitrogen includes an azo group [41].

Ionization of Amides. Amides in basic media may ionize by two
routes: attaching a hydroxyl group to a carbonyl group and detach-
ing a hydrogen atom from the nitrogen of the amino group (the latter route is not possible for tertiary amides):

The possibility of N-ionization is proved by the formation of salts of amides with alkali metals [42] and by H—D exchange in basic media [43]. The presence of BOH^- and $BOO^=$ tetrahedral ions is proven by the oxygen exchange between the amides and the medium [44]. The N-ionized form has not been observed in amides in alkali solutions by physical methods and, apparently, its concentration is low. The concentration of BOH^- and $BOO^=$ ions can be determined by spectral methods in the IR and UV regions [45]. Figure 2.4 shows the changes in the concentrations of the ionized and un-ionized forms of N-methylacetamide relative to the concentration of alkali.

The concentration of the ionized forms of the amides is found from the equations

$$c_{BOH^-} = \frac{c_0}{1 + \dfrac{b_0}{K_{BOO^=}\, a_{H_2O}} + \dfrac{K_{BOH^-}}{b_0}} \qquad (2.13)$$

$$c_{BOO^=} = \frac{c_0}{1 + \dfrac{K_{BOO^=}\, a_{H_2O}}{b_0} + \dfrac{K_{BOO^=} K_{BOH^-}\, a_{H_2O}}{b_0^2}} \qquad (2.14)$$

where b_0 is the alkalinity, which expresses the ionization of carbonyl-containing compounds, and K_{BOH^-} and $K_{BOO^=}$ the equilibrium constants of attachment of a hydroxyl group to the carbon of the carbonyl group.

<u>Molecularity of the Limiting Stage</u>. To establish the mechanism of decomposition of amides in basic media it is necessary to know:

which of the forms of the amide are reactive?

is a water molecule involved in the limiting stage of the process?

The N-ionized form is usually taken to be unreactive, inasmuch as it is difficult to imagine how an activated complex is formed from it [46].

Thus the decomposition of amides, in the general form, may be represented as shown by (III) and (II) on p. 38.

Analysis of the experimental data shows:

for amides which are practically in the un-ionized form in the reaction medium, keff changes proportionally with the product $b_0 a_{H_2O}$ (caprolactam, valerolactam, enantholactam [45]);

$$\sim\overset{-}{N}-\underset{\underset{O}{\|}}{C}\sim\ +\ H_2O$$

$$\downarrow\uparrow\ +OH^-$$

$$\underset{\underset{O}{\|}}{\sim\overset{\overset{H}{|}}{N}-C}\sim\ \underset{-}{\overset{+OH^-}{\rightleftharpoons}}\ \sim\overset{\overset{H}{|}}{N}-\underset{\underset{O^-}{|}}{\overset{\overset{OH}{|}}{C}}\sim\ -\ \begin{cases} \xrightarrow{\quad} M_3^{\neq} \xrightarrow{k_{act3}} & (III) \\[2em] \underset{-H_2O}{\overset{+OH^-}{\rightleftharpoons}}\ \sim\overset{\overset{H}{|}}{N}-\underset{\underset{O^-}{|}}{\overset{\overset{O^-}{|}}{C}}\sim\ \overset{+H_2O}{\rightleftharpoons}\ M_2^{\neq} \xrightarrow{k_{act\ BOO^=}} & (II) \end{cases}$$

$$\downarrow\uparrow +OH^- \qquad\qquad \downarrow\uparrow +H_2O$$
$$M_4^{\neq} \qquad\qquad\qquad M_1^{\neq}$$
$$\downarrow\ (IV) \qquad\qquad\quad \downarrow\ (I)$$
$$k_{act4} \qquad\qquad\quad k_{act\ BOH^-}$$

for amides which are in the ionized form, k_{eff} changes proportionally with the activity of the water (trifluoroacetamide [47]);

for amides whose ionization takes place to a considerable degree, k_{eff} changes proportionally with the activity of the water and the concentration of the ionized forms BOH^- and $BOO^=$ (butyrolactam, N-methylacetamide [45]).

We thus conclude that, for the decomposition of amides, routes I and II are the most probable, and in the following equation for k_{eff} the ratios of the activity coefficients $f_{BOH^-}/f_1^{\neq}$ and $f_{BOO^=}/f_2^{\neq}$ remain practically unchanged with the concentration of the base:

$$k_{eff} = \frac{k_{act\ BOH^-}K_{BOO^=}\dfrac{a_{H_2O}^2}{b_0}\dfrac{f_{BOH^-}}{f_1^{\neq}} + k_{act\ BOO^=}\,a_{H_2O}\dfrac{f_{BOO^=}}{f_2^{\neq}}}{1 + K_{BOO^=}\dfrac{a_{H_2O}}{b_0} + K_{BOH^-}K_{BOO^=}\dfrac{a_{H_2O}}{b_0}} \qquad (2.15)$$

For routes I and II, the exact mechanism of participation of water in the limiting stage is not known. Here two cases are possible:

the limiting stage is the reaction of the ionized form of the reagent with a water molecule;

with the ionized forms the water forms complexes whose decomposition or isomerization is itself a limiting stage.

In the first case, to determine the true bimolecular constant it is necessary to divide the pseudomolecular constant by the concentration of free water. In the second case, to find the true rate constant it is necessary to know the concentration of the complexes of the ionized form with a molecule of water.

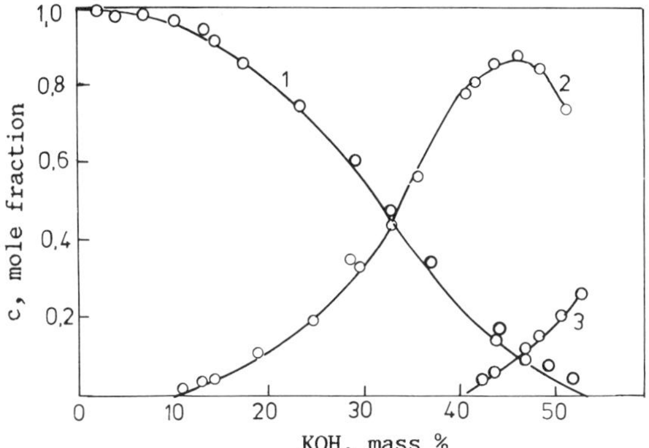

Fig. 2.4.   Dependence of the concentration of un-ionized
form (1) and of BOH⁻ (2) and BOO⁼ (3) of N-
methylacetamide on the concentration of potassium hydroxide [45].

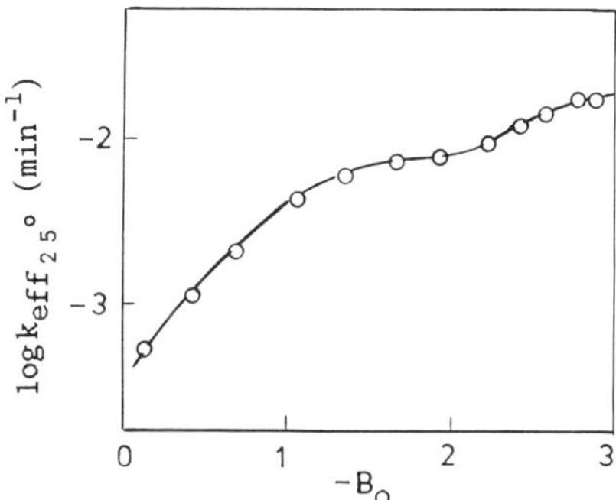

Fig. 2.5.   Dependence of log $k_{\mathrm{eff}\,25^\circ}$ on the alkalinity function of aqueous solu-
tions of potassium hydroxide in the
hydrolysis of N-methylacetamide.

Let us consider special cases of Eq. (2.15), taking as an example methylacetamide [48].

First case: $c_B \gg$ (c_{BOH^-} and $c_{BOO^=}$) and

$$k_{\mathrm{eff}} = (k_{\mathrm{act}\ BOH^-}/K_{BOH^-})\, b_0 a_{H_2O}\, \frac{f_{BOH^-}}{f_1^{\neq}} \tag{2.16}$$

This condition is satisfied for N-methylacetamide (Fig. 2.5)
right up to 18% KOH. The experimental data are given in Table 2.1.

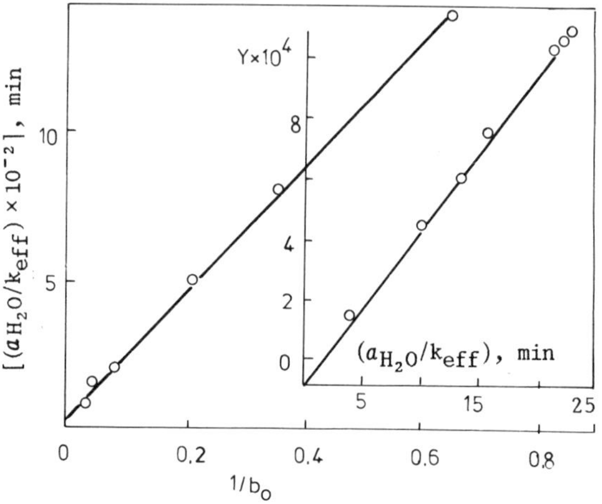

Fig. 2.6. Graphical determination of $k_{act\,BOH^-}$, $k_{act\,BOO^=}$, K_{BOH^-}, and $K_{BOO^=}$ from Eqs. (2.16) and (2.18) for the hydrolysis of N-methylacetamide:

$$y = -K_{BOH^-}\frac{a_{H_2O}}{b_0}\left(-\frac{1}{b_0} + \frac{k_{act\,BOH^-}}{K_{BOH^-}}\frac{a_{H_2O}}{k_{eff}}\right) + \frac{a_{H_2O}}{b_0}$$

TABLE 2.1. Kinetic Parameters of the Hydrolysis of N-Methylacetamide in Aqueous Solutions of Potassium Hydroxide

KOH, mass %	$-\log k_{eff\,25°C}$ [min^{-1}]	$-b_0$	$-\log a_{H_2O}$	$-\log \dfrac{k_{eff}}{b_0 a_{H_2O}}$
2.00	3.80	—0.56	—	3.24
4.61	3.46	—0.20	0.01	3.25
8.99	3.11	0.17	0.02	3.26
13.89	2.82	0.47	0.04	3.25
18.17	2.60	0.72	0.07	3.25

<u>Second case</u>: $c_B \sim c_{BOH^-}$, but $(c_B + c_{BOH^-}) \gg c_{BOO^=}$.

The concentration range for N-methylacetamide is from 18 to 44% KOH:

$$k_{eff} = \frac{k_{act\,BOH^-}a_{H_2O}}{1 + K_{BOH^-}/b_0}\frac{f_{BOH^-}}{f_1^{\neq}} \tag{2.17}$$

A graphical solution of the equation is given in Fig. 2.6.

At higher acid concentrations there is a higher rate constant as compared with the values given by Eq. (2.17). It may be assumed that the increase in k_{eff} is linked with the formation of a doubly ionized form, which is more reactive than the BOH^- form, in these solutions.

In this case, the experimental data can be described by the general equation (2.15), a graphical solution of which is given in Fig. 2.6.

<u>Third case.</u> $c_{BOO=} \gg (c_{BOH^-}$ and $c_B)$ and

$$k_{eff} = k_{act\,BOO^-}\, a_{H_2O}\, \frac{f_{BOO=}}{f_e^{\neq}} \tag{2.18}$$

A similar equation is obtained if $c_{BOH^-} \gg c_{BOO=}$ and c_B.

With decomposition of amides in weakly basic media there is a possibility of deviations from the above equations, caused by the accelerating action of acid salts [49] or with change in the conformation of the amides, e.g., for dipeptides [50].

Thus the decomposition of the majority of amides is according to mechanism B-2, and is described by Eq. (2.15).

2.1.3. Special Features of the Decomposition of Chemically Unstable Polyimides in Aggressive Media

For the scission of the polymer chain in a polyimide (PI), it is necessary to break two bonds, an imide and an o-carboxyamide bond. The simplest model compounds containing these bonds are N-phenylphthalimide (I) and N-phenylphthalamic acid (II):

<u>Acid Catalytic Decomposition.</u> At the present time the following scheme is accepted for the decomposition of o-carboxyamides [51]:

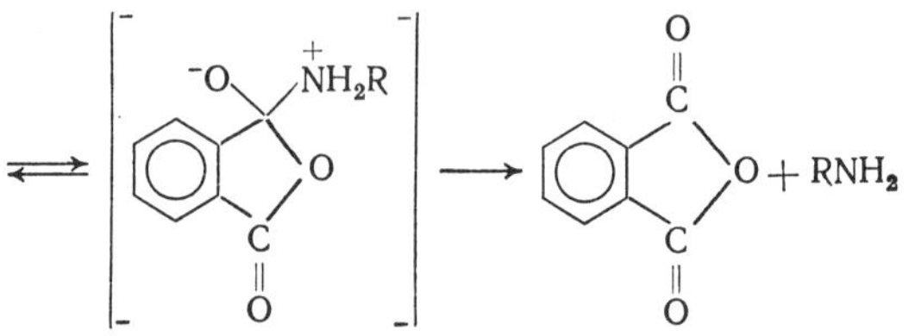

This scheme does not envisage participation of water in the decomposition of the o-carboxyamides. In actual fact, formation of an anhydride has been shown by a labeled atoms method [51], while in [52] an anhydride has been identified quantitatively in the decomposition of N-phenylphthalamic acid in concentrated acid solutions. A comparison of the kinetic data on the decomposition of substituted N-phenylphthalamic acids with results from quantum chemistry calculations (Fig. 2.7) shows that the decomposition of the o-carboxyamides passes through a stage of equilibrium transition of isomer I (the energetically stable form) to isomer II (the reactive form in the decomposition) [35]:

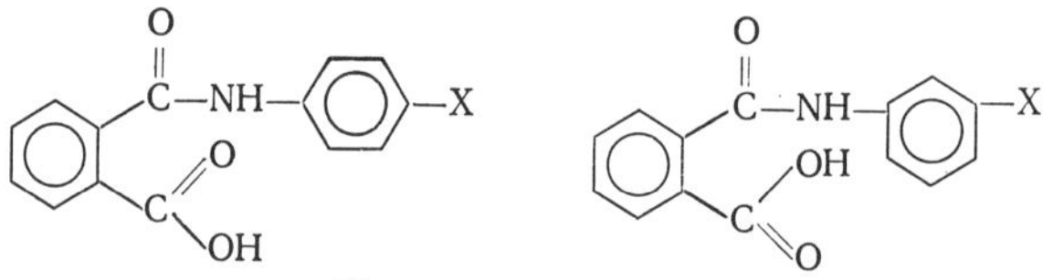

and that the equilibrium concentration of isomer II depends on the structure of the o-carboxyamides and on the experimental conditions.

For some o-carboxyamides (derivatives of maleic acid) the limiting stage is the stage of proton transfer with the formation of a zwitterion. In this case, the reaction conforms to general acid catalysis [53]. For most o-carboxyamides the limiting stage is irreversible decomposition of the zwitterion with formation of an anhydride and amide. In this case the reaction conforms to specific acid catalysis [52].

The decomposition by acid catalysis of N-phenylphthalamic acid proceeds by two routes (see Fig. 2.7):

in the range of moderately concentrated solutions of H_2SO_4 ($H_A > 1$) there is intramolecular acid catalysis (the un-ionized and nonprotonated form is reactive);

in the region of concentrated solutions of H_2SO_4 ($H_A < 1$) acid catalysis is effected by externally solvated protons.

The change in k_{eff} is described by the equation

$$k_{eff} = \frac{k_{act}\,c_{H_S^+} + k^0_{act}}{1 + h_A/K_{BH_0^+} + K_{diss}/h_A} \qquad (2.19)$$

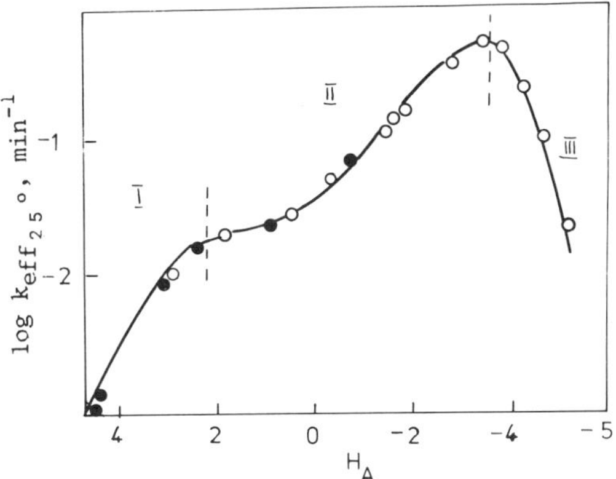

Fig. 2.7. Dependence of log $k_{eff_{25}°}$ on the acidity function of aqueous solutions of sulfuric acid for the decomposition of N-phenylphthalamic acid: ●) from data in [51]; ○) from data in [52]:

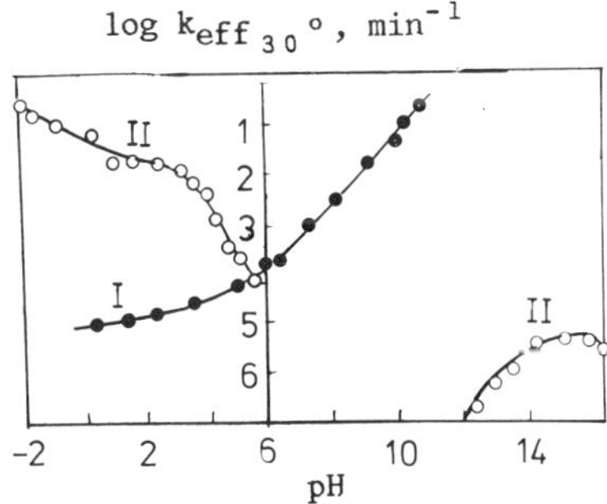

Fig. 2.8. Comparison of the reactivity of o-carboxyamide (N-phenylphthalamic acid) and imide (N-phenylphthalimide) groups at various pHs:

where k^0_{act} is the actual rate constant of the intramolecular decomposition, and K_{diss} the dissociation constant of the carboxyl group.

TABLE 2.2. Kinetic and Thermodynamic Parameters of the Decomposition of N-Phenylphthalamic Acid and N-Phenylphthalimide in Acidic and Basic Media at 25°C

Type of catalysis	k^0_{act}	k_{act}	$k_{BH^+_0}$
	N-phenylphthalamic acid		
Acidic	$(1.6\pm0.1)\cdot10^{-2}$, min^{-1}	$3.5\pm0.2\cdot10^{-2}$, liter/(mole·min^{-1})	$(1.7\pm0.2)\cdot10^3$
Basic	—	$(5.0\pm0.1)\cdot10^{-5}$, min^{-1}	$K_{BOH^-}=5.0\pm0.2$
	N-phenylphthalimide		
Acidic	—	$(6.3\pm0.2)\cdot10^{-5}$, liter/(mole·min^{-1})*	
Basic	$(4.0\pm0.3)\cdot10^{-5}$, min^{-1}	$(1.2\pm0.2)\cdot10^4$, min^{-1}	

*Found in aqueous acetic acid solutions of sulfuric acid (9 moles/liter CH_3COOH).

The decomposition of N-phenylphthalimide by acid catalysis proceeds by mechanism A-2. N-Phenylphthalimide is insoluble in concentrated aqueous acid solutions, and its decomposition was studied in concentrated aqueous solutions of sulfuric acid. The change in k_{eff} is described by Eq. (2.11). The value of k_{act} is $(6.3 \pm 0.2)\cdot10^{-5}$ liter/(mole·min), which is lower by more than two orders than k_{act} for the decomposition of N-phenylphthalamic acid under analogous conditions [52].

Catalysis Decomposition by Bases. The decomposition of N-phenylphthalamic acid in basic media proceeds by mechanism B-2 and is described by Eq. (2.17). The change in k_{eff} for the decomposition of N-phenylphthalimide in aqueous basic solutions is satisfactorily described by the equation [54]

$$k_{eff} = k_0 + \frac{k_{act}}{K_{BOH^-}}\, b_0 a_{H_2O} \qquad (2.20)$$

where k_0 is the rate constant of the uncatalyzed process.

Comparison of the Reactivity of N-Phenylphthalamic Acid and N-Phenylphthalimide in Acidic and Basic Media. The kinetic and thermodynamic parameters of the decomposition of N-phenylphthalamic acid and N-phenylphthalimide in acidic and basic media are shown in Table 2.2.

Comparison of the reactivity of the o-carboxyamide and imide bonds at various pH values (Fig. 2.8) shows that in basic media we get the most favorable conditions for the reaction of a PI to give an intermediate product for a polyamido acid (PAA). In acidic media the differing reactivity of imide and uncyclized c-carboxyamide groups made it possible to work out an absolute kinetic method for determining the degree of cyclization of soluble PIs [55].

2.1.4. Conversion of o-Substituted Benzanilides and Corresponding Hetero Rings in Aggressive Media

The decomposition of hetero chain polymers in aggressive media is a reversible—consecutive process, including hydrolysis of the hetero ring, cyclization, and hydrolysis of the amide. The scheme for model compounds is as follows:

$$
\underset{X}{\overset{N}{\bigcirc}}\!\!C-\bigcirc \;\underset{}{\overset{H_2O}{\rightleftharpoons}}\; \bigcirc\!\!\overset{NH-CO-\bigcirc}{\underset{XH}{}}\;\overset{H_2O}{\longrightarrow}\; \bigcirc\!\!\overset{NH_2}{\underset{XH}{}} + \bigcirc\!\!\overset{COOH}{}
$$

where X = NH (polybenzimidazoles), X = O (polybenzoxazoles), and X = COO (polybenzoxazinones).

Dissociation in Acidic Media. The mechanism of hydrolysis of the rings and their formation (cyclization of o-substituted benzanilides) are best considered together, since, according to the principle of "detail equilibrium," the intermediate compounds in this reaction are bound to be identical in structure. Since in the hydrolysis of the rings there can be dissociation both of the C=N bond and of the C—X bond, the reaction can follow two routes.

In a number of papers on the hydrolysis of Schiff's bases, benzimidates [56, 57], 2-methyloxazoline [58], and other compounds, it is suggested that the process proceeds by way of the formation of a carbinolamine, which may be one of the possible intermediate compounds.

In addition, it is accepted that the formation of hetero ring polymers from the appropriate polybenzanilides is by way of an amidol compound [59-61]. The possibility of amide—amidol tautomerism is suggested in [62-64].

Thus the hydrolysis of the rings and their formation may be by way of an amidol compound or cyclic carbinolamine (see p. 46).

Examination of the experimental data shows that, irrespective of the route of the process, the slow stage in the ring hydrolysis and in the cyclization of o-substituted benzamides is the formation of an intermediate compound (amidol or carbinolamine).

The hydrolysis of 2-phenylbenzoxazole and its formation are satisfactorily described by Eq. (2.4) [65].

2-Phenylbenzimidazole is resistant to hydrolysis with 2-10% concentrations of sulfuric acid at temperatures up to 140°C [66]. This may be explained by delocalization of π-electrons according to a conjugated bond system, as a result of which there is a powerful stabilizing effect [67]. In addition, for 2-phenylbenzimidazole pK_{BH^+} is 5.33 [68], and even in dilute acid solutions practically all the 2-phenylbenzimidazole is in the protonated unreactive form. The constant k_{eff} of the cyclization of o-aminobenzamide (or formation of 2-phenylbenzimidazole) is constant in the range of moderately concentrated acid solutions (0.1-20% H_2SO_4), and falls in more concentrated solutions [66]. In the present case the source of the protons which catalyze the cyclization of the o-aminobenzanilide is a protonated amino group, and k_{eff} does not depend on the concentration of hydrated protons in the solution; in concentrated solutions of sulfuric acid an unreactive o-protonated form of the amide is formed, and this leads to a reduction in k_{eff} of the cyclization. According to this mechanism,

$$k_{eff} = \frac{k_{act}}{1 + h_A/K_{BH_0^+}} \frac{f_B}{f^{\neq}} \tag{2.21}$$

with the limiting stage of the process evidently being the formation of carbinolamine [69], while the ratio $f_B/f^{\neq}$ is practically independent of the acid concentration.

2-Phenyl-4N-3,1-benzoxazinone readily hydrolyzes at room temperature, and the dependence of k_{eff} on acid concentration is of an extremal character (Fig. 2.9). With concentrations 1-58% H_2SO_4 the reaction is practically irreversible, and the sole product of the reaction is benzoylanthranilic acid, which very slowly decomposes to give o-aminobenzoic and benzoic acids. With 58-80% acid concentrations the reaction becomes reversible. The decomposition of

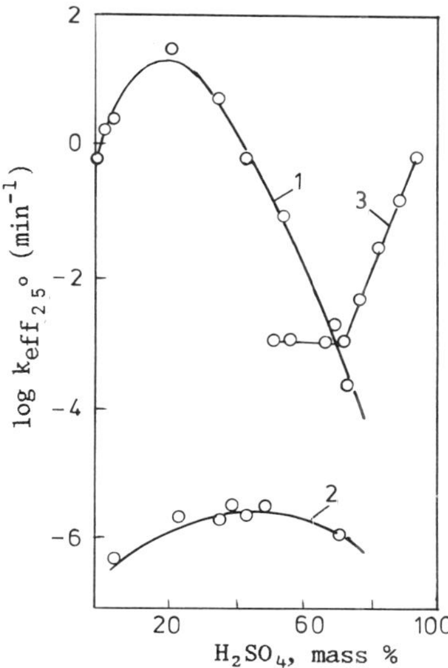

Fig. 2.9. Dependence of log keff_{25}° on concentration of sulfuric acid for the hydrolysis of 2-phenyl-4N-3,1-benzoxazinone (1) and the hydrolysis (2) and cyclization (3) of benzoylanthranilic acid [66].

2-phenyl-4N-3,1-benzoxazinone proceeds at the ester bond and the change in keff is expressed as

$$k_{eff} = \frac{k_{act}c_{H_S^+}}{1 + h_0/K_{BH_0^+}} \cdot \frac{f_B f_{H_S^+}}{f^{\neq}} \qquad (2.22)$$

which takes into account that the protonation of the ester bond in 2-phenyl-4N-3,1-benzoxazinone is expressed by the acidity function H_0 [66].

In contrast with these o-substituted anilides keff of the cyclization of benzoylanthranilic acid rises sharply in concentrated solutions of sulfuric acid (Fig. 2.9). Apparently, with the cyclization of benzoylanthranilic acid two forms are reactive: a form which is unprotonated at the amide bond and a dehydrated O-protonated form, the formation of which can be represented in highly concentrated acidic media with a low activity of the water, as shown on p. 48.

In concentrated solutions of sulfuric acid there becomes possible the protonation of the carboxyl group of benzoylanthranilic acid. It is known that the carboxyl group in substituted benzoic acids is practically fully protonated in ∿90% H_2SO_4 [70]. The pres-

$$\underset{\text{COOH}}{\text{NHCO}}-\bigcirc \;\overset{H_S^+}{\rightleftharpoons}\; \underset{\substack{\text{COOH}\\(BH^+)}}{\text{NH}-\overset{\overset{OH^+}{\|}}{C}\cdots}\bigcirc \;\rightleftharpoons\; \underset{\substack{\text{COOH}\\(B^+)}}{N=\overset{+}{C}}-\bigcirc \;+\;H_2O$$

ence of a positive charge on the amide bond of benzoylanthranilic acid in concentrated solutions of sulfuric acid lowers the basicity of the carboxyl group, and it may be noted that in concentrations up to 90% H_2SO_4 (Fig. 2.9), the carboxyl group will be in an ionized form.

The equation which describes the change in k_{eff} of cyclization of benzoylanthranilic acid has the form

$$k_{eff}=\frac{k_{act_B}\dfrac{K_{BH_0^+}}{K_{B^+}}\dfrac{a_{H_2O}}{h_A}\dfrac{f_B}{f^{\neq}}+k_{act_{B^+}}\dfrac{f_{B^+}}{f^{\neq}}}{1+\dfrac{K_{BH_0^+}}{K_{B^+}}\dfrac{a_{H_2O}}{h_A}+\dfrac{a_{H_2O}}{K_{B^+}}} \tag{2.23}$$

where k_{act_B} and $k_{act_{B^+}}$ are the actual rate constants of cyclization of forms B and B^+; K_{B^+} is the equilibrium constant of dehydration of form BH^+.

With sulfuric acid concentrations >75%, $c_{BH^+} \gg$ (c_B and c_{B^+}) and

$$k_{eff}=\frac{k_{act_{B^+}}\dfrac{f_B}{f^{\neq}}}{\dfrac{a_{H_2O}}{K_{B^+}}} \tag{2.24}$$

Figure 2.9 shows that k_{eff} is in fact inversely proportional to a_{H_2O}, i.e., in this case also $f_B/f^{\neq}$ remains practically unchanged with acid concentration.

The hydrolysis of o-substituted benzanilides is satisfactorily described by Eq. (2.4) [66].

Dissociation in Basic Media. In basic media, unlike in acidic media, o-substituted benzanilides practically do not cyclize. The hydrolysis of o-substituted benzanilides, and of the corresponding rings, in basic media, proceeds by mechanism B-2, and the change in k_{eff} of these reactions is expressed by Eq. (2.17).

Comparison of the Reactivity in Acidic and Basic Media. Table 2.3 gives the kinetic and thermodynamic parameters of the dissocia-

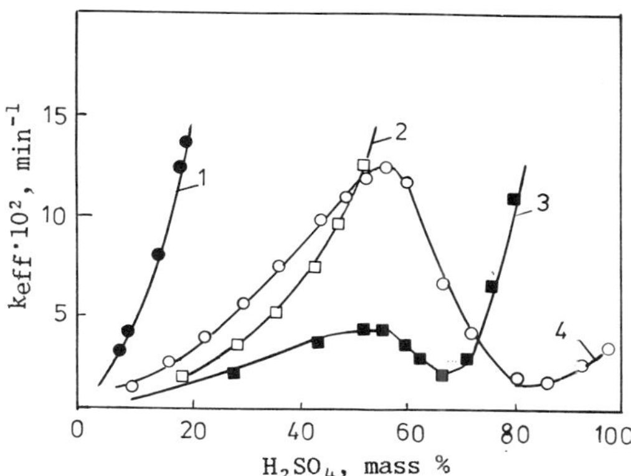

Fig. 2.10. Relationship between k_{eff} and $C_{H_2SO_4}$ for the
hydrolysis of various esters of acetic acid:
1) tert-butyl-, 2) phenyl-, 3) methyl-, 4)
sec-butyl acetate [82].

tion of o-substituted benzanilides and of the corresponding rings
in acidic and basic media. These parameters make it possible, with
the help of the corresponding equations, to calculate the hydrolytic
stability of cyclized and uncyclized fragments in hetero chain polymers.

2.2. COMPOUNDS WITH AN ESTER GROUP

Compounds with ester groups decompose both in acidic and in
basic media, in the same way as do amides.

2.2.1. Acid Catalytic Decomposition of Esters

Cryoscopic measurements in 100% H_2SO_4 have shown that in this
highly acidic medium esters attach to themselves a single proton
[71]. It has been found possible, using NMR, to establish the precise point of attachment of the proton: in 97% H_2SO_4 there is protonation of the carbonyl oxygen [72].

Unlike the case of amides, whose protonation is satisfactorily
described by the acidity function H_A, as found using amide indicators,
the mechanism of protonation remains unclear with esters.

The majority of investigators consider that the protonation
of esters is not expressed by the acidity function H_0 [73-76] since

$$d \log \frac{c_B}{c_{BH^+}} \bigg/ dH_0 \neq 1 \qquad (2.25)$$

TABLE 2.3. Kinetic and Thermodynamic Parameters of Corresponding Rings in Acidic and Basic Media

	Acid		
	k_{act}, min^{-1}		
X	hydrolysis of ring	hydrolysis of anilide	cyclization of anilide
—O—	$(6,0\pm0,5)\cdot10^{-6}$	$(4,0\pm0,3)\cdot10^{-6}$	$(6,0\pm0,3)\cdot10^{-7}$
—NH—	stable	$(2,5\pm0,2)\cdot10^{-5}$	$(7,7\pm0,3)\cdot10^{-6}$
—COO—	$8,8\pm0,2$	$(5,3\pm0,2)\cdot10^{-7}$	$(6,9\pm0,1)\cdot10^{-3}$
			$(1,4\pm0,2)\cdot10^{-4}$

*The value determined is k_{act}/K_{BOH^-}.

It should, however, be noted that in determining the degree of ionization of esters by UV spectroscopy there are frequently difficulties connected with the shift of absorption bands in the UV region of the spectrum when the concentration of strong acid in the solution is altered, and also connected with the differing degree of overlap of the absorption bands of the un-ionized and ionized forms (absence of isobestic points) in solutions of differing acidity.

Vinnik and Librovich [77] used IR spectroscopy to show that in a D_2SO_4–D_2O system the ionization of ethyl acetate is satisfactorily described by the acidity function D_0. However, unfortunately, these data were obtained only for a single ester.

There are at present two approaches to the mechanism of protonation of esters.

As with amides, the ionization of esters must be described by the acidity function H_A since in both cases a proton becomes attached to the carbonyl oxygen [78, 79]. In point of fact, the ionization of the majority of esters is formally expressed as H_A [75, 76].

The protonated form of the ester interacts with bases which are present in the acid solution [80]. For instance, in an aqueous solution of sulfuric acid the following reactions may take place:

$$BH^+ + (H_2O)_n \xrightleftharpoons{K_n} B \cdots H^+ (OH_2)_n$$

$$BH^+ + HSO_4^- \xrightleftharpoons{K_{HSO_4^-}} B \cdots H_2SO_4$$

Conversion of o-Substituted Benzanilides and of the
at 25°C

K_{BH^+}		Base		K_{BOH^-}
		k_{act_1}, min^{-1}		
of ring	of anilide	hydrolysis of ring	hydrolysis of anilide	of anilide
>0 — 14 ± 1	200 ± 10 72 ± 2 $(8\pm2)\cdot10^3$	— — >10	$(6,3\pm0,1)\cdot10^{-5}$* — $(1,3\pm0,1)\cdot10^{-1}$	— — 270 ± 5

The measured indicator ratio is given by the equation

$$I = \frac{c_{B\cdots H^+ (OH_2)n} + c_{B\cdots H_2SO_4}}{c_{BH^+}} = K_n a^h_{H_2O} + K_{HSO_4^-} c_{HSO_4^-} \tag{2.26}$$

A form protonated on the alkoxyl oxygen has not been observed experimentally (if indeed it is formed, this is only in low concentrations). Decomposition of the ester may take place either at the
acyl—oxygen bond or at the alkyl—oxygen bond, and scission may be
mono- or bimolecular, but one rarely meets decomposition of the acyl—
oxygen bond according to the monomolecular law AA1-1 or decomposition
of the alkyl—oxygen bond according to a bimolecular law. The AA1-1
mechanism applies in the decomposition of esters where the corresponding alcohols readily form carbonium ions. This type of decomposition is favored by the following factors [81]:

the weakly-defined nucleophilic character of any attacking
agent, i.e., the absence of strong bases;

the presence of an alcohol group which readily gives up an
electron to an oxygen atom which is combined with an alkyl group;

the presence of an acyl group which strongly attracts electrons.

This multiplicity of mechanisms gives rise to differing patterns of the relationship between k_{eff} and acid concentration (Fig.
2.10).

According to Yates [82], acetates in general may be divided
into four groups:

Group I (primary alkyl acetates). The value cf k_{eff} rises and
passes through a maximum in the region of 50-60% H_2SO_4, and the re-

TABLE 2.4.   Enthalpy and Entropy of Activation of the Hydrolysis
             of Ethyl Acetate in Aqueous Solutions of Sulfuric Acid
             [82]

Concentration of H_2SO_4, mass %	$\Delta H^{\neq}$, kJ/mole	$\Delta S^{\neq}$, cal/deg	Concentration of H_2SO_4, mass %	$\Delta H^{\neq}$, kJ/mole	$\Delta S^{\neq}$, cal/deg
40.2	71	—15.3	71.8	77	—12.4
55.9	69	—15.7	98.4	100	+2.4
62.5	68	—17.0			

action proceeds by mechanism A_{AC}-2.   With 80-90% H_2SO_4 the decomposition mechanism changes to A_{AC}-1.   This is confirmed by a sharp
rise in k_{eff}, a considerable reduction in the ratio of the rate of
hydrolysis and the rate of isotope exchange of the oxygen of the
carbonyl group of the ester, and by a change in the enthalpy and
entropy of activation (Table 2.4).

Group II (secondary alkyl acetates and benzyl and allyl acetates).   There is a similar pattern.   The change in k_{eff} with acid
concentration is similar to the change in k_{eff} in group II.   The
change in the mechanism of the reaction from A_{AC}-2 to A_{AC}-1 is in
the region 65-75% H_2SO_4.

Group III (phenyl acetates).   The value of k_{eff} rises over the
whole range of acid concentrations.   Probably the reaction proceeds
in parallel by the two mechanisms A_{AC}-2 and A_{AC}-1.

Group IV (tertiary alkyl acetates).   The value of k_{eff} rises considerably with the acid concentration, and the reaction is by mechanism A_{AC}-1.

Mechanisms A_{AC}-2 and A_{AC}-1.   In contrast with amides, there is,
in the case of esters, a lack of information on the relative reactivity of the forms protonated on the carbonyl and alkoxyl oxygen
and, accordingly, in the majority of investigations, the structure
of the protonated form of the ester is not specifically established.

There are two approaches to the explanation of the experimental
data on the hydrolysis of esters.   According to Khaldna and co-
workers [83], decomposition of esters in acidic media may follow
three parallel routes [see (I)-(III), p. 53].

Since the structure of the protonated form of the ester is not
known, this form may be designated BH^+.   As indicated above, the
protonated form of an ester can form unreactive particles with bases,
anions of the acid, and water molecules present in the solution.

$$
\begin{array}{c}
B{\cdots}H_2SO_4 \;\rightleftharpoons\; \longrightarrow\; M_1^{\neq} \xrightarrow{\;k_{act_1}\;} \\[2mm]
BH^+ \;\underset{}{\overset{+H_2O}{\rightleftharpoons}}\; M_2^{\neq} \xrightarrow{\;k_{act_2}\;} \\[2mm]
B{\cdots}H^+(OH_2)_n \;\underset{K_n}{\rightleftharpoons}\; \longrightarrow\; M_3^{\neq} \xrightarrow{\;k_{act_3}\;}
\end{array}
$$

(I)

(II)

(III)

The limiting stages of the process are the decomposition of
BH^+ (mechanism A_{AC}-1) and the decomposition of BH^+ with one or two
molecules of water (mechanism A_{AC}-2). In the latter case, Bunnett's
approach is used to explain the bimolecular reactions. The following equation, corresponding to this scheme,

$$
k_{eff} = \frac{k_{act_1} + k_{act_2}\, a_{H_2O} + k_{act_3}\, a_{H_2O}^2}{1 + K_{BH^+}/h_0 + K_{HSO_4^-}\, c_{HSO_4^-} + \displaystyle\sum_{n=4} K_n a_{H_2O}^n}
\tag{2.27}
$$

describes, for instance, the change in k_{eff} for the hydrolysis of
ethyl acetate in the range of concentration of H_2SO_4 from 1.39 to
99.8 mass %.

Apparently the protonation of ethyl acetate is described by
the acidity function H_0; the ratio of the activity coefficients does
not depend on the medium, and n can have only two values: 1 and
4 [83].

It must be noted that the scheme and the corresponding equation
are of formal character, but one virtue of the scheme is that the
change in the mechanism of the reaction in the range 80-90% H_2SO_4
is taken into account in Eq. (2.27). According to Vinnik and Librovich [84], in aqueous solutions of strong acids the hydrolysis of
esters proceeds by way of activated complexes formed from the un-
ionized form of the ester, a hydrated proton, and a nucleophilic reagent Y (water molecules and an anion of an undissociated acid).
H_S^+ becomes attached to an alkoxyl oxygen atom.

With the ionization of esters there are formed, in considerable
amounts, species of low reactivity: a complex protonated on a carboxyl group of the form

$$
c_{BH^+} = c_B\, \frac{h_0}{K_{BH^+}}
\tag{2.28}
$$

and a complex with a hydrated proton

$$
c_{BH_S^+} = c_B\, \frac{a_{H_S^+}}{K_d}
\tag{2.29}
$$

TABLE 2.5. Values of k_{eff} for the Hydrolysis of tert-Butyl Acetate in Aqueous Solutions of Hydrochloric Acid at 25°C [81]

Concentration of HCl, mass %	$-\log k_{eff}$, sec^{-1}	$-H_0$	$-\log k_{eff}$ $+H_0$	Concentration of HCl, mass %	$-\log k_{eff}$, sec^{-1}	$-H_0$	$-\log k_{eff}$ $+H_0$
2,2	3,9	—0,15	3,75	16,0	2,0	1,68	3,68
5,0	3,4	—0,35	3,75	19,5	1,5	2,13	3,63
10,6	2,6	1,02	3,62	22,7	1,1	2,55	3,65

The hydrated proton, in the opinion of the authors, attaches itself either to a carbonyl group or else to both atoms of oxygen in the ester molecule. The experimental data on the hydrolysis of a number of esters in aqueous solutions of strong acids (H_2SO_4, $HClO_4$, HCl, and HBr) are satisfactorily described by an equation which makes the assumption that the multiplier of the activity coefficients remains practically unchanged with the acid concentration:

$$k_{eff} = \frac{k_{act}^{} H_S^+ c_{H_S^+} + k_{act,Y}^{} H_S^+ c_{H_S^+} c_Y}{1 + a_{H_S^+}/K_d + h_0/K_{BH^+}} \qquad (2.30)$$

This equation explains the influence of neutral salts on the reaction rate of hydrolysis (rise in c_Y), but it is understood that in the presence of nucleophilic reagents there is a trimolecular reaction, which is of low probability in solutions.

Mechanism $AA1-1$. This mechanism is responsible for, e.g., the decomposition of tert-butyl acetate [85] or lactones [86]. Table 2.5 shows that the experimental data for tert-butyl acetate are satisfactorily described by Eq. (1.26).

It may be suggested that the form protonated on the alkoxyl oxygen is reactive and that the hydrolysis proceeds as follows:

$$\underset{R-O-C\sim}{\overset{O}{\|}} \underset{\rightleftharpoons}{\overset{H^+}{}} \underset{R-O-C\sim}{\overset{H^+\ O}{|\ \ \|}} \rightleftharpoons M^{\neq} \longrightarrow R^+ + \underset{HO-C\sim}{\overset{O}{\|}}$$

Thus, the acid hydrolysis of esters may proceed by a number of mechanisms, all of which are described by complex kinetic equations.

The decomposition of ester groups in commercial polyesters which have been prepared by the polycondensation of dicarboxylic acids and diols (or primary alcohols) proceeds by mechanism $AAC-2$ in dilute or moderately concentrated acid solutions. In these solutions, for the majority of esters k_{eff} for the hydrolysis rises proportionately with the concentration of hydrated protons.

2.2.2. The Base-Catalyzed Decomposition of Esters

The decomposition of esters in basic media proceeds very rapidly and, accordingly, the kinetics of these reactions are investigated only in dilute basic solutions, which makes it difficult to establish the true mechanism of decomposition.

It is now established that the decomposition reactions of esters, as of amides, proceed by way of tetrahedral intermediate compounds

$$\sim\!\overset{\text{O}}{\overset{\|}{\text{C}}}\!-\!\text{O}\!-\!\text{R} + \text{B}^- \underset{k_{-1}}{\overset{k_1}{\rightleftarrows}} \sim\!\underset{\text{B}}{\overset{\text{O}^-}{\overset{|}{\underset{|}{\text{C}}}}}\!-\!\text{O}\!-\!\text{R} \overset{+\text{H}_2\text{O}}{\rightleftarrows} \text{M}^{\neq} \overset{k_{\text{ac}^-}}{\longrightarrow}$$

the presence of which has been established by means of isotope exchange [87] and by identification of various products of combination, e.g., of a methylatoxyl in the case of ethyl trifluoroacetate [88]. The reactivity of esters in basic media is determined by three main factors:

1) the susceptibility of the carbonyl group of the ester to attack, an index of this being the value of k_1. There is a correlation between the frequency of deformational vibration of the C$=$O bond and the reactivity of the ester [89];

2) the nucleophilicity of B in relation to the carbonyl group. The relative nucleophilicities are determined from the kinetic data on the hydrolysis and alcoholysis of the esters (Table 2.6) [90];

3) the relative stability of B and OR in the tetrahedral intermediate compound.

Each of these three factors may depend both on the properties of the medium in which the decomposition proceeds and on the presence of certain cations.

Generally, k_{eff} for the hydrolysis of esters in basic media is found from the expression

$$k_{\text{eff}} = k_0 + \sum_i (k_{\text{B}} c_{\text{B}})\, i \tag{2.31}$$

where k_0 and k_{B} are the effective rate constants of the decomposition of the ester under the action of water and the bases.

TABLE 2.6. Nucleophilicity and Basicity of Various
            Oxygen-Containing Bases [90]

Base	Nucleophilicity in relation to C=O	pK_a	Base	Nucleophilicity in relation to C=O	pK_a
H_2O	—	−1.74	CH_3O-	1,6	15.0
$HO-$	1.0	15.8	C_2H_5O-	4,2	16.0

The decomposition of the esters in basic media proceeds by
mechanisms B_{AC}-2 and B_{A1}-2. Monomolecular mechanisms have not been
definitely found in the decomposition of esters in basic media.

2.3. COMPOUNDS CONTAINING AN ACETAL GROUP

The decomposition of compounds with an acetal group is catalyzed solely by acids. Their catalytic action is explained by the
possibility of coordination with an unshared pair of electrons on
the oxygen atom, resulting in the formation of an oxonium compound.
Such coordination is impossible with hydroxide ions, and this explains the stability of acetals in alkaline media. In spite of the
similar structure of acetals and ethers, there are significant differences in the rates of acid catalytic decomposition of these compounds under identical conditions. Ethers are less reactive, by 7-
8 orders, than the corresponding acetals [91]. The higher reactivity of acetals as compared with other oxygen-containing compounds
is due to the presence of an induction effect from the two adjacent
oxygen atoms:

$$\sim\ddot{O} \rightarrow \overset{\delta-}{CH_2}-O\sim$$

In addition, in the case of acetals the decomposition of an oxonium
ion is energetically more favorable since the carbonium ion which
is formed is resonance-stabilized as a mesomeric form [92]:

$$R\overset{+}{C}H_2 \longleftrightarrow RO^+{=}CH_2$$

The stability of such cations is confirmed by the fact that the decomposition of acetals is not accompanied by isomerization [93].
It is generally accepted that the hydrolysis of acetals proceeds
by mechanism A-1. The following factors are usually adduced to support mechanism A-1:

the kinetic isotope effect k_{D_2O}/k_{H_2O} = 2.5-3.0 [94];

positive values of the entropy of activation [95];

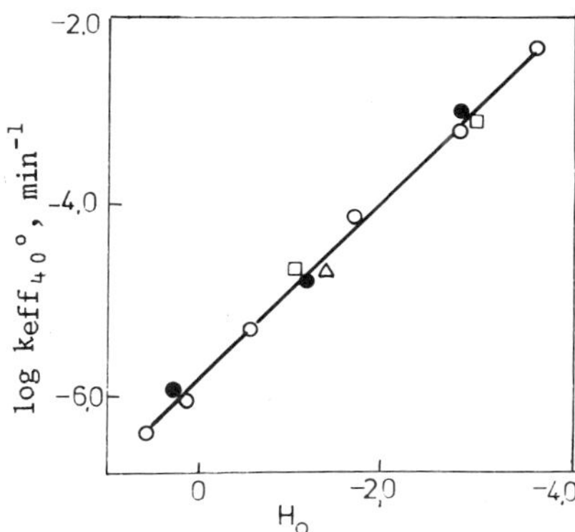

Fig. 2.11. Dependence of log $k_{\text{eff}_{40}}°$ on the
acidity function H_0 in aqueous solu-
tions of sulfuric ($\circ$), hydrochloric
($\bullet$), perchloric ($\square$), and phosphoric ($\triangle$) acids [92, 99] for the decomposition of trioxane.

reduction in the reaction rate with raising of the pressure
[96, 97];

the rate constant of hydrolysis of the acetals is independent
of the steric constant of the substituent [94, 98].

The hydrolysis of acetals was one of the first reactions to
which the acidity function H_0 was applied [92]. The acid catalytic
hydrolysis of trioxane is a classic example of a process according
to mechanism A-1. Figure 2.11 shows the linear relationship between
log k_{eff} and H_0 with a slope equal to unity for aqueous solutions
of hydrochloric, sulfuric, perchloric, and phosphoric acids [92,
99].

Thus, the change in k_{eff} is described by the following equation:

$$k_{\text{eff}} = \frac{k_{\text{act}}}{K_{\text{BH}^+}}\, h_0 \, \frac{f_{\text{BH}^+}}{f^{\neq}} \tag{2.32}$$

The ratio of the activity coefficients does not depend on the composition of the medium since the ionized form of acetals and the
activated complex are identical in charge, composition, and structure. It is this particular relationship which is adduced as the
main argument in support of the protonation of acetals proceeding
according to the Hammett indicator mechanism. In the general case
the hydrolysis of acetals is promoted by the orientation of an unshared pair of electrons of the unprotonated oxygen parallel to the
$\sim$C—O$\sim$ bond. For cyclic acetals such orientation is rendered diffi-

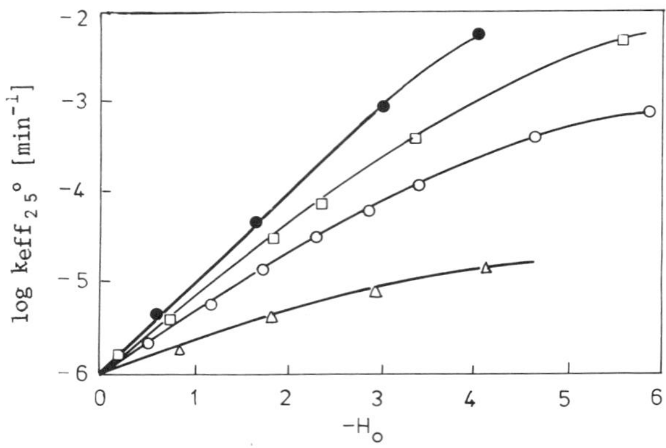

Fig. 2.12. Dependence of log $k_{eff_{25}°}$ for the hydrolysis
            of cellobiose on the acidity function H_0 in
            aqueous solutions of hydrochloric (●), per-
            chloric (□), sulfuric (○), and phosphoric
            (△) acids [104].

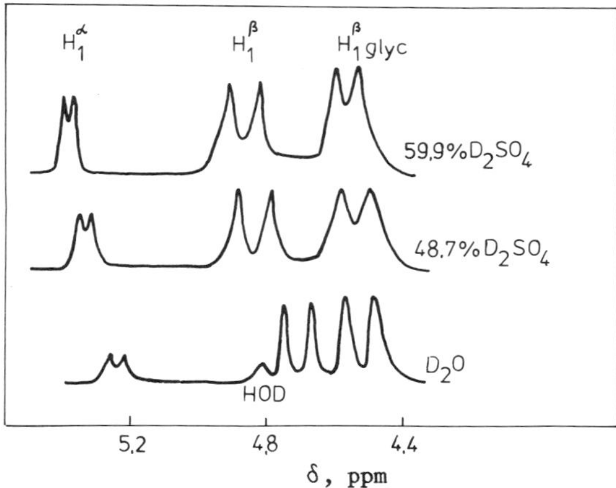

Fig. 2.13. PMR spectra of cellobiose in solutions of sul-
            furic acid of various concentrations [105].

cult and accordingly cyclic acetals are less reactive by several
orders than linear acetals (see Chap. 4). The mechanism of the acid-
catalyzed decomposition of oligomers of polyoxymethylene is likewise
described in Chap. 4.

2.3.1. Decomposition of Glycosides

Glycosides are a type of cyclic acetal of the general formula

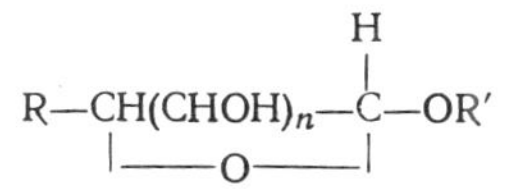

$$R-CH(CHOH)_n-\underset{\underset{\displaystyle |\!-\!-\!-\!O\!-\!-\!-|}{|}}{\overset{\overset{\displaystyle H}{|}}{C}}-OR'$$

The acid-catalyzed decomposition of glycosides is the subject of numerous investigations.

The protonation of glycosides in aqueous acid solutions has not been studied since, in strongly acidic media in which protonation of the oxygen is bound to take place to a considerable degree [100], there are fast decompositions and irreversible chemical dissociations [101]. Protonation at the glycoside and cyclic oxygen may lead to decomposition. Depending on protonation of one or the other oxygen, decomposition can proceed by two mechanisms, cyclic or acyclic. The cyclic mechanism, first proposed by Edward for pyranosides [102], is accepted by the majority of investigators:

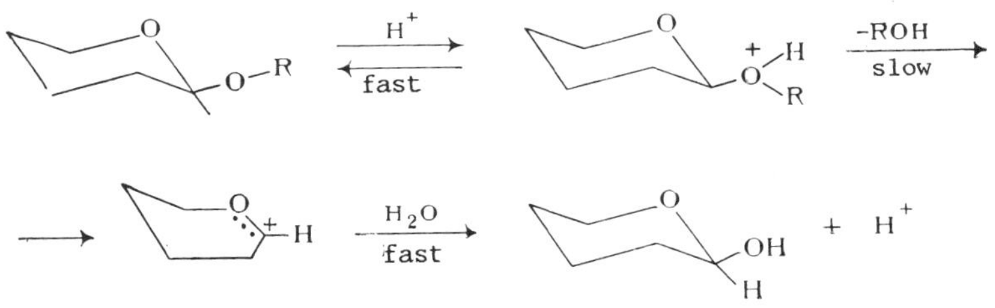

The process includes fast equilibrium protonation of the glycosside oxygen and slow heterolysis with the formation of a cyclic carbonium ion, which is in the half-chair conformation.

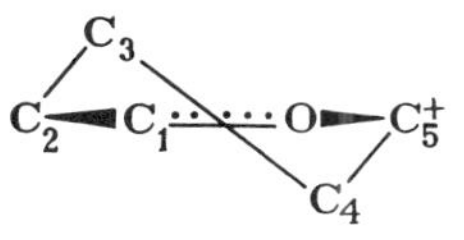

$$C_2\!-\!\!\overset{\displaystyle C_3}{C_1}\cdots O\!-\!C_5^{+} \quad (C_4)$$

The acyclic mechanism proposed by Armstrong and Glover [103] includes protonation of the cyclic oxygen and scission of the pyranose ring at the C_1-O bond with the formation of an acyclic carbonium ion. Data supporting this or the other mechanism are assessed in detail in a review [94]. The proofs adduced to support the cyclic mechanism are not exhaustive, and it is impossible to fully exclude the presence of an acyclic mechanism.

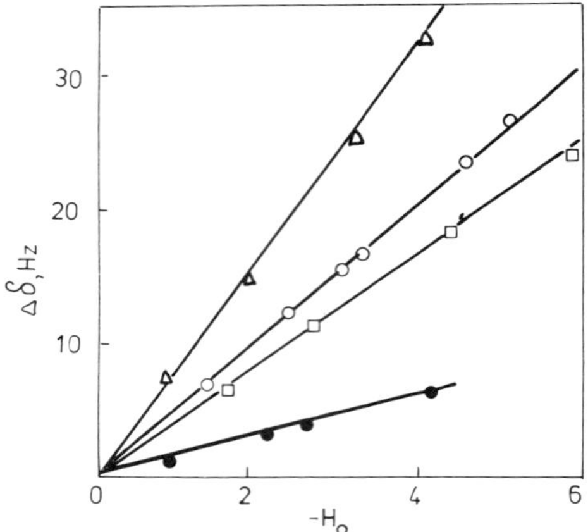

Fig. 2.14.   Dependence of the chemical shifts of the pro-
             ton of cellobiose at C_2 on the acidity func-
             tion H_0 for hydrochloric (●), perchloric
             (□), sulfuric (○), and phosphoric (∆) acids
             [105].

At the present time, mechanism A-1 is generally accepted for
the glycoside hydrolysis reaction. The main proof is a linear connection between log k_{eff} and H_0. Examination of the literature
shows that in the majority of cases (for all acids except hydrochloric) the slope of this relationship is less than unity and
changes with the acid concentration. Figure 2.12 shows log k_{eff} vs.
H_0 for the hydrolysis of cellobiose in aqueous solutions of HCl,
$HClO_4$, H_2SO_4, and H_3PO_4 [104].

To explain the mechanism of glycoside hydrolysis, Khalturinskii
[105] used PMR and the change in optical rotation to study the state
of these compounds in acidic media. With a rise in the acid concentration (Fig. 2.13) there is a shift of the α—β anomer equilibrium
toward the α form.

Protonation and conformational change of the glycoside (the
constants of spin-spin interaction remain constant in the various
media) do not occur to any noticeable extent; there is a symbatic
shift of the PMR spectral lines to a lower field, which differs according to the acid (Fig. 2.14). This latter effect is absent from
the EPR spectra of trioxane [105]. If we accept that the hydrolysis
of pyranosides proceeds according to the Edward mechanism [102],
the protonated form and the activated complex are bound to have different structures, and the ratio $f_{BH^+}/f^{\neq}$ will not be constant, but
will depend on the interaction of the pyranoside molecule with the
medium. As a first approximation we can accept that $f_{BH^+}/f^{\neq} \approx \rho\Delta\delta$, where
$\Delta\delta$ is the chemical shift of the protons of the glycoside molecules
in the PMR spectra with a change in the medium and ρ a coefficient

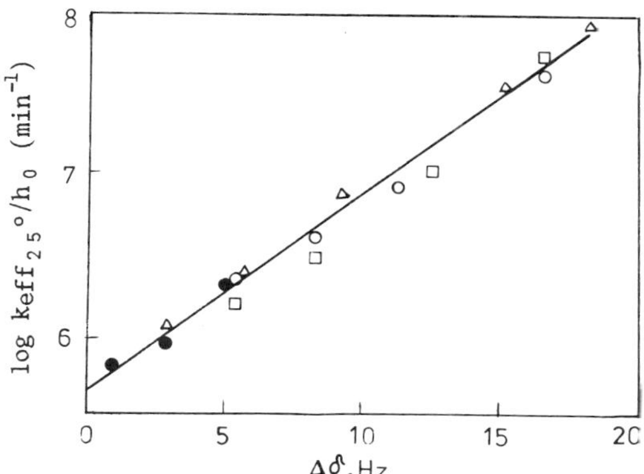

Fig. 2.15. Graphical solution of Eq. (2.33) for the hydrolysis of cellobiose in aqueous solutions of hydrochloric (●), perchloric (□), sulfuric (○), and phosphoric (△) acids at 25°C [105].

of proportionality. As a reference state we take the state in aqueous solution. In this case, the equation may be written as follows:

$$\log \frac{k_{eff}}{h_0} \approx \rho\Delta\delta + \log \frac{k_{act}}{K_{BH^+}} \tag{2.33}$$

The interaction of the glycosides with the medium proceeds by way of the hydroxyl groups. In this connection it is interesting to clarify in what positions the hydroxyl groups have a significant influence on the course of hydrolysis.

We now quote values for $\log k_{act}/K_{BH^+}$ for cellobiose, methyl-α-D-glucopyranoside [105], and a group of derivatives of methyl-α-D-glucopyranosides in which the hydroxyl groups in various positions are successively replaced by hydrogen atoms:

Compound	$\log \left(\dfrac{k_{act}}{K_{BH^+}}\right)_{25\,°C}$
Cellobiose	6.1
Methyl-α-D-glucopyranoside	7.0
Methyl-α-D-2-deoxyglucopyranoside	3.1*
Methyl-α-D-3-deoxyglucopyranoside	6.6*
Methyl-α-D-xylopyranoside	6.3*

*Data from [194].

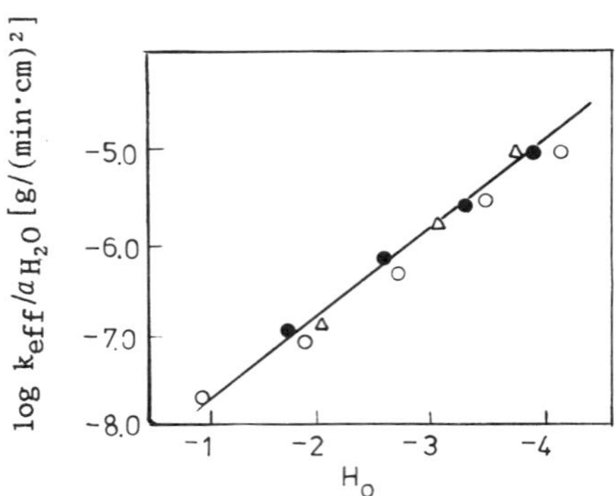

Fig. 2.16. Dependence of log k_{eff}/a_{H_2O} on H_0
for the decomposition of D_4 in aque-
ous solutions of hydrochloric (●),
sulfuric (○), and phosphoric (△)
acids [118].

It may be seen that the replacement of a hydroxyl group by a
hydrogen atom at C_2 leads to a rise in the rate constant by approximately four orders, whereas analogous replacement at other carbon
atoms has a negligible effect on the rate constant.

Thus, when hydrolysis of pyranosides takes place, the process
is significantly influenced by interaction of the glycoside molecule
with the medium by way of a hydroxyl group at C_2. Such a conclusion nevertheless does not exclude influence of steric and induction effects from the hydroxyl group on the reactivity of pyranosides.

Equation (2.33) satisfactorily describes the experimental data
on the acid-catalyzed hydrolysis of cellobiose (Fig. 2.15) and of
other glycosides [105].

Thus, the acid decomposition of glycosides proceeds by mechanism A-1, but the limiting stage of the hydrolysis of cellobiose is
transition from the chair conformation (the energetically favorable
form in solutions) into that of a half-chair (the cyclic carbonium
ion which it forms), this stage being significantly influenced by
the interaction of the glycoside molecule with the medium; this interaction may be assessed from the chemical shifts of the protons.

2.4. COMPOUNDS CONTAINING A SILOXANE GROUP

Study of the mechanism of decomposition of organosiloxanes in
acidic and basic media is of great importance for the interpretation

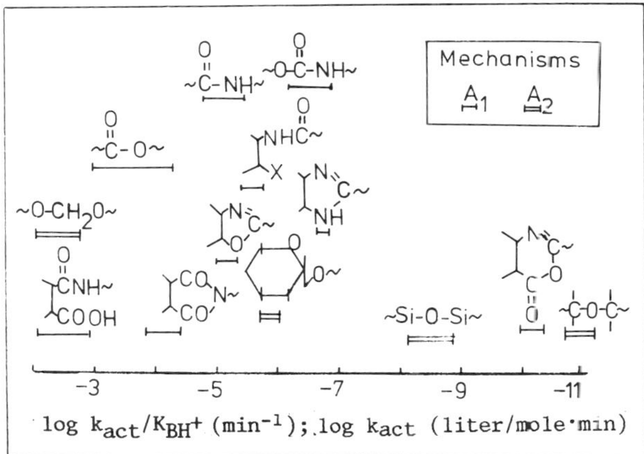

Fig. 2.17. Kinetic parameters of the decomposition of the simplest model compounds with various chemically unstable groups in aqueous acid solutions: for mechanism A_1 there is given log k_{act}/K_{BH^+}; for mechanism A_2, log k_{act} [liters/(mole·min)].

of data on the chemical degradation of polyorganosiloxanes, and also on the polycondensation and catalytic regrouping of these compounds. Most investigations relate to the decomposition of organosiloxanes in acidic media.

There are at present a number of points of view as to the mechanism of acid decomposition of organosiloxanes. It has been suggested [106-108] that the scission of organosiloxanes is effected according to the S_N-1 mechanism, according to which there is fast and equilibrium protonation of the oxygen of the siloxane group with subsequent slow decomposition of the protonated form into a silanol and a silyl cation, the latter reacting rapidly with a nucleophilic reagent, e.g., an alcohol:

$$\equiv Si\text{—}O\text{—}Si\equiv \; \underset{}{\overset{H^+}{\rightleftharpoons}} \; \equiv Si\overset{\overset{H}{|}}{\underset{+}{\text{—}O}}\text{—}Si\equiv \qquad \text{(fast)}$$

$$\equiv Si\overset{\overset{H}{|}}{\underset{+}{\text{—}O}}\text{—}Si\equiv \; \rightleftharpoons \; \equiv Si\text{—}OH + \overset{+}{Si}\equiv \qquad \text{(slow)}$$

$$\equiv \overset{+}{Si} + HOR \; \rightleftharpoons \; \equiv SiOR + H^+ \qquad \text{(fast)}$$

There has also been suggested [109, 110] a bimolecular mechanism, S_N-2, according to which the limiting stage is interaction of a nucleophilic reagent with the protonated form of the organosiloxane:

$$\equiv Si-O-Si\equiv \; \overset{H^+}{\rightleftharpoons} \; \equiv Si-\overset{\overset{\displaystyle H}{|}}{\underset{+}{O}}-Si\equiv \qquad \text{(fast)}$$

$$\equiv Si-\overset{\overset{\displaystyle H}{|}}{\underset{+}{O}}-Si\equiv + ROH \; \rightleftharpoons \; \equiv Si-OH + \equiv Si-\overset{\overset{\displaystyle H}{|}}{\underset{+}{O}}R \qquad \text{(slow)}$$

$$\equiv Si-\overset{\overset{\displaystyle H}{|}}{\underset{+}{O}}-R \; \rightleftharpoons \; \equiv Si-O-R + H^+ \qquad \text{(fast)}$$

According to [111-115], scission of the siloxane group takes place by way of six-membered activated complexes, in which there participates either an acid molecule (oxygen-containing acids) alone or an acid molecule and a water molecule (hydrohalogen acids) or a water molecule and a hydrated proton:

$$\equiv Si-O-Si\equiv \; + H_2SO_4 \; \rightleftharpoons \; [\text{complex}] \; \longrightarrow \; \equiv SiOSO_2OH + \equiv SiOH$$

$$\equiv Si-O-Si\equiv \; + HX + H_2O \; \rightleftharpoons \; [\text{complex}] \; \longrightarrow \; \equiv SiX + \equiv SiOH + H_2O$$

$$\equiv Si-O-Si\equiv \; + H_2O + H_S^+ \; \rightleftharpoons \; [\text{complex}] \; \longrightarrow \; 2\equiv SiOH + H_S^+$$

The kinetics of decomposition of octamethylcyclotetrasiloxane (D_4) in aqueous solutions of strong acids have been investigated under heterogeneous conditions, since organosiloxanes have a very low solubility in aqueous solutions of strong acids [116]. The decomposition of D_4 took place with vigorous stirring under conditions where the reaction rate did not depend on the intensity of stirring. The constant k_{eff} for decomposition was found from the equation

$$-\ln \frac{m}{m_0} = k_{eff} \frac{S_0}{m_0} t \qquad (2.34)$$

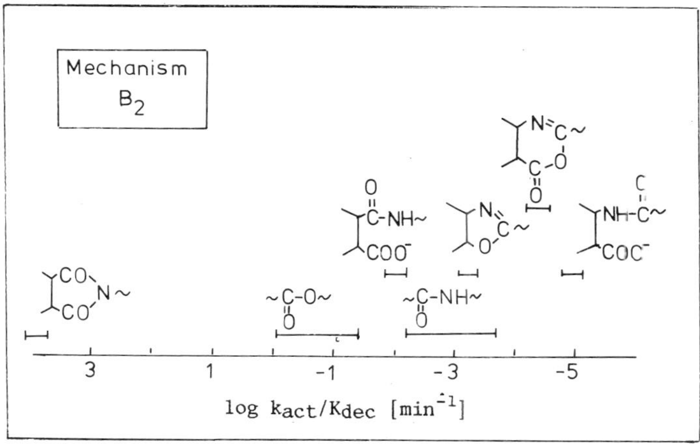

Fig. 2.18. Kinetic parameters of the decomposition of the simplest model compounds with various chemically unstable groups in aqueous basic solutions.

TABLE 2.7. Decomposition of the Principal Chemical Groups

Chemical group	Acid catalysis	Base catalysis
$\sim\underset{\substack{\| \\ O}}{C}\underset{\substack{\| \\ H}}{N}\sim$	$\log K_{eff}$, $E \approx 84$ J/mole, Conc. of acid	$\log K_{eff}$, $E \approx 68$ J/mole, Conc. of base
$\sim\underset{\substack{\| \\ O}}{C}\text{—}O\sim$	$E \approx 75$ J/mole	$E \approx 52$ J/mole
$\sim\underset{\substack{\| \\ O}}{C}\diagdown_{\underset{\substack{\| \\ O}}{C}}N\sim$	$E \approx 84$ J/mole	$E \approx 68$ J/mole
$\sim O\text{—}\underset{\substack{\| \\ H}}{\overset{\substack{H \\ \|}}{C}}\text{—}O\sim$	$E \approx 100$ J/mole	
CH_2OH ring with OH, $O\sim$	HCl, HClO$_4$, H$_2$SO$_4$, H$_3$PO$_4$	
$\sim O\text{—}\overset{\substack{\| }}{\underset{\substack{\|}}{Si}}\text{—}O\sim$	$E \approx 50$ J/mole	

where m_0 and m are the initial and current masses and S_0 the total surface of the drops of D_4 which are formed when the solution is stirred.

The decomposition of D_4 takes place in several stages. The first is scission of a siloxane bond with the formation of a linear siloxane, and is slow; the second, the further dissociation of the linear siloxane into a polymer, is fast. It has been suggested that the protonation of siloxanes, like silanes [117], is described by the acidity function H_0. The experimental data are satisfactorily described by Eq. (2.18) where $r = 1$. Figure 2.16 shows the linear dependence of $\log\,(k_{eff}/\alpha_{H_2O})$ on H_0 with slope 0.9 ± 0.1 (correlation coefficient 0.967) for aqueous solutions of HCl, H_2SO_4, and H_3PO_4. The alcoholysis of cyclic and linear organosiloxanes in weakly acidic media likewise proceeds by the S_N-2 mechanism [118, 119]. The nature of the alcohol has a significant influence on the kinetic parameters of the process.

Thus, examination of the data shows two patterns in the decomposition of compounds with chemically unstable groups in aggressive media. When the un-ionized form of the compound is reactive (decomposition of carbonyl-containing compounds by mechanism A-2) protonation of the carbonyl group reduces the concentration of the reactive form, and this leads to a considerable reduction in k_{eff} (or rise in chemical resistance) in these media.

When the ionized form is reactive (decomposition of the compounds by mechanisms A-1 and B-2), k_{eff}, as a rule, rises with a rise in the concentration of the aggressive medium; certain reactions proceeding by mechanism B-2 constitute an exception.

Table 2.7 shows the principal chemically unstable groups, the patterns of change in k_{eff} as a function of the concentration of aggressive medium, and typical activation energies.

Figures 2.17 and 2.18 show the kinetic parameters of decomposition of the simplest model compounds with chemically unstable groups in aqueous acidic media (mechanisms A-1 and A-2) and basic media (mechanism B-2).

REFERENCES

1. R. N. Lacey, J. Chem. Soc., No. 4, 1633 (1960).
2. G. Fraenkel and C. Franconi, J. Am. Chem. Soc., 82, No. 17, 4478 (1960).
3. R. J. Gillespie and T. Birchall, Can. J. Chem., 41, No. 1, 148 (1963).
4. C. O'Connor, Quart. Rev., 24, No. 4, 553 (1970).
5. J. T. Edward, J. B. Leane, and J. C. Wang, Can. J. Chem., 40, No. 8, 1521 (1962).

6. Yu. V. Moiseev, G. I. Batyukov, and M. I. Vinnik, Izv. Akad. Nauk SSSR, Ser. Fiz., 20, No. 10, 1306 (1962).
7. K. Yates, J. B. Stevens, and A. R. Katritzky, Can. J. Chem., 42, No. 8, 1957 (1964).
8. K. Yates and J. B. Stevens, Can. J. Chem., 43, No. 3, 529 (1965).
9. C. A. Bunton, B. N. Figgis, and B. Nayak, Adv. Mol. Spectrosc., 3, 1209 (1962).
10. A. J. Kresge, P. H. Fitzgerald, and Y. Chiang, J. Am. Chem. Soc., 96, No. 14, 4698 (1974).
11. R. S. de Lockerente, O. B. Nagy, and A. Bruylants, Org. Magn. Reson., 2, No. 2, 179 (1970).
12. A. R. Katritzky, A. J. Waring, and K. Yates, Tetrahedron, 19, No. 3, 465 (1963).
13. K. Yates and J. C. Riordan, Can. J. Chem., 43, No. 8, 2328 (1965).
14. E. Tamme, N. L. Khaldna, and Kh. I. Kuura, Reakts. Sposobn. Org. Soedin., 8, No. 4, 1131 (1971); 9, No. 3, 617 (1972).
15. P. D. Bolton and J. R. Wilson, Tetrahedron Lett., No. 13, 847 (1963).
16. A. R. Osborn and E. Whalley, Can. J. Chem., 39, No. 3, 597 (1961).
17. J. A. Leisten, J. Chem. Soc., No. 2, 765 (1959).
18. F. A. Long and M. A. Paul, Chem. Rev., 57, No. 5, 935 (1957).
19. M. L. Bender, Chem. Rev., 60, No. 1, 53 (1960).
20. J. K. Koskikallio, Acta Chem. Scand., 18, No. 8, 1831 (1964).
21. R. J. L. Martin, Aust. J. Chem., 18, No. 6, 807 (1965).
22. K. J. Laidler and P. A. Landskroener, Trans. Faraday Soc., 52, No. 2, 200 (1956).
23. P. D. Bolton and T. Henshall, J. Chem. Soc., No. 9, 3369 (1962).
24. E. M. Kosower, Molecular Biochemistry, McGraw-Hill, New York (1962).
25. A. J. Hall and D. P. N. Satchell, Chem. Ind. (London), No. 13, 527 (1974).
26. H. Pracejus, Chem. Ber., 92, No. 4, 988 (1959).
27. V. F. Klages and E. Zange, Ann. Chem., 607, 35 (1957).
28. C. R. Smith and K. Yates, Can. J. Chem., 50, No. 5, 771 (1972).
29. A. R. Fersht, J. Am. Chem. Soc., 93, No. 14, 3504 (1971).
30. H. Benderley and K. J. Rosenheck, J. Chem. Soc., D, No. 3, 179 (1972).
31. R. B. Martin and W. C. Hutton, J. Am. Chem. Soc., 95, No. 4, 4752 (1973).
32. C. A. Bunton, C. J. O'Connor, and T. A. Turney, Chem. Ind. (London), 1835 (1976).
33. C. J. Giffney and C. J. O'Connor, Aust. J. Chem., 29, No. 2, 307 (1976).
34. P. P. Nechaev, Yu. V. Moiseev, and G. E. Zaikov, Vysokomol. Soedin., A, 14, No. 5, 1048 (1972).
35. W. P. Jencks, Catalysis in Chemistry and Enzymology, McGraw-Hill, New York (1969).

36. Yu. V. Moiseev, V. S. Markin, and G. E. Zaikov, Usp. Khim.,
    $\underline{45}$, No. 3, 510 (1976).
37. M. I. Vinnik and I. M. Medvetskaya, Zh. Fiz. Khim., $\underline{41}$, No.
    7, 1775 (1967).
38. I. G. Orlov, Candidate's Dissertation, Inst. Khim. Fiz. Akad.
    Nauk SSSR, Moscow (1975).
39. J. W. Barnett and C. O'Connor, J. Chem. Soc., D, No. 9, 525
    (1972).
40. A. Signor and E. Bordignon, J. Org. Chem., $\underline{30}$, No. 10, 3447
    (1965).
41. F. H. Stodola, J. Org. Chem., $\underline{37}$, No. 2, 178 (1972).
42. J. Tafel and O. Wassmuth, Ber. Dtsch. Chem. Ges., $\underline{3}$, 2831
    (1907).
43. A. Berger, A. Loewenistein, and S. Meiboom, J. Am. Chem. Soc.,
    $\underline{81}$, No. 1, 62 (1959).
44. M. L. Bender and R. D. Ginger, J. Am. Chem. Soc., $\underline{77}$, No. 2,
    348 (1955).
45. M. I. Vinnik and Yu. V. Moiseev, Tetrahedron, $\underline{19}$, 1441 (1963).
46. S. S. Biechler and R. W. Taft, J. Am. Chem. Soc., $\underline{79}$, No. 18,
    4927 (1957).
47. Yu. V. Moiseev, Candidate's Dissertation, Inst. Khim. Fiz.
    Akad. Nauk SSSR, Moscow (1963).
48. Yu. V. Moiseev, E. Ya. Bakhrakh, and M. I. Vinnik, Zh. Fiz.
    Khim., $\underline{37}$, No. 4, 784 (1963).
49. U. Meresaar and L. Bratt, Acta Chem. Scand., A, $\underline{28}$, No. 7, 715
    (1974).
50. I. G. Orlov, et al., Reakts. Sposobn. Org. Soedin., $\underline{2}$, No. 4,
    180 (1965).
51. M. L. Bender, Y. L. Chow, and F. Chloupek, J. Am. Chem. Soc.,
    $\underline{80}$, No. 20, 5380 (1958).
52. P. P. Nechaev et al., Int. J. Chem. Kinet., $\underline{6}$, No. 2, 245
    (1974).
53. M. F. Aldersley et al., J. Chem. Soc., B, $\underline{12}$, 1487 (1974).
54. M. L. Ernst and G. L. Schmir, J. Am. Chem. Soc., $\underline{88}$, No. 21,
    5001 (1966).
55. P. P. Nechaev et al., Vysokomol. Soedin., A, $\underline{15}$, No. 3, 702
    (1973).
56. J. T. Edward and S. C. R. Meacock, J. Chem. Soc., No. 5, 2009
    (1957).
57. R. H. de Wolfe and F. B. Augustine, J. Org. Chem., $\underline{30}$, No. 3,
    699 (1965).
58. R. Greenhalgh, R. M. Heggie, and M. A. Weinberger, Can. J.
    Chem., $\underline{41}$, No. 7, 1662 (1963).
59. V. V. Korshak, G. M. Tseitlin, and A. I. Pavlov, Dokl. Akad.
    Nauk SSSR, $\underline{163}$, No. 1, 116 (1965).
60. V. S. Yakubovich et al., Dokl. Akad. Nauk SSSR, $\underline{159}$, No. 3,
    630 (1964).
61. T. Kubota and R. Nakanihi, J. Polym. Sci., B, $\underline{2}$, No. 6, 655
    (1964).
62. A. R. Katritzky et al., Chem. Ind. (London), No. 18, 711 (1966).

63. C. A. Grob and B. Fisher, Helv. Chim. Acta, 38, No. 7, 1794 (1955).
64. E. H. Cordes and W. P. Jencks, J. Am. Chem. Soc., 85, No. 18, 2843 (1963).
65. V. N. Kulagin et al., Izv. Akad. Nauk SSSR, Ser. Khim., No. 8, 1770 (1974).
66. A. I. Donskikh et al., Izv. Akad. Nauk SSSR, No. 3, 562 (1978).
67. H. Zimmermann and H. Geisenfelder, Z. Elektrochem., 65, No. 4, 368 (1961).
68. O. J. Rabiger and M. M. Joullie, J. Org. Chem., 29, No. 3, 476 (1964).
69. K. J. Morgan and A. M. Turner, Tetrahedron, 25, No. 4, 915 (1969).
70. W. M. Schubert, J. Donohue, and J. D. Gardner, J. Am. Chem. Soc., 76, No. 1, 9 (1954).
71. R. J. Gillespie and E. A. Robinson, Nonaqueous Solvent Systems, T. C. Waddington (ed.), Academic Press, London and New York (1965).
72. G. E. Maciel and D. D. Traficante, J. Phys. Chem., 69, No. 3, 1030 (1965).
73. K. Yates and R. A. McClelland, J. Am. Chem. Soc., 89, No. 11, 2686 (1967).
74. C. A. Lane, M. F. Cheung, and G. F. Dorsey, J. Am. Chem. Soc., 90, No. 23, 6492 (1968).
75. Yu. P. Siigur and Yu. L. Khaldna, Reakts. Sposobn. Org. Soedin., 7, No. 1, 211 (1970).
76. Yu. P. Siigur et al., Reakts. Sposobn. Org. Soedin., 7, No. 2, 412 (1970).
77. M. I. Vinnik and N. B. Librovich, Reakts. Sposobn. Org. Soedin., 7, No. 4, 1221 (1970).
78. M. Liler, J. Chem. Soc., D, No. 2, 115 (1971).
79. T. Birchall and R. J. Gillespie, Can. J. Chem., 41, No. 10, 2642 (1963).
80. Yu. L. Khaldna and T. K. Rodima, Reakts. Sposobn. Org. Soedin., 10, No. 3, 719 (1973).
81. A. G. Davies and J. Kenyon, Quart. Rev., 9, No. 3, 203 (1955).
82. K. Yates, Acc. Chem. Res., 4, No. 4, 136 (1971).
83. Yu. R. Siigur and Yu. L. Khaldna, Reakts. Sposobn. Org. Soedin., 7, No. 2, 431 (1970).
84. M. I. Vinnik and N. B. Librovich, Izv. Akad. Nauk SSSR, Ser. Khim., 10, 2211 (1975).
85. P. Salomoe, Suomen Kemistil., No. 1, 145 (1959).
86. F. A. Long and M. Purchase, J. Am. Chem. Soc., 72, No. 7, 3267 (1950).
87. M. L. Bender, J. Am. Chem. Soc., 73, No. 4, 1626 (1951).
88. J. E. Leffler, Reactive Intermediates of Organic Chemistry, Wiley-Interscience, New York (1956).
89. W. P. Jencks et al., Arch. Biochem. Biophys., 88, No. 2, 193 (1960).

90. M. L. Bender, Mechanisms of Catalysis of Nucleophilic Reactions of Carboxylic Acid Derivatives [Russian translation], Mir, Moscow (1964).

91. J. Mathieu and A. Allais, Principles of Organic Synthesis [Russian translation], Izd. Inostr. Lit., Moscow (1962).

92. W.Kern and V. Jaacks, J. Polym. Sci., $\underline{48}$, No. 150, 399 (1960).

93. M. M. Kreevoy, C. R. Morgan, and R. W. Taft, J. Am. Chem. Soc., $\underline{82}$, No. 12, 3064 (1960).

94. J. N. Be Miller, Adv. Carbohydr. Chem., $\underline{22}$, 25 (1967).

95. W. G. Overend, C. W. Rees, and J. S. Sequeira, J. Chem. Soc., No. 9, 3429 (1962).

96. R. J. Withey and E. Whalley, Can. J. Chem., $\underline{41}$, No. 7, 1849 (1963).

97. E. Whalley, Trans. Faraday Soc., $\underline{55}$, No. 5, 798 (1959).

98. M. M. Kreevoy and R. W. Taft, J. Am. Chem. Soc., $\underline{77}$, No. 21, 5590 (1955).

99. M. A. Paul, J. Am. Chem. Soc., $\underline{74}$, No. 1, 141 (1952).

100. E. M. Arnett, in: Contemporary Problems in Physical Organic Chemistry, M. E. M. Vol'pin (ed.) [Russian translation], Mir, Moscow (1967), p. 195.

101. H. Leamaire and H. J. Lucas, J. Am. Chem. Soc., $\underline{73}$, No. 11, 5198 (1951).

102. J. T. Edward, Chem. Ind. (London), 1102 (1955).

103. H. E. Armstrong and W. H. Glover, Proc. Roy. Soc., B, $\underline{80}$, 312 (1908).

104. N. A. Khalturinskii et al., Dokl. Akad. Nauk SSSR, $\underline{198}$, No. 1, 149 (1971).

105. N. A. Khalturinskii, Candidate's Dissertation, Inst. Khim. Fiz. Akad. Nauk SSSR, Moscow (1972).

106. D. P. N. Satchell, J. Chem. Soc., No. 4, 1752 (1960); No. 5, 1894 (1962).

107. K. Thinius and W. Saupe, Plaste Kautsch., $\underline{16}$, No. 4, 252 (1969).

108. K. Damm, D. Golitz, and W. Noll, Angew. Chem., $\underline{76}$, No. 6, 273 (1964).

109. K. Damm, D. Golitz, and W. Noll, Z. Anorg. Chem., $\underline{340}$, No. 1, 1 (1965).

110. K. A. Andrianov and S. E. Yakushkina, Vysokomol. Soedin., $\underline{7}$, No. 4, 613 (1965).

111. Z. Lasocki, J. Kulpinski, and W. Gador, Polim. Tworz. Wielkoczast., $\underline{12}$, 442 (1970).

112. A. G. Kuznetsova, V. I. Ivanova, and S. A. Golubtsov, Plast. Massy, No. 9, 19 (1969); translated in Soviet Plast., No. 9 (1969).

113. M. G. Voronkov, Doctoral Dissertation, Institute of Petrochemical Synthesis, Academy of Sciences, USSR, Moscow (1961).

114. M. G. Voronkov and L. A. Zhagata, Zh. Obshch. Khim., $\underline{38}$, No. 10, 2327 (1968).

115. M. G. Voronkov and L. A. Zhalata, Izv. Akad. Nauk Latv. SSR, Ser. Khim., No. 3, 507 (1964).

116. L. P. Razumovskii et al., Izv. Akad. Nauk SSSR, Ser. Khim.,
     No. 11, 2448 (1973).
117. L. H. Sommer, Stereochemistry, Mechanism and Silicon, McGraw-
     Hill, New York (1966).
118. L. P. Razumovskii et al., Izv. Akad. Nauk SSSR, Ser. Khim.,
     No. 7, 1502 (1975).

Chapter 3

SPECIAL FEATURES
OF THE
CHEMICAL DEGRADATION
OF POLYMERS

The study of the kinetics of chemical degradation is an exceptionally complex problem, both with regard to obtaining accurate experimental data and with regard to finding the kinetic parameters of the individual elementary processes. Nowadays progress achieved in physical organic chemistry (the introduction of acidity and alkalinity functions, the establishment of the connection between the rate of acid—base reactions and their functions, and the use of correlation equations in chemistry) makes it possible not only to determine the kinetic parameters of the individual processes, but also to predict their changes as a function of the thermodynamic parameters of the reaction medium and of the structure of the reagents.

In this connection, nevertheless, new problems arise.

Can the basic postulates developed in the field of physical organic chemistry for low-molecular compounds be applied to the chemical degradation of polymers?

To what extent can one model these processes by using low-molecular analogs?

In this chapter we endeavor to answer these questions.

Flory [1, p. 4] formulated the basic principles according to which the reactivity of functional groups in oligomers and polymers is independent of the molecular mass (MM) for various types of reactions. Functional groups in a polymer should have the same reactivity as in a low-molecular model compound if the following provisions are observed:

the reaction takes place in a homogeneous medium, and the initial reagents and the reaction products are soluble in this medium;

in each elementary process there participates a single type of functional group of the polymers, and the molecules of all the other reagents are small and comparatively labile;

an appropriate low-molecular analog is selected taking into account the steric hindrances which arise with change in the conformation of the macromolecule in the reaction.

As pointed out by Alfrey [2, p. 9], identical reactivity of functional groups of a polymer and a low-molecular analog may not be the case when one of these provisions is not observed.

Today, however, a large number of polymers are known where the reactivity of various groups in the polymer is of anomalous character, and this is explained by the specific influence of the polymeric state [3, p. 16; 4, p. 250; 5]. Let us consider the catalytic degradation reactions of polymers in a homogeneous medium. For simplicity we take a catalytic reaction proceeding as follows:

$$\sim R + cat \underset{k_2}{\overset{k_1}{\rightleftharpoons}} \sim X \qquad (a)$$

$$\sim X + R' \xrightarrow{k_3} P + cat \qquad (b)$$

where R is the functional group of the polymer and R' the reagent, e.g., a molecule of water in hydrolysis.

What will be the difference in this reaction if the functional group belongs to a polymer and not to a low-molecular analog? It is assumed that the low-molecular analog is selected taking into account the steric hindrances which arise with change in the conformation of the macromolecule in the course of the reaction.

Solutions of polymers may have a viscosity several orders higher than solutions of their low-molecular analogs. What influence will the macroviscosity of the solution have on the course of the above reaction?

In addition, functional groups in polymers are as a rule located close to each other, and so it is interesting to ascertain the influence of interaction of the functional groups on the degradation reaction.

3.1. QUANTITATIVE ASSESSMENT OF THE INFLUENCE
OF MACROVISCOSITY ON THE KINETICS OF DEGRADATION

The macroviscosity of the reaction medium determines the rate of diffusion of the substances dissolved in it and, consequently, it influences the rate of the chemical reactions. The question of the influence of diffusion factors on the kientics of homogeneous chemical reactions of small molecules is considered in [6-8, p. 17]. When monomolecular reactions occur, diffusion does not play a significant role. Let us consider the influence of diffusion on the kinetics of bimolecular reactions. If, for instance, the rate of the reaction of R with cat is high in relation to the rate of diffusion of R and cat, then in the direct vicinity of R the concentration of the catalyst may fall, which will lead to a concentration differential causing diffusional flow of the particles in the direction toward R. Under steady-state conditions, the rate of this flow becomes equal to the reaction rate.

According to [6] we get

$$k_{eff} = \frac{4\pi\delta D k_{act}\beta}{k_{act} + 4\pi\delta D\beta} \tag{3.1}$$

where k_{eff} and k_{act} are, respectively, the effective and actual rate constants; δ is the sum of the radii of the reacting particles; β is a coefficient of proportionality, which is 1 molecule^{-1}; and D is the sum of the diffusion coefficients of cat and reagent R' in the medium.

It must be noted that even when the functional group in the polymer is completely lacking in lability, the frequency of collisions does not necessarily change significantly. For sufficiently flexible macromolecules the lability of a functional group in the polymer does not differ greatly from the lability of the small molecules.

From Eq. (3.1) it is easy to find the conditions for the reaction to be kinetically or diffusion-controlled:

a) if $k_{act} \ll 4\pi\delta D\beta$, then

$$k_{eff} = k_{act} \tag{3.2}$$

i.e., there is equilibrium-statistical distribution of the particles in the solution (the reaction is kinetically controlled);

b) if $k_{act} \gg 4\pi\delta D\beta$, then

$$k_{eff} = 4\pi\delta D\beta \tag{3.3}$$

TABLE 3.1. Diffusion Coefficients of Acids in Water at
            25°C

Acid	$D \cdot 10^6$, cm^2/sec	Reference
HCl	30.0	[11. p. 789]
H_2SO_4	12.0	[10]
H_3PO_4	8.7	[12]

i.e., there is abrupt breakdown of equilibrium-statistical distribution of the particles in the solution (the reaction is diffusion-controlled).

At ordinary temperatures in nonviscous media the values of the diffusion coefficients of the catalyst lie within the range 10^{-6} to 10^{-5} cm^2/sec^{-1} (Table 3.1), while the values of δ for typical functional groups and catalysts are of a few angstroms. Consequently, the value of $4\pi\delta D\beta$ is 10^{-12}-10^{-11} cm^3/(molec·sec) or 10^9-10^{10} liters × (mole·sec)$^{-1}$.

Thus it is evident that those reactions for which $k_{act} > 10^8$ liters/(mole·sec) are diffusion controlled. It is in fact under these conditions that there takes place interaction of a prcton, hydroxide ion [9], or enzymes [10] with functional groups of reagents, i.e., stage (a) of the degradation of polymers.

Reactions for which $k_{act} < 10^8$ liters/(mole·sec) are not diffusion-controlled and take place under kinetic conditions. These include the decomposition reactions of functional groups and of the majority of enzyme—substrate complexes, with the formation of reaction products, i.e., stage (b) of the degradation of polymers. In the majority of cases this stage limits the rate of the entire degradation process. When we look at viscous systems, the value of δ remains practically unchanged, but the diffusion coefficient changes inversely with the viscosity of the system in accordance with the Stokes—Einstein equation:

$$D = \frac{k_B T}{6\pi r \eta} \qquad (3.4)$$

where r is the radius of the diffusing particle and η the viscosity of the medium.

If, as a first approximation, we consider the molecules as spherical particles of constant specific gravity ρ, then

$$D = \frac{k_{\mathrm{B}}T}{6\eta}\left(\frac{3\pi^2}{4\rho N_{\mathrm{A}}}\right)^{-1/3} M^{-1/3} \tag{3.5}$$

i.e., the diffusion coefficient must not depend very strongly on the molecular mass M of the diffusing substance [9, p. 262].

Where the intermolecular interactions are considerable we use the Debye equation

$$k_{\mathrm{eff}} = \frac{k_{\mathrm{act}}}{1 + \dfrac{k_{\mathrm{act}}\displaystyle\int_0^\infty \exp\left(U k_{\mathrm{B}}^{-1}T^{-1}\right)\dfrac{dr}{r^2}}{4\pi D}} \tag{3.6}$$

The value of k_{eff} depends on the specific form of the function u = u(r). For instance, the interaction of two ions is expressed by the Coulomb equation

$$U = \frac{z_A z_B e^2}{\varepsilon r} \tag{3.7}$$

where z_A and z_B are the charges on the interacting ions, ε the permittivity of the medium, and r the interion distance (r > δ).

The expression for k_{eff} takes the form

$$k_{\mathrm{eff}} = \frac{k_{\mathrm{act}}}{1 + \dfrac{k_{\mathrm{act}}\,\varepsilon k_{\mathrm{B}}T\,\{[\exp\,(z_A z_B e^2/\varepsilon\delta k_{\mathrm{B}}T)] - 1\}}{4\pi D\beta z_A z_B e^2}} \tag{3.8}$$

If even one of the reagents is not an ion, then $z_A z_B$ = 0 and Eq. (3.8) changes into (3.1).

The value of D for electrolyte solutions depends on the viscosity in accordance with the equation in [10]:

$$D = D^0\left(\frac{\eta^0}{\eta}\right)\left(1 - \frac{e^2 f}{4\varepsilon k_{\mathrm{B}}T}\right) \tag{3.9}$$

where D^0 and η^0 are the diffusion and viscosity coefficients, respectively, at infinite dilution, and f a variable depending on the concentration of electrolyte.

It must be noted that Eq. (3.9) is inapplicable to very viscous media.

From the above it may be concluded that even in very viscous media stage (b) is never diffusion-controlled, i.e., this stage of

TABLE 3.2. Kinetic Parameters of Degradation Reactions of Polymers and Their Low-Molecular Analogs

Polymer and its low-molecular analog	Solvent and catalyst	$k_{act\ 25\ °C}$, sec^{-1}	E, J/mole	$\log A$, sec^{-1}	Reference
Polycaproamide	H_2O—H_2SO_4	$8.3 \cdot 10^{-8}$	84 ± 4	7,2	[13]
Cyclic trimer of caprolactam	The same	$8.3 \cdot 10^{-8}$	84 ± 4	7,4	[13]
Cellulose	H_2O—HCl H_2O—H_2SO_4 H_2O—$HClO_4$	$1.1 \cdot 10^{-8}$	126 ± 4	14,2	[13]
Cellobiose	H_2O—H_3PO_4	$1,6 \cdot 10^{-8}$	126 ± 4	14,0	[13]
Poly(N,N-diethylacrylamide)	H_2O—buffer	$2.5 \cdot 10^{-7}$	92	9,5	[14]
		$3,1 \cdot 10^{-9}$	103	9,5	[15[
N,N-Diethylisobutyramide	The same	$2,0 \cdot 10^{-7}$	91	9,7	]14]
		$8,0 \cdot 10^{-9}$	99	9,5	[15]
Polyacrylamide	''	$2,0 \cdot 10^{-5}$	60	5.6	[16]
Isobutyramide	''	$2,0 \cdot 10^{-5}$	60	5,6	[16]
Poly(N-vinyl-2-pyrrolidone	''	$3,1 \cdot 10^{-7}$	105 ± 4	11,7	[17]
N-Isopropyl-5-methyl-2-pyrrolidone	''	$4,0 \cdot 10^{-8}$	113 ± 4	12,3	[17]

catalytic degradation always takes place under kinetic conditions (we have in mind a microkinetic approach as compared with macrokinetic; cf. Chap. 6).

Thus the macroviscosity of the reaction medium has a significant influence on the rate of reaction of catalysts with functional groups in the reagents, but has practically no influence on the rate of decomposition of complexes of functional groups with the catalysts leading to the formation of products. In the majority of cases the latter reaction is the limiting stage of the whole degradation process.

3.2. THE INFLUENCE OF THE INTERACTION OF FUNCTIONAL GROUPS ON THEIR REACTIVITY

If functional groups in polymer molecules are at a distance apart, which for practical purposes rules out reciprocal influence, then the kinetic parameters of the degradation reactions of the polymers and of their low-molecular analogs are closely similar, i.e., the principle that the reactivity is independent of the degree of polymerization holds.

In Table 3.2 we give the kinetic parameters of the degradation reactions of polymers and their low-molecular analogs. These parameters are closely similar within the limits of experimental error.

It is interesting to note that in the reaction of H—D exchange the reactivity (or k_{eff}) of polymers of the acrylic series is ap-

TABLE 3.3. Kinetic Parameters of the Reaction of H—D Exchange in Polymers and Their Low-Molecular Analogs

Polymer and its low-molecular analog	Solvent and catalyst	k^*_{act} 25°C, sec^{-1}	E, J/mole	log A, [sec^{-1}]	Reference
Poly(isopropylpropion-amide)	D_2O—DCl	$5 \cdot 10^{-5}$	24	10.3	[18]
N-Isopropylpropion-amide	D_2O—KOD	$2.5 \cdot 10^{-3}$	84	12.0	
Poly(aminomethacryl-2-lysine)	D_2O—DCl	$6.6 \cdot 10^{-5}$	80	9.8	[19]
ε-Aminoisobutyryl-L-lysine	D_2O—KOD	$2.5 \cdot 10^{-3}$	67	9.1	
Poly(methylpropion-amide)	D_2O—KOD	$4 \cdot 10^{-4}$	76	9.7	[20]
N-Methylacetamide	D_2O—KOD	$1 \cdot 10^{-2}$	72	10.4	

*Rate constant at minimum point of k_{eff} vs. pD.

TABLE 3.4. Kinetic Parameters of Dehydrochlorination Reactions of Model Compounds of PVC in the Gaseous Phase [22]

Model compound	Temperature range	$k_{177 °C} \cdot 10^{11}$, sec^{-1}	E, J/mole	log A, [sec^{-1}]
2-Chlorobutane	347—394	3.6	206	13
2,4-Dichloropentane	347—397	3.3	206	13
2,4,6-Trichloroheptane	346—386	3.4	214	14
3-Chloro-1-pentene	319—380	8.9	130	15
4-Chloro-2-pentene	296—345	859	172	13
6-Chloro-2,4-heptadiene	362—347	45900	147	18

TABLE 3.5. Kinetic Parameters of Dehydrochlorination Reactions of Model Compounds of PVC in the Liquid Phase [23]

Model compound	Temperature range	$k_{177 °C} \cdot 10^{7}$, sec^{-1}	E, J/mole	log A, [sec^{-1}]
8-Chlorohexadecane	236—284	1	151	10.5
8-Chloro-6-tridecene	165—196	10^3	105	8.1
4-Chloro-2-dodecene	156—180	$8 \cdot 10^2$	125	6.9
7-Chloro-3,5-nonadiene	70—96	$4 \cdot 10^5$	81	8.0
6-Chloro-2,4-octadiene	87—113	$3,2 \cdot 10^5$	75	7.1

proximately two orders lower than for the corresponding low-molecular analogs (Table 3.3). This, in the opinion of Scarpa et al. [18],
is due to the reduction in the dissociation constant for water in
the vicinity of the amide groups as compared with the solution, i.e.,
the reduction in the reaction rate of H—D exchange in the polymers
is due not to the structure of the polymer molecules themselves,
but to the specific state of the second reaction (the molecule of
D_2O).

Frequently, however, the functional groups are located close
to each other (~ 10 Å). It is this that brings about the anomalous
reactivity of the "polymeric" functional groups. At the present
time we distinguish between the following types of anomalous reactivity in polymers.

Effect of an Adjacent Group. This may be either of long-range
order (known in the literature as a "chain effect"), or of short-
range order (the influence of groups located directly around a particular functional group).

Conformational Effects. These are brought about by a change
in reactivity as a consequence of a change in the conformation of
the macromolecule in a particular reaction medium or else in the
course of reaction.

Apparently these effects can be explained on the basis of the
polar, resonance, and steric effects, well known in organic chemistry [21, p. 562], in exactly the same way as the mechanism of the
hydrolytic effect of a number of ferments has been explained by
means of clear chemical analogies. Let us consider some examples
of these effects.

3.2.1. Effect of an Adjacent Group
of Long-Range Order

This may be illustrated by comparing the reactivities of model
compounds of poly(vinylchloride) (PVC) with varying numbers of
double bonds.

It is now established that the dehydrochlorination process is
stepwise, with the double bonds which form activating the next process of dehydrochlorination (allyl activation).

Tables 3.4 and 3.5 show the kinetic parameters of the dehydrochlorination reaction of model compounds of PVC in gaseous and liquid phases.

When we compare the rate constants of dehydrochlorination,
whether in the gaseous or liquid phase, it is clear that the addi-

tion of each double bond to the model compound raises the value of
k by approximately two orders.

Thus the formation of individual double bonds and the growth
of a system of conjugated bonds due to allyl activation can be simulated by the dehydrochlorination reactions of alkyl chlorides and
chloroalkenes, respectively.

3.2.2. Effect of An Adjacent Group
of Short-Range Order

These effects become apparent in that a substituent present
on the same carbon atom as a functional group either favors or dis-
favors the formation of an activated complex. For instance, the
hydrolysis of a copolymer of acrylic acid with p-nitrophenyl methacrylate proceeds more rapidly than that of p-nitrophenyl acetate
since in the former case there is readily formed a six-membered
activated complex with an attack of the carboxylate ion on the ester
group [24]:

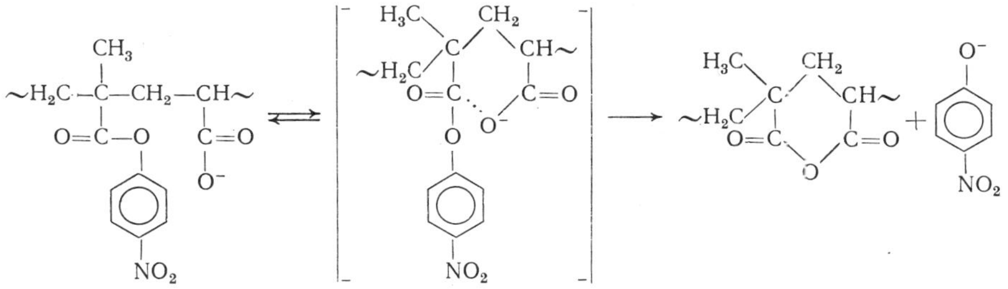

The following are the rate constants of the intramolecular hydrolysis of a copolymer of acrylic acid and of the molecular interaction of a model compound with an acetate ion at 0°C and pH 6.0:

Copolymer of acrylic acid with p-nitrophenyl meth-acrylate	$2 \cdot 10^{-3}$ sec^{-1}
p-Nitrophenyl acetate + acetate ion	$6.5 \cdot 10^{7}$ liters $\times$ (mole·sec)$^{-1}$*

*Found by extrapolation ($E = 66$ J/mole).

The influence of adjacent groups on a particular functional
group is considered in [25-31]. In a polymer molecule there may,
at low degrees of dissociation, be functional groups of three types:

those whose nearest neighbors have not reacted;

those which have a single neighbor which has reacted;

those which have two neighbors which have both reacted.

The rates of dissociation of functional groups under conditions of excess concentration of catalyst are determined as follows:

$$- \frac{dX}{dt} = k_0 N_0 + k_1 N_1 + k_2 N_2 \tag{3.10}$$

where k_0, k_1, and k_2 are the rate constants of functional groups of types 1, 2, and 3, respectively; and N_0, N_1, and N_2 are the molar fractions of the respective types of functional group.

The solution of the system of corresponding differential equations linking the change in N_n sequences of n reacted groups with the respective concentrations of functional groups of various types, the rate constants, and the time leads to the following expressions [25]:

$$N_0 = \exp \left\{ - \left(2 \frac{k_1}{k_0} + 1 \right) k_0 t - 2 \left(\frac{k_1}{k_0} - 1 \right) [(\exp - k_0 t) - 1] \right\} \tag{3.11}$$

$$N_1 = [(\exp - k_0 t) - 1] \exp \left\{ -- \left(2 \frac{k_1}{k_0} + 1 \right) k_0 t - 2 \left(\frac{k_1}{k_0} - 1 \right) [(\exp - k_0 t) - 1] \right\} \tag{3.12}$$

For N_2 the solution has a complex form, e.g., for $k_1/k_0 = 1$,

$$N_2 = \frac{2}{(k_2/k_0 - 2)(k_2/k_0 - 3)} \left[\left(\frac{k_2}{k_0} - 3 \right) \exp(-2k_0 t) \right.$$

$$\left. - \left(\frac{k_2}{k_0} - 2 \right) \exp(-3k_0 t) + \exp(-k_2 t) \right] \tag{3.13}$$

Figure 3.1 shows the dependence of the degree of dissociation of the functional groups in the polymer on time for various values of the constants [27]:

$$k_0 = k_1 = k_2 \qquad \text{(linear relationship)}$$
$$k_0 = 1, \ k_1 < 1, \ k_2 < 1 \qquad \text{(autoinhibition)}$$
$$k_0 = 1, \ k_1 > 1, \ k_2 > 1 \qquad \text{(autoacceleration)}$$

We most frequently encounter the second case, where degradation as a result of autoinhibition does not run to the end, i.e., the process runs to a particular limiting degree of dissociation. In [27-29] a probability approach to the solution of this problem is adopted.

The following are the limiting degrees of dissociation of functional groups in the polymer for various ratios of k_0 and k_1 [31]:

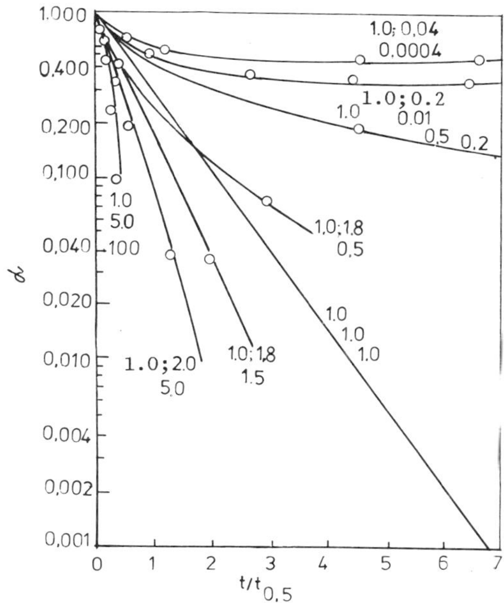

Fig. 3.1. Dependence of the degree of dissociation of the
functional groups in the polymer on time. The
numbers against the curves give the various
values of the reaction rate constants k_0, k_1,
and k_2 [26].

α	0.667	0.632	0.606	0.600	0.586	0.577	0.570
k_1/k_0	1	0.5	0.25	0.20	0.10	0.05	0.01

In the same paper, a careful study is made of the kinetics of
alkaline hydrolysis of polyacrylamide. To describe the experimental data the authors of the paper made the scheme more complex, assuming that the carboxylate group can influence not only the adjacent but also the next amide group.

The following data show the influence of carboxylate groups
on the reactivity of the amide groups in the alkaline hydrolysis of polyacrylamide at 60°C (A is the amide group, C the carboxylate group):

	—A—A—A—	—A—A—A—C—A—	—A—C—A—A—C—	—A—C—A—C—A—
$k \cdot 10^4$ liters/(mole·sec)	10.8	3	2	0.65
$\Delta F^{\neq}$ J/mole	0	3.1	4.2	12.6

TABLE 3.6.   Rate Constants of Hydrolysis of Copolymers of Methacrylic
Acid and Its Esters Having Various Microtacticities
(MMA — methyl methacrylate; MAA — methacrylic acid;
DPMMA — diphenylmethyl methacrylate)

Copolymer	Medium	$k_0 \cdot 10^4$, min^{-1}	$k_1 \cdot 10^4$, min^{-1}	$k_2 \cdot 10^4$, min^{-1}
MMA→MAA (iso)	0.2 N KOH	90	35	35
MMA→MAA (syndio)		5.8	1.2	0.3
DPMMA→MAA (iso)	Pyridine—water (95:5)	0.3	7	33
DPMMA→MAA (syndio)		0.5	0.5	0.5

The increase in the Helmholtz energy (or free energy) of activation as a consequence of the influence of adjacent carboxylate
groups was determined by the formula

$$\Delta F^{\neq} = \frac{eR}{\varepsilon r k_{\mathrm{B}}}$$ (3.14)

where r is the average distance between the amide group and the carboxylate group, taken as 3.5 Å.

Plate and Litmanovich [32] investigated in detail the kinetics
of hydrolysis of copolymers of methacrylic acid and its esters of
various microtacticities in basic media.   The kinetic characteristics of the process are given in Table 3.6.

In the case of MMA—MAA there is autoinhibition as a consequence
of the repulsion of the hydroxide ions by carboxylate ions.   The rate
constants are higher for a copolymer of isotactic structure than
for copolymers of a different structure, since in this instance a
six-membered activated complex can easily form.

In a less basic medium (pyridine—water) there is autoaccelera-
tion in the case of isotactic DPMMA—MAA.   In this medium some
of the carboxyl groups are in the dissociated state, the other in
the undissociated state, and this leads to bifunctional catalysis
by a mechanism of anchimeric cooperation [33].

In a syndiotactic copolymer such catalysis does not take place, and accordingly there is linear dependence of the degree of dissociation of the ester groups with time. The hydrolysis of ester bonds in copolymers of the acrylic and methacrylic series, as already shown, frequently takes place by an intramolecular mechanism of catalysis.

In a number of surveys [2-5], similar reactions are presented as examples of processes characteristic only of polymers. At the present time there is a wealth of experimental material showing the powerful influence of structure on the reactivity of monophenolic esters of dicarboxylic acids in the intramolecular reaction of hydrolysis in an alkaline medium [34, 35].

$$
\begin{array}{lll}
H_2C\!\!\begin{array}{l} \diagup CH_2\!-\!COOR \\ \diagdown CH_2\!-\!COO^- \end{array} & \text{(I)} & 1{,}0 \\[2ex]
\begin{array}{l} H_3C \\ C \\ H_3C \end{array}\!\!\begin{array}{l} \diagup CH_2\!-\!COOR \\ \diagdown CH_2\!-\!COO^- \end{array} & \text{(II)} & 20 \\[2ex]
\begin{array}{l} CH_2\!-\!COOR \\ | \\ CH_2\!-\!COO^- \end{array} & \text{(III)} & 30 \\[2ex]
\begin{array}{l} CH\!-\!COOR \\ \| \\ CH\!-\!COO^- \end{array} & \text{(IV)} & 10\,000 \\[2ex]
\begin{array}{l} CH\!-\!CH_2\!-\!COOR \\ HC \diagup\ \diagdown \\ \| O \\ HC \diagdown\ \diagup \\ CH\!-\!CH_2\!-\!COO^- \end{array} & \text{(V)} & 53\,000
\end{array}
$$

where $R = C_6H_5$ or $n = BrC_6H_4$

The above data show that the bicyclic compound (V) reacts 53,000 times as rapidly as the monophenyl glutarate (I).

The acceleration of the intramolecular reactions is brought about by various factors.

It is shown in [36, 37] that such acceleration cannot be brought about:

by local increase in the concentration of the reagents around the group being hydrolyzed with the occurrence of an intramolecular reaction; for instance the effective concentration of carboxylate groups around an ester group in the hydrolysis of a copolymer of acrylic acid and p-nitrophenyl methacrylate can be assessed from the ratio K (intramolecular)/K (intermolecular); it is approximately $3 \cdot 10^3$ moles/liter, which is considerably higher than the concentrations actually achieved;

by favorable relative orientation of the reacting groups since the kinetic differences between intramolecular and intermolecular reactions do not amount merely to an entropy effect.

A considerable contribution to the acceleration of various intramolecular and, evidently, polymeric reactions may be made by changes in the solvation.

For the formation of an activated complex in the intramolecular reaction of hydrolysis of monophenyl esters of dicarboxylic acids it is necessary for the water molecules which hydrate the charged carboxylate group to be removed. Consideration of molecular models shows that this group in ester (V) cannot be hydrated by water from the direction from which the attack on the ester group takes place. Accordingly, with the occurrence of this particular intramolecular reaction we do not require energy to remove water in this position from the hydrate shell.

It is interesting to present the following analogy. The catalytic activity of the proton, which is expressed by the acidity function, increases proportionally as a proton of the water molecules is removed from the hydrate shell. In nonaqueous solvents, e.g., acetic acid, nitromethane and sulfolane, the acidity function is, for identical concentrations of sulfuric acid, several orders higher than in an aqueous solution [38]. Probably it is specifically these factors which bring about the high rates of intramolecular reactions whether in polymers or in their low-molecular analogs.

3.2.3.   Conformational Effects

Depending on the temperature and composition of the solution, macromolecules may exist in differing conformations, in which the functional groups differ in reactivity. The functional groups may be at differing distances from each other and, in addition, their accessibility to an external reagent may differ. Such is the case with alkaline hydrolysis of polyacrylamide, where the formation of carboxyl groups leads to transition of the macromolecule from the globular to the unfolded shape. Unfortunately, there is no kinetic assessment of this effect. Similar effects are characteristic of polymer molecules with marked secondary structure (polypeptides, polyorganosiloxanes, etc.).

Maslova et al. [39] investigated the reaction of $^1H \rightarrow {}^3H$ exchange in polyriboadenylic acid (poly-A) and in its low-molecular fragment adenosine-5-monophosphate (AMP) at pH 7.1. The molecules of poly-A in neutral solutions at temperatures below 0°C have the conformation of single-thread helices. The stabilization of such helices is determined mainly by the stacking interaction, which takes place due to the π-electrons of the purine rings and hydro-

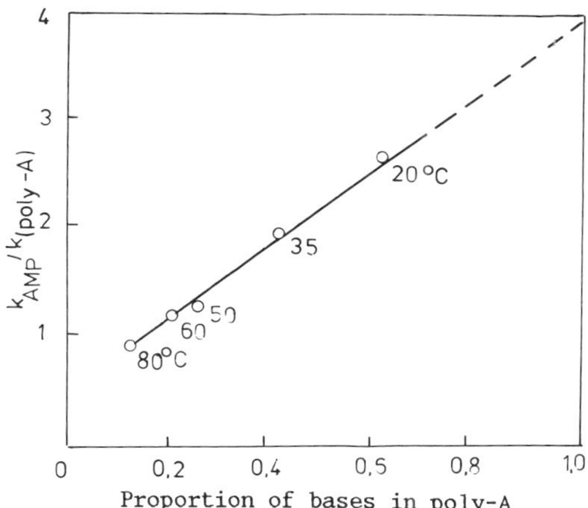

Fig. 3.2. Dependence of the ratio of the rate constants
of the reaction of $^1H \rightarrow {}^3H$ exchange for aden-
osine-5-monophosphate (AMP) and polyriboaden-
ylic acid (poly-A) on the proportion of bases
in poly-A taking part in the stacking interaction [39].

phobic forces. As the temperature is raised, the interaction between the bases weakens, and the molecules of the polymer become
more labile. The AMP molecules retained their structure at all temperatures.

Figure 3.2 shows the dependence of the ratio of the rate constants of the reaction of $^1H \rightarrow {}^3H$ exchange for AMP and poly-A on
the proportion of bases in poly-A taking part in the stacking interaction. The latter value was determined by an independent method.
The proportionality observed between k and this proportion may be
regarded as an indication that the stacking interaction is the reason for the decrease of the exchange of H atoms in the C—H groups
of the adenylic residues.

The reactivity of functional groups combined with crosslinked
polymer gels may vary greatly, even in the case of gels which have
swollen identically, this being due to the differing topological
structure of the gel [40].

Thus, the anomalous reactivity of "polymeric" functional groups,
which is brought about by the primary structure (the effects of an
adjacent group) can, evidently, be determined and simulated
by means of low-molecular analogs. The anomalous reactivity which
is brought about by the secondary structure (conformational effects)
is a specific feature of the polymeric state.

REFERENCES

1. P. J. Flory, Principles of Polymer Chemistry, Cornell University Press, New York (1953).
2. T. Alfrey, Jr., in: Chemical Reactions of Polymers, Vol. 19, Wiley-Interscience, New York (1964).
3. H. Morawetz, in: Chemical Reactions of Polymers, Vol. 19, Wiley-Interscience, New York (1964).
4. N. A. Platé, in: Kinetics and Mechanism of Formation and Conversion of Macromolecules, Nauka, Moscow (1968), p. 250.
5. I. A. Tutorskii, S. A. Novikov, and B. A. Dogadkin, Usp. Khim., 35, No. 1, 191 (1966).
6. R. M. Noyes, in: Progress in Reaction Kinetics, Vol. 1, Oxford University Press, London (1961), p. 129.
7. A. M. North, The Collision Theory of Chemical Reaction in Liquids, Wiley, New York (1964).
8. S. G. Entelis and G. P. Tiger, Kinetics of Reactions in the Liquid Phase, Khimiya, Moscow (1973).
9. I. V. Berezin and A. A. Klesov, Practical Course in Chemical and Enzyme Kinetics, Moscow State University (1977).
10. E. A. Moelwyn-Hughes, Physical Chemistry, 2nd edn., Pergamon Press, New York (1964).
11. O. W. Edwards and E. O. Hoffmann, J. Phys. Chem., 63, 1830 (1959).
12. Handbook of Chemistry and Physics, Cleveland (1955), p. 2026.
13. Yu. V. Moiseev, V. S. Markin, and G. E. Zaikov, Usp. Khim., 45, No. 3, 510 (1976).
14. I. Moens and G. Smets, J. Polym. Sci., 7, 931 (1947).
15. G. Smets and W. van Humbeeck, J. Polym. Sci., A, 1, 1227 (1963).
16. G. Smets, Makromol. Chem., 34, 190 (1959).
17. A. Conix and G. Smets, J. Polym. Sci., 15, 221 (1955).
18. J. S. Scarpa, D. D. Mueller, and J. M. Klotz, J. Am. Chem. Soc., 89, No. 24, 6024 (1967).
19. Y. Kakuda, N. Perry, and D. D. Mueller, J. Am. Chem. Soc., 93, No. 23, 5992 (1971).
20. A. Hvidt and R. Corrett, J. Am. Chem. Soc., 92, No. 19, 5546 (1970).
21. R. W. Taft, in: Three-Dimensional Effects in Organic Chemistry [Russian translation], Izd. Inostr. Lit., Moscow (1960), p. 562.
22. V. Chyty, B. Obereigner, and D. Lim, Conference on Chemical Transformation of Polymers, Bratislava (1958), p. 15.
23. Z. Mayer, B. Obereigner, and D. Lim, J. Polym. Sci., C, No. 33, 289 (1971).
24. M. L. Bender and M. S. Nevew, J. Am. Chem. Soc., 80, No. 20, 5388 (1958).
25. T. Alfrey, Jr. and W. G. Leoyd, J. Chem. Phys., 38, 318 (1963).
26. J. B. Keller, J. Chem. Phys., 37, 2584 (1962); 38, 325 (1963).
27. C. B. Arends, J. Chem. Phys., 38, 322 (1963).
28. L. Lasare, J. Chem. Phys., 39, 727 (1963).

29. D. A. McQuarrie, J. P. McTague, and R. Reise, Biopolymers, 3, No. 6, 657 (1965).
30. L. V. Noa et al., Vysokomol. Soedin., A, 15, No. 4, 877 (1973).
31. M. Higuchi and R. Senju, Polym. J., 3, No. 3, 370 (1972).
32. A. D. Litmanovich, N. A. Platé, L. M. Postnikov, and L. N. Valuev, Vysokomol. Soedin., A, 17, No. 5, 1112 (1975).
33. H. Morawetz and T. Oreskes, J. Am. Chem. Soc., 80, No. 10, 2591 (1958).
34. T. C. Bruice and U. K. Pandit, J. Am. Chem. Soc., 82, No. 22, 5858 (1960).
35. T. C. Bruice and U. K. Pandit, Proc. Natl. Acad. Sci. USA, 46, 402 (1960).
36. M. L. Bender, Mechanisms of Catalysis of Nucleophilic Reactions [Russian translation], Mir, Moscow (1964).
37. W. P. Jencks, Catalysis in Chemistry and Enzymology, McGraw-Hill, New York (1969).
38. L. Hammett, Physical Organic Chemistry, McGraw-Hill, New York (1970).
39. R. N. Maslova, E. A. Lesnik, and Ya. V. Varshavskii, Mol. Biol. (Moscow), 3, 728 (1969).
40. H. Morawetz, J. Polym. Sci., Polym. Symp., No. 62, 271-285 (1978).

Chapter 4

PRINCIPAL TYPES
OF DECOMPOSITION
OF POLYMER MOLECULES

The structure of polymer molecules determines the character
of their decomposition. On the one hand, chemically unstable bonds
can have identical reactivity and, on the other hand, terminal
bonds can have relatively high reactivity, which complicates the
process of chemical degradation. In addition, in the polymers
there may be bonds with relatively high reactivity, so-called "weak
bonds."

In considering the kinetics of degradation, it is necessary
to distinguish between decomposition at a terminal bond and decomposition at any bond in the main chain. In the former there is
depolymerization at the terminal bonds, and in the latter random
decomposition.

4.1. DEPOLYMERIZATION AT THE TERMINAL BONDS

In [1], Berlin et al. consider the theoretical kinetic behavior (the change in the number of molecules and the mass of a
polymer) in the decomposition of polymers of this type.

Under the action of a catalyst an inactive molecule is converted into an active one with the same chain length. According
to the ratio of the rates of initiation and depolymerization, two
limiting cases are possible:

the rate of initiation is much lower than the rate of depolymerization; in this case there is instantaneous depolymerization, and
the number-average ($\bar{P}_n$) and mass-average degree of polymerization
($\bar{P}_m$) remain unchanged during the course of degradation;

the rate of initiation is much higher than the rate of depolymerization; in this case there is stepwise depolymerization.

For a monodisperse molecular mass distribution (MMD)

$$\frac{\overline{P}_n}{\overline{P}_{n_0}} = \frac{\overline{P}_m}{\overline{P}_{m_0}} = 1 - \alpha \qquad (4.1)$$

i.e., the degree of polymerization decreases linearly with increase in the degree of dissociation of the mass of the polymer. For the most probable MMD $\overline{P}_n$ and $\overline{P}_m$ remain constant, but for a broad MMD they increase with increase in the degree of dissociation of the mass of the polymer.

Depolymerization at the terminal bonds in its pure form occurs only rarely in chemical degradation. Usually, decomposition of this type proceeds together with random decomposition of the polymer chain. One example of depolymerization at the terminal bond is the degradation of polyoxymethylene with hydroxyl terminal groups (POM-OH) in aqueous basic solutions. In these solutions there is no random decomposition and degradation proceeds as follows:

$$\sim CH_2OCH_2OCH_2OH + OH^- \rightleftharpoons \sim CH_2OCH_2OCH_2O^- + H_2O$$
$$\sim CH_2OCH_2 - O \cdots CH_2O^- \longrightarrow \sim CH_2OCH_2OH + HOCH_2O^-$$
$$\vdots \qquad \vdots$$
$$H \cdots O - H$$

The detachment of a proton from aliphatic alcohols by an ion takes place with a rate constant of around 10^8 liters/(mole·min) [2], while the depolymerization of POM-OH with $M_v = 7000$ in aqueous basic solutions takes place with a rate constant of $(1.0 \pm 0.2) \cdot 10^2$ liters × (mole·min)$^{-1}$ [3]. This polymer has a broad MMD [4] and it can be expected that in this case there is stepwise depolymerization. In point of fact, during the course of degradation M_v rises somewhat.

4.2. RANDOM DECOMPOSITION

An examination of the kinetic behavior in the random decomposition of polymers has been presented by Kuhn [5] and Simha [6]. With decomposition of this type, the reactivity of the chemically unstable units in the polymer chain is identical. After the scission of the chain there are formed two inactive fragments (the depolymerization reaction does not take place) and, moreover, the decomposition can take place in any part of the macromolecules.

The value of $\overline{P}_n$ changes during the course of the degradation, as follows:

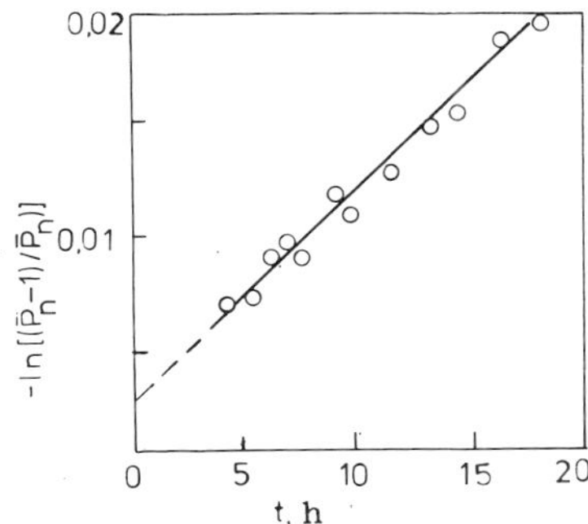

Fig. 4.1.   Dependence of the number-average de-
            gree of polymerization on time for the
            hydrolysis of methylcellulose in 35%
            hydrochloric acid at 0°C on the coor-
            dinates of Eq. (4.2); see [8].

$$-\ln \frac{\bar{P}_n - 1}{\bar{P}_n} = -\ln \frac{\bar{P}_{n_0} - 1}{\bar{P}_{n_0}} + k_{\text{rand}}\, t \tag{4.2}$$

For fairly low degrees of dissociation of chemically unstable
bonds (<0.05) it is possible to use a simpler equation:

$$\frac{1}{\bar{P}_n} = \frac{1}{\bar{P}_{n_0}} + k_{\text{rand}}\, t \tag{4.3}$$

The concentration of macromolecules with degree of polymerization
$\bar{P}_x$ changes with time, as follows:

$$c_x = c^0\,[1 - \exp(-k_{\text{rand}}t)]\exp-[(\bar{P}_x - 1)\,k_{\text{rand}}t]\,\{2 + (\bar{P}_n - \bar{P}_x - 1)\,[1 - \exp(k_{\text{rand}}t)]\} \tag{4.4}$$

where c^0 is the initial concentration of the polymer.

With this type of decomposition the amount of polymer decreases
due to chain scissions near the end of the chain, and the degree of
dissociation with respect to the mass of the monomer which is forming is linked with the number-average degree of polymerization by
the relationship

$$\left(\frac{\bar{P}_{n_0}}{\bar{P}_n}\right)^2 - 1 \approx \frac{(\bar{P}_{n_0})^2}{2}\,\alpha \tag{4.5}$$

i.e., in the initial period of degradation there is a sharp reduction in the number-average degree of polymerization without any significant reduction in the mass of the polymer.

Random decomposition is frequently met with in processes of chemical degradation.

In cellulose the hemiacetal terminal bonds are somewhat more reactive than the nonterminal bonds. This may be seen from a comparison of the rate constants of acid-catalyzed hydrolysis of cellulose and cellobiose, and also from the data in [7], where it is shown that in cellotriose the reactivity of the two hemiacetal bonds differs 1.5-fold. However, the difference in the reactivity of terminal and nonterminal bonds is insignificant and, accordingly, the process of acid-catalyzed hydrolysis of cellulose under homogeneous conditions may as a first approximation be regarded as random.

Figure 4.1 shows experimental data on the hydrolysis of methylcellulose in about 35% HCl at 0°C [8] drawn on the coordinates of Eq. (4.2).

4.3. MIXED TYPE OF DECOMPOSITION

Frequently the random decomposition of macromolecules is accompanied by depolymerization of the fragments which form, i.e., a mixed type of decomposition takes place. The kinetic behavior for this type of decomposition is considered in [1]. The following cases are possible.

1. One fragment is stable, while the other decomposes completely (instantaneous depolymerization). In this case, irrespective of the type of initial MMD, the number-average degree of polymerization changes, as follows:

$$\frac{\overline{P}_n}{\overline{P}_{n_0}} = 1 - \alpha \tag{4.6}$$

up to quite high degrees of dissociation since the scissions of the macromolecules around the ends, with depolymerization of a large part of the molecule, begin to influence the process only at very low values of the degree of polymerization ($\overline{P}_n < 10$).

2. Both fragments decompose completely. In this case,

$$\frac{\overline{P}_n}{\overline{P}_{n_0}} = \frac{n}{n_0} (1 - \alpha) \tag{4.7}$$

where n_0 and n are the initial and current numbers of polymer molecules present.

For the most probable distribution the change in $\overline{P}_n$ and $\overline{P}_m$ proceeds as follows:

TABLE 4.1. Values of the Induction Constants of the Substituents
            and of the Rate Constants of Degradation of $ROCH_2OR'$
            Oligomers in Aqueous Solutions of Sulfuric Acid [9]

n	R	R'	$\Sigma\sigma^*$	$-\log\left(\dfrac{k_{act}}{K_{BH^+}}\right)_{25\,°C}$ [min^{-1}]
1	CH_3	CH_3	0	2.82
2	CH_3	$CH_2\!-\!OCH_3$	0.52	2.18
3	CH_3	$(CH_2O)_2CH_3$	0.15	2.64
4	CH_3	$(CH_2O)_3CH_3$	0.05	2.74

$$\frac{\overline{P}_n}{\overline{P}_{n_0}} = \frac{\overline{P}_m}{\overline{P}_{m_0}} = \sqrt{1-\alpha} \tag{4.8}$$

3. The length of the kinetic depolymerization chain is much
less than the number-average degree of polymerization. This, the
most complex case, occurs with the acid-catalyzed degradation of
polyoxymethylene, where chain scission is accompanied by depolymerization of the fragments with hemiacetal bonds being formed. In
addition, the presence of an induction effect of two adjacent oxygen atoms is responsible for the differing reactivity of the terminal and nonterminal bonds, and it may be expected that the chain scission will not proceed randomly. It is easy to verify this last suggestion in an investigation of the process of acid-catalyzed degradation of oligomers of polyoxymethylene with methoxy terminal
groups (POM-OCH$_3$), the presence of which excludes depolymerization
of the initial compounds.

In principle, the decomposition of oligomers of POM-OCH$_3$ can
take place by the three following mechanisms:

random decomposition with differing reactivity of the terminal
and nonterminal acetal bonds;

random decomposition with identical reactivity of the acetal
bonds;

decomposition of one bond with rapid subsequent decomposition
of the intermediate products which form.

As already indicated (Chap. 2), in the case of oligomers of
POM-OCH$_3$ there is decomposition at one acetal bond, while the hemiacetal forming decomposes rapidly, i.e., the oligomers of POM-OCH$_3$
decompose by the third mechanism.

Nevertheless, for oligomers with $\bar{P}_n > 1$ it is necessary to know the site of scission of the first bond and, accordingly, we can distinguish two cases within the suggested third mechanism.

If we assume that any given bond may decompose, then with increase in chain length the rate constant is bound to increase. If, however, decomposition takes place only at the terminal bonds and these bonds are identical in reactivity in the various oligomers, then the rate constant need not depend on the chain length. Table 4.1 shows that oligomers of POM-OCH$_3$ have differing reactivities [9].

The most probable explanation of these experimental data may be the differing reactivities of the acetal bonds.

In point of fact, the relative basicity of the oxygen atoms in the methoxy terminal groups is higher than in the oxymethylene groups [10, 11]. And although these data are obtained from the shift of the absorption band of the phenol γ_{OH}, which forms a complex with the oligomers in nonaqueous solvents, it is difficult to imagine inversion of the order of basicity of these bonds in aqueous solutions.

For a quantitative explanation of the data given in Table 4.1, we may use the Taft equation, which links the reactivity of compounds with their structure. For oligomers of POM-OCH$_3$, in which there is no hyperconjugation, the Taft equation takes the form

$$\log k_{eff} = \log k^0_{eff} + \rho\sigma^* \qquad (4.9)$$

where $k_{eff} = k_{act}/k_{BH^+}$, and as the reaction site we take the group $-OCH_2O$, which expresses the specific character of the acetal bond. As already noted, when decomposition takes place by the third mechanism, two cases are possible: 1) decomposition taking place only at a terminal acetal bond (for oligomers with n = 3 and 4 the constant k_{eff} has to be divided by 2, assuming that decomposition of the terminal acetal bonds can proceed in parallel with equal probability), and 2) decomposition of the oligomers of POM-OCH$_3$ taking place with scission of one of the acetal bonds (in contrast with random decomposition this is decomposition where the possibility is unequal). In each oligomer, by reason of the induction influence of the terminal bond, differing possibilities are created for the scission of the nonterminal acetal bonds. In the former case the overall induction constant $\Sigma\sigma^*$ is equal to the sum of the induction constants of the substituents $-CH_3$ and $-CH_3(OCH_2)_n-$ (Table 4.1). In the second case, $\Sigma\sigma^*$ is obtained by the summation of the induction constants for all the bonds and division of the found value by the total number of bonds.

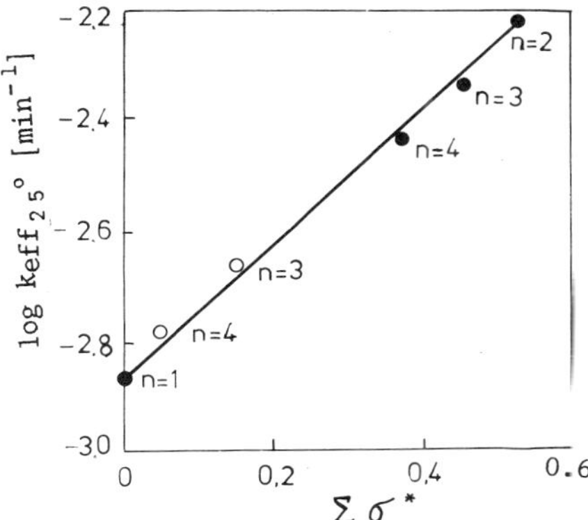

Fig. 4.2. Dependence of log k_{eff} on the decomposition of oligomers of polyoxymethylene with methoxy terminal groups (POM-OCH$_3$) at 25°C on the sum of the induction constants of the substituents [9]: o) decomposition at a terminal acetal bond; ●) decomposition at any acetal bond.

Figure 4.2 shows that the correlation relationship between log k_{eff} and $\Sigma\sigma^*$ is identical for the two cases: $\rho = 1.25 \pm 0.05$ and log $k^0_{eff} = 2.82$. Thus it may be concluded that the terminal acetal bonds have the higher reactivity. Unfortunately, in aqueous solutions under homogeneous conditions it is not possible to investigate the acid-catalyzed degradation of an oligomer with $n > 4$. Nevertheless, such experiments can be carried out in hexafluoroacetone (HFA), which is a common solvent for POM and inorganic acids at room temperature [12].

As in aqueous acid solutions, the dimer of POM-OCH$_3$ has a very high reactivity, and then with an increase in the number of acetal bonds k_{eff} decreases. The following are the values of k_{eff} for the degradation of oligomers and polymers of POM in a solution of $4.4 \cdot 10^{-3}$ mole/liter HCl and 8.99 mole/liter H$_2$O in HFA at 30°C [12]:

Terminal group	M_v	$k_{eff} \cdot 10^2$, min^{-1}	Terminal group	M_v	$k_{eff} \cdot 10^2$, min^{-1}
—OCH$_3$	76	6.5	—OCH$_3$	3500	6.1
	106	9.2		10500	6.3
	136	8.7		19500	6.6
	166	8.0	—OOCCH$_3$ Delrin 500		5.8
	226	7.1		100 000	5.8
	256	6.0	—OH	7 000	3.3
	1500	6.5		220 000	5.9

TABLE 4.2.   Value of k_{act}/K_{BH^+} of the Decomposition of Terminal and Nonterminal Acetal Bonds in the Depolymerization of POM-OCH_3 in H_2O—HFA Solutions of Various Compositions at 30°C

Concentration of water, mole/liter	k_{act}/K_{BH^+}, liter/(mole·min)		
	Decomposition of terminal bonds	Decomposition of nonterminal bonds	Depolymerization
8,99	$(1,1\pm0,2)\cdot10$	$2\pm0,4$	$(3\pm1)10$
55,5	$(3,1\pm0,2)\cdot10^{-3}$	$(3\pm1)\cdot10^{-4}$	$6\pm1^*$

*Value from [14].

The polymers have practically identical reactivity. This, at first sight, convincing result may be explained by the scission of the POM chain with the formation of one or two molecules of formaldehyde, with subsequent closure of the chain (such a mechanism was suggested by Grassie [13] for the thermal degradation of POM) and scission of the POM chain with instantaneous depolymerization of the fragments which are being formed.

However, the considerable change in the molecular mass (MM) as a function of the degree of dissociation, which is not expressed either by Eq. (4.6) or by Eq. (4.8), indicates that in this case there is chain scission with subsequent stepwise depolymerization. If we assume that scission of the POM chain takes place basically at random, and that with stepwise depolymerization with exponential MMDs the MM remains practically unchanged, we get a system of equations for the change in mass m of the polymer and the number of molecules with hemiacetal bonds (N) [12]:

$$\frac{dm}{dt} = -2k_{rand}N - k_{dep}N \tag{4.10}$$

$$\frac{dN}{dt} = k_{rand}\,m - 3k_{rand}\,N - \frac{2k_{dep}N^2}{m} \tag{4.11}$$

Under the initial conditions m = m_0 and N = 0 with t = 0, Eqs. (4.10) and (4.11) have the solution

$$m = m_0 \exp\left\{\left(\frac{2k_{dep}}{k_{rand}}+1\right)[\ln(2 - \exp - k_{rand}\,t) - k_{rand}\,t]\right\} \tag{4.12}$$

which does not include the original degree of polymerization of the POM. The value of k_{rand} can be calculated from Eq. (4.3) for low degrees of polymerization from independent experiments, and the value of k_{dep} from Eq. (4.12). For instance, the experimental data given

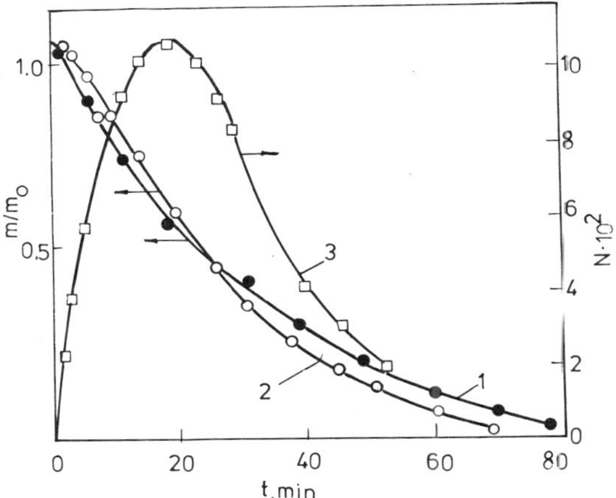

Fig. 4.3. Kinetic curves of the change in mass of the
           polymers and in the number of molecules with
           hemiacetal bonds in the degradation of POM-
           OCH_3 with $\bar{M}_v = 19,500$ at 30°C in a hexafluoro-
           acetone solution containing $4.4 \cdot 10^{-3}$ mole/
           liter HCl and 8.9 mole/liter H_2O [12]: 1)
           experimental curve; 2, 3) curves calculated
           from Eqs. (4.12) and (4.11), respectively.

in Fig. 4.3 are satisfactorily described by Eq. (4.12), if k_{dep} falls
within the limits 0.1-0.2 min^{-1}. The discrepancy observed between
the theoretical and experimental kinetic curves with high degrees
of dissociation is brought about by a marked contribution of chain
scission at the more reactive terminal acetal bonds.

The practically identical reactivity of POMs of different MMs
may be explained as follows. In the initial period of degradation
of the POM there are formed, due to random scission, a considerable
number of POM-OH molecules, the depolymerization of which leads
to the evolution of formaldehyde, the amount of POM-OH being practically independent of the degree of polymerization of the original
polymer.

In the case of a POM-OH molecule with $\bar{M}_v = 7000$, there are initially
a large number of sites of depolymerization, which ultimately yields
a higher value of k_{eff}.

Table 4.2 gives the ratios of the actual rate constants and
basicity constants for the decomposition of terminal and nonterminal
acetal bonds in the depolymerization of POM-OCH_3 in H_2O—HFA solutions of various compositions at 30°C.

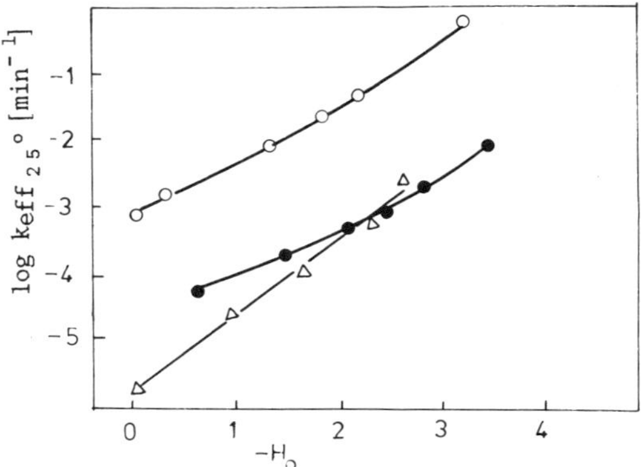

Fig. 4.4.   Relationship of log k_{eff} vs. H_0 for
the degradation of POM-OH with $\bar{M}_v$ =
220,000 (●) or 7000 (○) and POM-
OCH_3 with M_v = 6500 (Δ) in aqueous
solutions of sulfuric acid [15].

The ratio of the rate constants of acetal terminal and nonterminal bonds is higher in an aqueous solution than in an H_2O—HFA solution, i.e., in the aqueous solution the oligomers of POM-OCH_3 decompose preferentially at the terminal acetal bonds, while in the
H_2O—HFA solution the acetal terminal and nonterminal bonds are closely similar in reactivity. It is this which explains the weak extremal dependence of the reactivity of the oligomers of POM-OCH_3 in
the H_2O—HFA mixture (see above).

A change in the medium leads to a change in the ratio between
the rate constants of random decomposition and of depolymerization.
In water, decomposition of one of the acetal bonds leads to instantaneous depolymerization of the fragments which form, while in H_2O—
HFA solutions depolymerization proceeds stepwise.

The ratio of k_{rand} and k_{dep} is also influenced by the concentration of catalyst. Chain scission proceeds by the mechanism of specific acid catalysis, A-1, and the change in k_{rand} is described by
Eq. (1.28); the depolymerization reaction of molecules with hemiacetal bonds is by the mechanism of general acid catalysis, and the
change in k_{dep} is described by Eq. (1.16).

Figure 4.4 shows the dependence of log k_{eff} of the degradation
of POM-OH powders of various $\bar{M}_v$ on the acidity function of aqueous
solutions of H_2SO_4 [3]. The general equation which links the change
in k_{eff} of the degradation of POM-OH with the parameters of the medium h_0 and $c_{H_S^+}$ takes the form

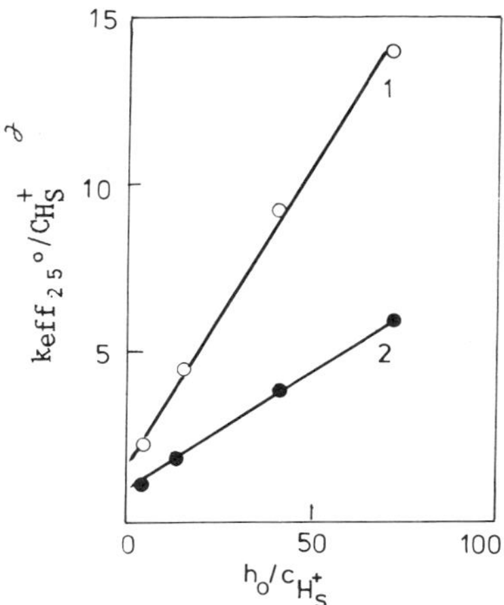

Fig. 4.5. Graphical solution of Eq. (4.13) for the degradation of POM-OH with $\bar{M}_v = 7000$ ($\circ$) and $\bar{M}_v = 220{,}000$ ($\bullet$) in aqueous solutions of sulfuric acid [15]: 1) $(k_{eff_{25°}}/c_{HS}^+ \cdot 10^3$; 2) $(k_{eff_{25°}}/c_{HS}^+) \cdot 10^4$.

$$k_{eff} = k_{dep(act)} c_{HS}^+ \left(\frac{1}{P_{n_0}} + \frac{k_{rand(act)} h_0}{K_{BH^+} \cdot k_{\ell(act)} \cdot c_{HS}^+} \right) \tag{4.13}$$

where $k_{\ell(act)}$ is the actual rate constant of the loss of active depolymerization sites.

The experimental data form a straight line using the coordinates k_{eff}/c_{HS}^+ vs. h_0/c_{HS}^+ (Fig. 4.5), giving one and the same value for $k_{dep(act)} = (1.3 \pm 0.3) \cdot 10^{-1}$ liter/(mole·min).

In dilute solutions of H_2SO_4 ($h_0 \sim c_{HS}^+$) for a low-molecular POM-OH ($\bar{M}_v = 7000$)

$$\frac{1}{P_{n_0}} \gg \frac{k_{rand(act)} h_0}{K_{BH^+} \cdot k_{\ell(act)} \cdot c_{HS}^+}$$

i.e., there takes place practically only the depolymerization reaction brought about by the initial concentration of hemiacetal bonds in the POM-OH. In highly concentrated solutions of H_2SO_4 ($h_0 \gg c_{HS}^+$) for any particular POM-OHs,

$$\frac{1}{P_{n_0}} \ll \frac{k_{rand(act)} \cdot h_0}{K_{BH^+} \cdot k_{\ell(act)} \cdot c_{HS}^+}$$

$$k_{eff} = \frac{k_{rand(act)} \cdot k_{dep(act)} \cdot h_0}{K_{BH^+} \cdot k_{\ell(act)}} \tag{4.14}$$

i.e., the depolymerization reaction is governed mainly by the concentration of hemiacetal bonds forming as a result of chain scission.

Equation (4.14) is applicable for POM-OCH$_3$s of various MMs with any particular concentration of H$_2$SO$_4$.

Thus, acid-catalyzed degradation of POM is an example of a complex type of decomposition, where the ratio of random decomposition and depolymerization at the terminal bonds depends on the nature of the terminal bonds, the degree of polymerization, the medium, and the concentration of catalyst.

4.4. DEGRADATION OF THE POLYMER CHAIN WITH PRIMARY DECOMPOSITION OF "WEAK BONDS"

The concept of "weak bonds" was first proposed by Schulz and Husemann [15], who suggested that these are located regularly every 200-250 units in the cellulose molecule. In [16] it was shown that within the interval of time in which one or two scissions take place per molecule, a broad MMD becomes very probable. This suggests support for random decomposition since "weak bonds" are most probably distributed statistically along the chain.

4.5. ANOMALIES IN THE DEGRADATION OF A POLYMER CHAIN IN THE SOLID STATE

The decomposition of molecules in a solid polymer may differ in nature from that in solution. For instance, according to one model, the cellulose molecules have a folded conformation in the protofibrils, each fold containing about eight elementary units [17]. On the basis of such a structure, Manley et al. [18] suggested a relatively high reactivity of the glycosidic bonds in the tips of the folds. To confirm this, experiments have been carried out on the acid-catalyzed degradation of cellulose of various types under heterogeneous conditions.

The MMDs of the specimens (acid-soluble degradation products, and insoluble residue) were determined after differing intervals of time up to 70% dissociation. During the course of degradation the cellulose molecules decomposed into fragments with an average degree of polymerization of about eight, irrespective of the type of polymer, the MMD of the unreacted part of the cellulose remaining unchanged. The results confirm the presence in the cellulose molecule of at least two types of bonds, differing considerably in their reactivity: the glycosidic bonds in the tips of the folds have higher reactivity than those in the sides of the folds. If the degradation of the polymer takes place under heterogeneous conditions, but nevertheless in the kinetic region and in the external solution oligomers with degree of polymerization $\bar{P}_x$ are dissolved, then the degree of polymerization of the rest of the polymer, which decomposes randomly, will change, as shown by the equation in [19]:

$$\bar{P}_n = \frac{\bar{P}_n + (\bar{P}_{n_0} - \bar{P}_x)(\bar{P}_x - 1)[1 - \exp(-kt)]}{1 + (\bar{P}_{n_0} - P_x)[1 - \exp(-kt)]} \tag{4.15}$$

The degree of dissociation with respect to the mass, and the rate of change with time, are respectively:

$$\alpha = 1 - \left\{\exp - [(P_x - 1)/kt]\ \bar{P}_{n_0}^{-1}\right\} \left\{\bar{P}_{n_0} + [1 - \exp(-kt)](\bar{P}_{n_0}\bar{P}_x)(\bar{P}_x - 1)\right\} \tag{4.16}$$

$$\frac{d\alpha}{dt} = k\left\{(\bar{P}_x - 1)(1 - \alpha)\left[\exp - (kt\bar{P}_x)(\bar{P}_{n_0} - \bar{P}_x)(\bar{P} - 1)\cdot\bar{P}_{n_0}^{-1}\right]\right\} \tag{4.17}$$

The initial rate of the degradation process is

$$\left(\frac{d\alpha}{dt}\right)_{t\to 0} = \frac{(\bar{P}_x - 1)\bar{P}_x}{\bar{P}_{n_0}}\,k \tag{4.18}$$

Thus the decomposition of macromolecules whose chemically unstable groups are separated by distances which for practical purposes exclude reciprocal influence takes place at random. One example is the homogeneous degradation of polyamides, polyesters, or cellulose and its derivatives. In degradation under heterogeneous conditions, polymers with a marked folded conformation may have an anomalously high reactivity of the bonds in the tips of the folds.

In macromolecules where the chemically unstable bonds are at distances allowing considerable reciprocal influence, e.g., polyoxymethylene, there is a complex type of decomposition:

the terminal acetal bonds are more reactive than the nonterminal bonds;

the hemiacetal bonds, which have high reactivity, are responsible for depolymerization.

The ratio of the reaction rates of chain scission and depolymerization depends on the structure of the polymer (the MM and the nature of the terminal bonds) and on the external conditions (the temperature, the medium, and the concentration of the catalyst).

REFERENCES

1. A. A. Berlin and N. S. Enikolopyan, Vysokomol. Soedin., A, <u>10</u>, 1475 (1968).

2. E. F. Caldin, Fast Reactions in Solution, Wiley, New York (1966).
3. L. V. Ivanova, Yu. V. Moiseev, and G. E. Zaikov, Vysokomol. Soedin., A, $\underline{14}$, No. 5, 1057 (1972).
4. V. Jaacks and H. Penneweis, Makromol. Chem., $\underline{124}$, 59 (1969).
5. W. Kuhn, Ber., $\underline{63}$, 1503 (1930).
6. R. Simha, J. Appl. Phys., 12, 569 (1941).
7. M. S. Feather and J. F. Harris, J. Am. Chem. Soc., $\underline{89}$, 5661 (1967).
8. M. L. Wolfrom, J. C. Sowden, and E. A. Metcalf, J. Am. Chem. Soc., $\underline{63}$, 1688 (1941).
9. L. V. Ivanova, Yu. V. Moiseev, and G. E. Zaikov, Izv. Akad. Nauk SSSR, Ser. Khim., No. 11, 2501 (1970).
10. N. I. Vasil'ev et al., Kinet. Katal., $\underline{11}$, No. 3, 579 (1970).
11. A. G. Gruznov, Candidate's Dissertation, Moscow State University (1971).
12. L. V. Ivanova et al., Vysokomol. Soedin., A, $\underline{16}$, No. 8, 1831 (1974).
13. N. Grassie and R. S. Roche, Makromol. Chem., $\underline{112}$, 16 (1968).
14. J. Lobering and V. Rank, Chem. Ber., $\underline{90}$, 2331 (1957).
15. G. V. Schulz and E. Husemann, Z. Phys. Chem. (Leipzig), B, $\underline{52}$, 23 (1942).
16. L. V. Ivanova, Summaries of Reports, Third Conference on Chemistry and Physical Chemistry of Polyacetals, Frunze (1971), p. 5.
17. R. St. J. Manley, J. Polym. Sci., A-2, $\underline{9}$, 1025 (1971).
18. M. Chang, T. C. Pound, and R. St. J. Manley, J. Polym. Sci., Polym. Phys. Ed., $\underline{11}$, 399 (1973).
19. L. A. Wall, in: Analytical Chemistry of Polymers (High Polymers, Vol. 12), G. M. Kline (ed.), Wiley-Interscience, New York (1962).

Chapter 5

DIFFUSION
OF AGGRESSIVE MEDIA
IN POLYMERS

To establish the mechanism of degradation of polymers in aggressive media it is necessary to know how the diffusion of each of the aggressive media proceeds and what is its state in the polymer, i.e., to what extent, for instance, the electrolytes dissociate into ions and the latter are hydrated with water.

Since the diffusion of aggressive media in polymers is not merely of theoretical, but also of considerable practical interest, we shall consider in this chapter the general rules governing the diffusion of aggressive media in polymers. Since the most common aggressive media are electrolyte solutions, it is to these that the most attention is given.

5.1. PRINCIPAL EQUATIONS OF DIFFUSION

Diffusion is the redistribution of a substance in space and time as a result of thermal migration of particles (molecules and ions).

This definition, on the one hand, indicates that diffusion is a macroprocess which is described by an equation of macroscopic physics. On the other hand, since the principal process is thermal migration of particles, diffusion characterizes the microscopic (or molecular-kinetic) properties of the medium. The principal macroscopic equation of isothermal diffusion, obtained in the framework of the theory of thermodynamics of irreversible processes [1, p. 295 of Russian translation; 2, p. 169 of Russian translation] is as follows:

$$I_i = L_i \,\mathrm{grad}\,\mu_i \tag{5.1}$$

which links the flow of the diffusion substance i with the kinetic coefficient L_i, which characterizes the mobility of the diffusing substance in the polymer and the conduction of the medium with respect to the diffusing substance. The chemical potential gradient provides the motive force for the diffusion.

On substituting in (5.1) the expression for the chemical potential, we get the following equation:

$$I_i = D_i^* \left(1 + \frac{d \ln y_i}{d \ln c_i}\right) \tag{5.2}$$

where y_i is the molar activity coefficient and $D_i^* = L_i RT/c_i$ the autodiffusion coefficient of the substance i.

Equation (5.2) is equivalent to the formulation of Fick's first law.

A statement of Fick's second law is obtained from Eq. (5.2) in its joint solution with the equation of continuity of the medium [1, p. 295 of Russian translation; 2, p. 169 of Russian translation]:

$$\frac{\partial c_i}{\partial t} = \mathrm{div}\left[D_i \left(1 + \frac{d \ln y_i}{d \ln c_i}\right) \mathrm{grad}\,c_i\right] \tag{5.3}$$

$\left(1 + \frac{d \ln y_i}{d \ln c_i}\right)$ is called the thermodynamic factor of the diffusion coefficient; in ideal systems its value is unity.

The solutions of Eqs. (5.2) and (5.3) are obtained in the analytical form only for a binary system. Solutions of special cases of Eq. (5.3) with constant diffusion coefficients and differing initial and boundary conditions are given in [3-6, p. 11 of Russian translation].

Various approaches are at present known for describing isothermal diffusion in multicomponent systems [7, 8]. Usually a number of simplifications are employed to deal with the ratios of the different diffusion coefficients for the components and the distribution of their concentrations in the system. For instance, in expressing the steady-state flow of an electrolyte in a polymer membrane swollen to equilibrium in water it is possible to exclude the influence of the chemical potential differential of the water if the difference in the activities of the water on either side of the membrane is slight, i.e., the electrolyte concentrations in the solution are low. The expressions for the flow of the components after the appropriate simplifications and transformations are brought to a form analogous with the Fick equation.

The molecular-kinetic view of diffusion is based on the assumption that diffusion is the result of successive jumps of diffusing particles from one equilibrium position to another. The possibility of such movement is usually connected with the presence in the polymers of free volume — a combination of molecular spaces, which, as a result of thermal motion in the polymer at temperatures above the glass transition temperature, are constantly disappearing and reappearing in such a way that their statistical-average distribution by size and concentration remains constant [9].

All the present theories of diffusion may be divided conventionally into two groups.

The first group (theory of random processes [1C], theory of transition state [11, p. 494 of Russian translation; 12], and theory of activation zones [13]) is based on a hole activation model, which looks at the movement of a molecule in the elementary process as a transition through an energy barrier which divides an equilibrium position, i.e., the diffusion is an activation process.

The second group includes various modifications of the free volume theory [14-16], the basis of which is the assumption that the formation of a microcavity close to a diffusing molecule does not require any expenditure of energy since the free volume within the polymer is being continuously redistributed between various regions of the space.

The conclusions which follow from the activation and activationless models make it possible to describe the diffusion behavior of a broad class of systems of polymer and low-molecular substances which are at a temperature above the glass transition temperature.

The diffusion coefficients which characterize the diffusional flows of the individual components are calculated in relation to the cross section, in relation to which the flow of the substance is being determined. The selection of the cross section depends on the special features of the system of polymer and low-molecular substance and on the method of investigation.

Usually three types of cross section are involved:

V — the volumes of the diffusing components on the two sides of the cross section remain constant, and the change in volume during the movement is zero;

M — the mass of the system on one side of the cross section remains unchanged during the diffusion, and the cross section moves along the diffusion coordinate;

1 — the location of the cross section is fixed on one side of the components.

The diffusion coefficients are designated D_V, D_M, and D_1 according to the cross section selected; the first two are called the coefficients of interdiffusion, and D_1 the relative diffusion coefficient.

These diffusion coefficients are linked by the relationships [17, p. 229 of Russian translation]:

$$D_V = D_1 \varphi_2 \tag{5.4}$$

$$D_M = D_V (\varphi_2/w_2)^2 \tag{5.5}$$

where φ_2 and w_2 are, respectively, the volume and mass fractions of the component.

With concentrations of the diffusant close to zero, the values of D_V, D_M, and D_1 coincide and are designated as D_0.

Solution of Eqs. (5.2) and (5.3) is possible in an analytical form only for a diffusion coefficient not dependent on the coordinates and concentrations. For the concentration dependence of the diffusion coefficient, which applies in the majority of cases, there is an approximation based on the employment of the integral diffusion coefficient $\bar{D}$. This coefficient is by definition linked with the diffusion coefficient as follows:

$$\bar{D} = \frac{1}{c_1 - c_2} \int_{c_1}^{c_2} D(c)\, dc \tag{5.6}$$

where c_1 and c_2 are the limiting values of the concentration range.

Determination of the diffusion coefficient for a number of successive concentration ranges makes it possible to explain the concentration-dependence of the differential diffusion coefficient $\bar{D}$. The corrections that should be taken into account in using Eq. (5.6) are given in [3].

5.2. METHODS OF INVESTIGATION OF THE DIFFUSION OF ELECTROLYTES IN POLYMERS

The methods of investigation of the diffusion of electrolytes in polymers do not differ fundamentally from those used to investigate the transport of other low-molecular substances, but they are to a greater or lesser degree related to the peculiarities of the electrolytes, and it is therefore desirable to consider these methods in detail.

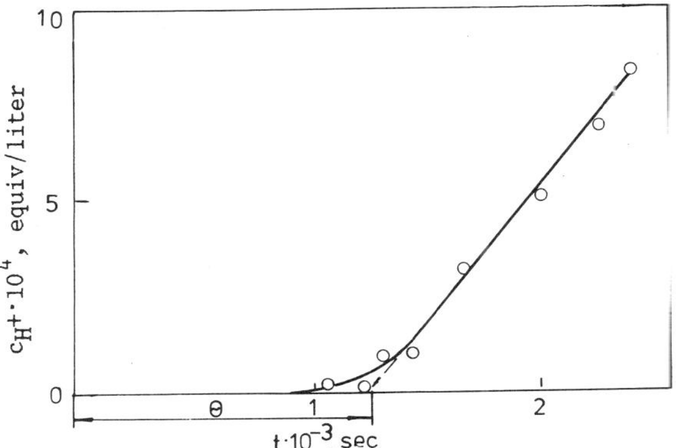

Fig. 5.1. Kinetics of change in hydrogen ion concentration in the solvent chamber with the passage
of phosphoric acid (pH 2.12) through a PVA
film saturated with water, at 27°C [19].

The present methods of determining the diffusion coefficients
of electrolyte solutions in polymers can be divided into the three
following major groups:

methods based on the measurement of the permeability in the
polymers;

methods based on the measurement of the physicochemical properties of the polymers;

methods based on recording the moving boundary of the diffusant
in the polymer.

5.2.1. Methods Based on the Measurement
of the Permeability in the Polymers

Determination of the diffusion parameters by means of measurement of permeability is based on recording the amount of electrolyte
transported in unit time through unit surface of a polymer film.

The typical kinetic curve expressing the change in concentration of electrolyte in a chamber of solvent with the passage of a
diffusant through the polymer has two characteristic sections (Fig.
5.1): linear, in which the transport process is of a steady-state
character; and curvilinear, in which it is under nonsteady-state
conditions.

The values of $\bar{D}$ can be calculated in two ways.

1. On the basis of data on the permeability and the solubility of the electrolyte, from a formula obtained from Fick's first law:

$$\overline{D} = \frac{q\,(c_{\mathrm{p}})\,l}{c^0 - c_{\mathrm{p}}}\,\frac{V}{s} \qquad (5.7)$$

where $q(c_{\mathrm{p}}) = \lim\,(dc_{\mathrm{p}}/dt)$; V is the volume of the chamber of the cell; s and l, respectively, the surface area and the thickness of the polymer film; and c^0 and c_{p} the concentrations, respectively, of the electrolyte in the surface layer of the polymer on the side of the electrolyte solution of constant concentration (the solubility of the electrolyte in the polymer) and on the side of the solution being measured.

Under the conditions of the experiment usually $c^0 \gg c_{\mathrm{p}}$ and, accordingly, Eq. (5.7) is simplified. The values of c^0 are determined from the data on the sorption of electrolyte solutions in the polymers.

2. On the basis of the retardation time θ (Fig. 5.1), from Frisch's equation [18]:

$$\frac{\theta\,(c^0)}{l^2} = \frac{\displaystyle\int_0^{c^0} \overline{D}_c\left[\int_c^{c^0} D\,(u)\,du\right] c\,dc}{\left[\displaystyle\int_0^{c^0} D\,(c)\,dc\right]^3} \qquad (5.8)$$

This equation makes it possible to calculate $\overline{D}$ from data on the retardation time for polymer—diffusant systems, for which the functional dependence $D(c)$ is known or can be found.

Under actual conditions θ and P are directly measurable parameters. In [19] it is attempted to use these data for the calculation of diffusion coefficients without measurement of the sorption.

For the solution of this problem there were used Eq. (5.8) and an equation for the permeability

$$p = \int_0^c D\,(c)\,dc \qquad (5.9)$$

After a series of transforms we get the following equation for $\overline{D}$:

$$\overline{D} = l^2\left[6\theta\,(c^0) + 6\,\frac{d\theta\,(c^0)}{dP\,(c^0)}\,P\,(c^0) + \frac{d^2\theta\,(c^0)}{dP^2\,(c^0)}\,P^2\,(c^0)\right]^{-1} \qquad (5.10)$$

One advantage of the resulting equation as compared with (5.8) is that it enables us to determine the diffusion coefficients without direct measurement of c^0.

Let us consider two special cases which emerge from Eq. (5.10).

1. The retardation time does not depend on the concentration of the diffusant. In this case we get the well-known Daynes—Barrer equation for the diffusion coefficient, which does not depend on the concentration:

$$\overline{D} = \frac{l^2}{6\theta} \qquad (5.11)$$

2. The diffusion coefficient depends on the concentration of diffusant in the polymer linearly

$$\overline{D} = D_0\,(1 + bc^0) \qquad (5.12)$$

where b is a constant.

In this case, the retardation time depends on the concentration, as follows:

$$\theta\,(c^0) \approx \frac{l^2}{12D_0}\,\left(\frac{4 + bc^0}{2 + bc^0}\right) \qquad (5.13)$$

At temperatures exceeding the glass transition temperature of the polymer, the diffusion coefficients as determined from the steady-state and nonsteady-state sections of the kinetic curve coincide satisfactorily over a wide range of concentrations. At temperatures near to and below the glass transition temperature these diffusion coefficients differ from each other (usually $D_{steady-state} > D_{nonsteady-state}$), this being due to the presence of relaxation processes in the polymer matrix [20].

To measure P, we used a variety of cells, consisting most frequently of two chambers separated by the polymer under investigation. In one of the chambers we put the solution under investigation, in the other the pure solvent. Also in the latter chamber we put a sensor which records the inflow of the electrolyte through the polymer.

The recording may be effected by using electrical conductivity measurements [21, 22], pH measurements [23-25], radioactive isotopes [26-29], or chemical [30-33] or optical methods [34].

Some methods make it possible to measure directly the concentration gradient of the electrolyte in the polymer, which has become established in the course of diffusion.

These methods include the removal of layers [35-37] by cutting the specimen into a number of layers in planes perpendicular to the coordinate of diffusion, followed by the determination of the average concentration of diffusant in each layer. Sometimes a multiple-film method is used [38, 39] in which the specimens are piled into a stack, pressed to improve the contact between the layers, and subjected to the action of the electrolyte solution. When steady-state transfer has been established, the stack is broken down into its constituent elements. Knowing the concentration gradient and the flow of the electrolyte, it is possible to calculate the diffusion coefficient from Eq. (5.2).

Nevertheless, in such measurements one cannot be sure that the boundary layers of the constituent elements do not distort the true diffusion process.

Methods based on permeability measurements are sufficiently accurate and convenient for determining diffusion parameters, but various complications arise.

First, the interaction of the flow of diffusing substances (e.g., when there is counterflow diffusion of water and electrolyte the permeability of the latter changes and there is even formed a new phase within the polymer [6, p. 205; 40]).

Second, finding different diffusion coefficients on the non-steady-state and steady-state sections of the kinetic curve of permeability; this is due to diffusion taking place in a nonswollen and unrelaxed matrix. This is encountered mostly in the diffusion of electrolytes in hydrophilic glassy polymers. To avoid this complication it is best to employ a not-too-large concentration differential or, where the concentrations are equal on the two sides of the diaphragm, to set up a concentration differential using a radioactive isotope.

The specimens must not have microcracks or macropores or there will be phase flow which distorts the diffusion process.

5.2.2. Methods Based on the Measurement of the Physicochemical Properties of the Polymers

Determination of the diffusion parameters by these methods is based on direct or indirect recording, on a time scale, of the amount of electrolyte and other components of the aggressive medium in the polymer.

For a parallelepiped where one of the sides is much smaller than the others (a thin film), the amount of substances which has diffused in time $t(M_t)$ can be found from two equations [3], obtained by solving (5.3) for boundary $c = c^0$ when $x = 0$ and $x = \ell$ and initial conditions $c = 0$ when $t = 0$ and $0 \leqq x \geqq \ell$

$$\frac{M_t}{M_\infty} = 4\left(\frac{\overline{D}t}{l^2}\right)^{1/2}\left[\frac{1}{\pi^{1/2}} + 2\sum_{m=0}^{\infty}(-1)^{m}i\,\text{erf}\,\frac{ml}{2\,(\overline{D}\,t)^{1/2}}\right] \tag{5.14}$$

[where M_∞ is the limiting amount of substance sorbed when $t \to \infty$, corresponding to concentration c^0; erf $(\ell/2(Dt)^{1/2})$ is the Gaussian error integral]

$$\frac{M_t}{M_\infty} = 1 - \frac{8}{\pi^2}\sum_{m=0}^{\infty}\frac{1}{(2m+1)^2}\,\exp\left[-\frac{\overline{D}\,(2m+1)^2\pi^2t}{l^2}\right] \tag{5.15}$$

In Eqs. (5.14) and (5.15) we often restrict ourselves to the first terms of the expression

$$\frac{M_t}{M_\infty} = \frac{4}{\pi^{1/2}}\left(\frac{\overline{D}t}{l^2}\right)^{1/2} \tag{5.16}$$

$$\frac{M_t}{M_\infty} = 1 - \frac{8}{\pi^2}\exp\left(-\frac{\overline{D}\pi^2t}{l^2}\right) \tag{5.17}$$

Equation (5.16) best describes the initial section of the kinetic curve of sorption ($M_t/M_\infty < 0.4$), and Eq. (5.17) best describes the middle and final sections. Analogous equations for bodies of different geometrical shapes are given in Table 5.1.

To maintain constant boundary conditions in the sorption process, the amount of diffusant in the solution must considerably exceed the value M_∞ in the polymer. In the opposite case the boundary conditions are not upheld, and we must use an equation from [41]

$$\frac{M_t}{M_\infty} = 1 - \sum_{n=1}^{\infty}\frac{2\alpha\,(1+\alpha)}{1+\alpha+\alpha^2q_n^2}\,\exp\left(-\frac{4\overline{D}q_n^2t}{l^2}\right) \tag{5.18}$$

where α is the ratio of the volumes of solution and film divided by the distribution coefficient of the diffusant.

The diffusion coefficient can be calculated by comparing the experimental curve (M_t/M_∞) vs. t with the family of theoretical curves for differing α given in [3]. Analogous equations are known for bodies of different geometrical shapes [3].

To determine the integral diffusion coefficients, we used the following methods based on changes in the physicochemical properties of the polymer.

TABLE 5.1. Diffusion Equations for Bodies of Different Geometrical Shapes

Shape of body	General equation	Equation for the initial section of the kinetic curve of sorption
Parallelepiped with dimensions of sides $2a$, $2b$, $2l$	$$\frac{M_t}{M_\infty} = 1 - \frac{1}{8}\left(\frac{64}{\pi^3}\right)^2 \sum_{n=0}^{\infty}\sum_{m=0}^{\infty}\sum_{k=0}^{\infty} \frac{\exp(-\alpha t)}{(2n+1)^2(2m+1)^2(2k+1)^2}$$ where $\alpha = \frac{\pi^2\overline{D}}{4}\left[\frac{(2n+1)^2}{a} + \frac{(2m+1)^2}{b} + \frac{(2k+1)^2}{l}\right]$	$$\frac{M_t}{M_\infty} = 2\left(\frac{Dt}{\pi}\right)^{1/2}\frac{s}{v}$$ where $s = 8(ab + al + bl)$; v is the volume
Cylinder with $h \gg r$	$$\frac{M_t}{M_\infty} = 1 - \frac{4}{r^2}\sum_{m=1}^{\infty}\frac{1}{\alpha_m^2}\exp(-\alpha_m^2 Dt)$$ where α_m indicates the positive roots of the equation $I_0(r\alpha_m) = 0$ The first four roots are equal to $\alpha_1 = 2,41/r$, $\alpha_2 = 5,52/r$, $\alpha_3 = 8,65/r$, $\alpha_4 = 11,79/r$	$$\frac{M_t}{M_\infty} = \frac{4}{r}\left(\frac{Dt}{\pi}\right)^{1/2}$$
Sphere	$$\frac{M_t}{M_\infty} = 1 - \frac{6}{\pi^2}\sum_{p=1}^{\infty}\frac{1}{p^2}e^{-p^2\pi^2 Dt/r^2}$$	$$\frac{M_t}{M_\infty} = \frac{6}{r}\left(\frac{Dt}{\pi}\right)^{1/2}$$

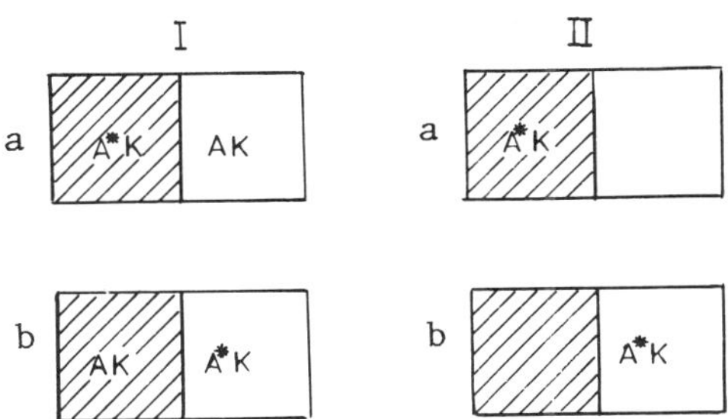

Fig. 5.2. Measurement of diffusion of an electrolyte AK
in a polymer (for explanation, see text).

Gravimetric Method. This is best used solely for a single-component diffusant. For multicomponent mixtures there is a possibility of sorption of the different components at considerably differing rates, for instance with the diffusion of nonvolatile electrolyte solutions into hydrophobic polymers. In the latter case the diffusion parameters can be determined from the change in concentration of the diffusing substance in the solution with sorption or desorption from the polymer. For recording the concentrations we used the methods cited above. When this method is used with desorption there may be complications linked, for instance, with the interaction of the flow of electrolyte with the flow of the solvent which is diffusing into the polymer. The most satisfactory results are obtained in the investigation of volatile electrolytes, whose concentration may be determined by desorption into a gaseous phase in a stream of inert carrier gas [42].

Spectral Methods. The concentration of diffusant in the polymer can be determined by various methods: optical spectroscopy [43], NMR [44], EPR (where the diffusant has paramagnetic properties) [45, 46], and spin echo [41, 48].

Radioactive Isotope Methods. These are very convenient not only for determining $\bar{D}$, but also for measuring the autodiffusion coefficients of individual molecules and ions.

Figure 5.2 shows the measurement of diffusion of an electrolyte AK in a polymer (shaded portion) by means of the labeled ion A*. This scheme includes carrying out the experiment under conditions of sorption (Ib and IIb) and desorption (Ia and IIa); in Ia and Ib we are measuring the autodiffusion coefficients and in IIa and IIb the integral diffusion coefficients of individual ions.

It is of interest to determine the diffusion parameters by isotope exchange, with the reaction taking place in the diffusion region. The course of the reaction, e.g., H—D exchange, may be conveniently recorded by IR spectroscopy [49].

<u>Electrical Conductivity Method</u>.   In an electrical field in a
polymer four different types of conduction may occur: ionic,
protonic, electronic, and induced.   In the diffusion of aggressive
media, specifically of electrolyte solutions, the first two are the
most important.

A satisfactory analysis of the process of electrical conductivity can be carried out only after determination of the degree of
dissociation of the electrolyte in the polymer, its hydration, and
its interaction with the polymer.

The task of finding such data is obvious, and, in fact, the
interrelationship of the diffusion coefficients and electrical conductivity is not an unambiguous one, i.e., correct under all experimental conditions.   For instance, with a reduction in the water content of the polymers there is a change from the ionic mechanism of
charge transfer to the electronic one [50].   The temperature has
a considerable influence on the mechanism of conduction [51].

For polymers with a high content of uncombined water (hydrogels) we may expect Kohlrausch's law on the independence of ion
movement to be satisfied:

$$\chi = \sum z_i \bar{c}_i u_i \tag{5.19}$$

wher<u>e</u> χ is the specific electrical conductivity; u_i the mobility;
and c_i the concentration of the ion i in the polymer.

If (5.19) is satisfied, the integral diffusion coefficient of
the ion in the polymer may be found from the equation in [52]:

$$\frac{1}{R_{el}} = \frac{z_i^2 F^2}{RT} c_i \left(\frac{2l_1}{\bar{D}_{solv\,i}} + \frac{l_1}{K_{(distr)}D_i} \right)^{-1} \tag{5.20}$$

where R_{el} is the electrical resistance of a specimen of unit cross
section; c_i, the ion concentration of the solution; l_1, the thickness of the stationary Nernst layer; $\bar{D}_{solv\,i}$, the diffusion coefficient of the ion in the Nernst layer; K_{distr}, the distribution coefficient of the ion between the polymer and the solution.

If we can disregard the resistance of the stationary Nernst
layer, Eq. (5.20) takes a simpler form:

$$\frac{1}{R_{el}} = \frac{z_i^2 F^2}{RT} \frac{\bar{c}_i \bar{D}_i}{l} \tag{5.21}$$

The voltage, and also the frequency of the alternating current, have
a significant influence on the conductivity.

When the voltage is raised, the ion transfer process is limited to a greater extent by the surface, stationary, layers than by
the bulk of the diaphragm, since the conduction of the diaphragm
rises sharply under these conditions.

At low frequencies of the alternating current the conduction
of electricity is determined by diffusion polarization in the stationary layers [52].

At high frequencies of the alternating current the link between
the diffusion parameters and the conductivity becomes more complex
since under these conditions the ions are not transported from one
boundary of the diaphragm to the other, but merely oscillate within
the limits of a particular section.

The determination of ionic mobility from electrical conductivity data is considered in [52, 53]. Correct results are obtained
for dilute electrolyte solutions in polyelectrolytes [54] and porous polymers [55].

With rise in the electrolyte concentration the concentration
dependence of the conductivity becomes nonlinear as a result of an
increase in the potential at the polymer—solution interface [56],
an increase in the electroosmotic flow of solvent, and of an increase
in the relaxation effect.

Measurement of the conductivity of polymer diaphragms is effected by means of three types of electrodes [51]: those not adjoining
the diaphragm (I), platinum (II), and mercury (III). The mercury
electrodes are the best; those of the second group give excessively
low results, and those of the first excessively high. Methods of
measuring conductivity are described in [58, 59].

Other Methods. Of the other methods of determining the diffusion parameters we must note the measurement of dielectric permittivity or of the microhardness of the polymer [60]. In both cases
it is necessary to plot calibration curves relating the property
of the polymer to the concentration of diffusant in the polymer.

5.2.3. Methods Based on Recording the Movement of the Boundary of the Diffusant in the Polymer

Determination of the diffusion parameters by means of these
methods is based on recording the movement of the boundary of the
diffusant in the polymer in unit time.

For a semi-infinite polymeric body bounded only by one plane
with $x = 0$, the solution of Eq. (5.3) with initial $c = 0$ at $t = 0$
and $x \geq 0$ and the boundary condition $c = c^0$ at $x = 0$ takes the form

$$c = c^0 \left(1 - \mathrm{erf}\ \frac{x}{2\,(Dt)^{1/2}} \right) \qquad (5.22)$$

The solution of this equation may be presented in the form of a series [3]. Limiting ourselves to the first term of the decomposition, we get an expression for the rate of movement of a plane of a given concentration c_x along the diffusion coordinate

$$x_t = \lambda t^{1/2} \qquad (5.23)$$

where x_t is the depth of penetration of the boundary of a diffusant of constant concentration c_x over the time of diffusion t; and λ the constant of penetration, equal to $(\pi D)^{1/2}(1 - (c_x/c^0))$.

The depth of penetration of the boundary of the diffusant in the polymer can be determined by autoradiography [61], optical methods [62], luminescent analysis [63], or acid—base indicators [62].

The first three methods make it possible to record directly the boundary of diffusant from the changes in the physical properties of the matrix of the polymer during the course of diffusion. When the latter method is used, the polymer article is cut along the coordinate of diffusion and the section is treated with a solution of any suitable acid—base indicator. On the sections there shows up a boundary between the colored part of the polymer, into which the acid or base has penetrated, and the uncolored part.

The acid—base indicator method has found wide use and, accordingly, here we shall deal in more detail with its advantages and disadvantages.

The change in color of the indicators is due to the occurrence of chemical reactions whose equilibrium depends on the pH of the medium. As the concentration of the indicator is much less than that of the acid or base in the system, the following equation holds:

$$\mathrm{pH} = \mathrm{p}K + \log \frac{\alpha}{1-\alpha} \qquad (5.24)$$

where K is the equilibrium constant of the reaction and α the degree of dissociation of the indicator.

The color change is found to occur with pH values in the range pK ± 1. This range is considerably influenced by temperature, the presence of salts, the nature of the solvent, and other factors. The mechanism of dissociation of the indicators in aqueous solutions of acids or bases is known in the majority of cases (the pK values are determined), and this makes it possible, using a set of in-

dicators, to assess approximately the pH of the solution, or else
to determine the value of α spectrophotometrically and from Eq.
(5.24) calculate the pH value.

In organic solvents the mechanisms of dissociation of the indicators are not known (the pK values are not determined). In addition, on changing from aqueous to nonaqueous media there is frequently inversion of the relative values of pK. For instance, the range
of color change of methyl red in water is at higher pH values than
for bromophenol blue; in alcohol solutions the reverse applies.

Since acids and bases in polymers may to a first approximation
be regarded as mixed media, all these uncertainties remain when indicators are used in nonaqueous media.

Thus, the acid—base indicator method does not, with rare excep-
tions (hydrogels), allow us to assess the concentration of acid or
base in a polymer matrix; it can merely be stated that the concentration of acid or base is higher in the colored portion of the
polymer than in the uncolored.

When indicators with differing pK values are used in water, we
usually get two different results.

The color change of indicators (i.e., one indicator in a series
of polymers) takes place at practically one and the same x_t. This
may be connected with the presence of anomalous distribution of the
diffusant (a sharp front) in the polymer. But, at the same time,
it may be brought about by closely similar values of pK, at which
the color change of the indicators in the matrix of the polymer takes
place.

The color change of different indicators occurs at different
values of x_t. From these data we can determine the actual distribution of the diffusant in the polymer only when we establish by independent experiments the relationship between the pK values for
different indicators in water and in the matrix of the polymer.

Generally, the diffusion coefficient cannot be calculated by
means of Eq. (5.23) since, as indicated already, we do not know the
value of c_x.

However, we can use the following method [64] assuming that
the color change of the indicator depends solely on the concentration of acid or base in the polymer (c_x) and does not depend on the
concentration of solvent, if this changes within slight limits:

$$c_1^0\left(1 - \frac{x_{t_1}}{(\pi D)^{1/2}\, t_1^{1/2}}\right) = c_2^0\left(1 - \frac{x_{t_2}}{(\pi D)^{1/2}\, t_2^{1/2}}\right) \qquad (5.25)$$

The relationship in (5.25) makes it possible to calculate the value of D where the diffusion coefficient does not depend on concentration.

Thus, although the acid—base indicator method does not generally make it possible to determine the diffusion parameters of aggressive media, nevertheless, in certain cases (the use of solutions with differing concentration of acid or base), it is possible to estimate these parameters with sufficient accuracy.

Methods making it possible to obtain spectra of microvolumes of polymers are of great interest.

Electron probe x-ray analysis is based on recording the characteristic x-ray emission which is excited by a locally focused electron beam. By means of this method it is possible to analyze 10^{-15} to 10^{-16} g of elements in the range of microvolumes from 3 to 10 μm^3.

The methods of spectral analysis in the visible UV range, and also IR spectroscopy, make it possible to record spectra in the 1-2 μm^2 range.

These methods are being used to detect substances diffusing through polymers with variable density [65].

5.3. ANOMALOUS DIFFUSION IN POLYMERS

For polymers at temperatures close to and above the glass transition temperature, diffusion processes often cannot be described within the framework of the Fick equations, with constant boundary conditions and with a diffusion coefficient depending only on the concentration. In such cases we designate the diffusion processes as anomalous. In the opinion of Frisch and Hoffenberg [66], anomalous diffusion processes are indicated by the following signs:

the sorption curves on the coordinates (M_t/M_∞) vs. $(t^{1/2}/\ell)$ are not monotonic but have characteristic inflection points (they are multistage or S-shaped);

the kinetic curves of sorption depend on the thickness of the specimen;

the permeability curves may have extrema on the nonsteady-state portions, and the retardation time does not obey the Frisch equation (5.8);

the speed of the moving boundary of the electrolyte solution is not proportional to $t^{1/2}$ (5.23), but is proportional to t^n (where $\frac{1}{2} < n \leq 1$).

To describe anomalous diffusion in polymers, there are several theories, but they can all be reduced basically to two models: molecular relaxation and diffusion-convective models [67, p. 219].

The molecular relaxation model takes into account the molecular restructuring of the macromolecules which occurs in the distribution of the diffusing substance within the polymer matrix [68, 69]. At temperatures close to and below the glass transition temperature of the polymer, the relaxation times of the molecular segments are comparable with or exceed the times of the diffusion process. Diffusion in which the rate of transport is limited only by relaxation of the macromolecules is often called Type II diffusion, in contrast with Fick diffusion, which is called Type I. One criterion of Type I diffusion is movement of the boundary of the diffusant in accordance with (5.23) [70].

It is assumed that, as a consequence of molecular relaxation, either the concentration of diffusant in the surface layer [69] or the diffusion coefficient [68] changes with time according to a first-order equation. In [67] an attempt is made to reconcile these two mechanisms by means of Eq. (5.26), which describes the change in concentration of diffusant not only on the surface but also within the polymer, as a consequence of diffusional transport and relaxation:

$$\frac{\partial c}{\partial t} = \frac{\partial c_0}{\partial a}\,\frac{\partial a}{\partial t} + \frac{1}{\tau}\,(c^0 - c_t) \qquad (5.26)$$

where a is the activity of the diffusing substance within the polymer, c_0 and c_t, respectively, are the initial and surface concentration of the diffusing substance, and τ a coefficient of proportionality characterizing the average relaxation time of the macromolecules.

Apparently the relaxation processes have the greatest influence on the initial stages of sorption, and retardation of the relaxation processes affects the change in the surface concentration with time first.

The change in surface concentration ($\overline{c_t}$) with time is expressed as follows [71, 72]:

$$\overline{c_t} = \overline{c_0} + (c^0 - \overline{c_0})\left(1 - \exp -\frac{t}{\tau}\right) \qquad (5.27)$$

where c_0 is the surface concentration of diffusant as $t \to 0$.

The solution of Eq. (5.3) with asymmetric boundary conditions, where the surface concentration changes as given by (5.25), for steady-state transport leads to the equation in [73]:

$$\frac{\theta}{l^2} = \frac{1}{6\overline{D}} + \tau \left(1 - \frac{c_0}{c_\infty} \right) \frac{1}{l^2} \qquad\qquad (5.28)$$

Equation (5.28) explains a number of experimental facts:

the dependence of the diffusion coefficient, as found from the nonsteady-state section of the permeability curve, on the thickness of the film;

the difference between the diffusion coefficients determined under steady-state and nonsteady-state conditions;

the reason for the experimentally determined retardation time in the relaxation systems being the sum of the retardation time brought about by Fick diffusion and the time necessary to reach equilibrium concentration in the surface layer and characterized by the average polymer—diffusant relaxation time.

Processes of molecular relaxation play a significant role in the diffusion of electrolyte solutions in hydrophilic polymers since the change in conformation of the macromolecules in the polymer is in the first place brought about by interaction between the diffusant and the polymer. In hydrophobic polymers the interaction of the diffusant molecules predominates over polymer—diffusant interaction, and the equations given above become ineffective for describing the diffusion processes.

One consequence of the molecular interaction of the diffusing substance is the formation of clusters [74, 75], which in the case of Fick diffusion leads to a reduction in the activity coefficient with increase in the concentration of diffusing substance. With anomalous diffusion the situation becomes more complex. The relaxation processes which take place in the matrix as the entry of diffusant progresses render the structure of the polymer more accessible for the movement of the molecules of the low-molecular substance and, consequently, increase its concentration. In turn, the increase in concentrations leads to more vigorous formation of clusters. Thus, the presence of two factors with opposite influence on the diffusion coefficient may lead either to this coefficient not varying with time [76], or to its decreasing with time [77].

Although the molecular relaxation model explains a wide range of experimental results, when we describe the diffusion process within the framework of this model we have to use a series of empirical parameters, the physical meaning of which is not always as clear as could be. Accordingly, for a strict explanation of the effects occurring with anomalous diffusion, we require additional information obtained by independent methods.

The diffusion-convective model envisages heavily swollen glassy polymers. In the limiting case there is for such polymers a constant rate of movement of the sharp front which divides the swollen outside layer from the nonswollen inside layer [70, 78]. Moreover, the movement of the volume of the liquid as a whole under the action of the various forces leads to transport of the electrolyte dissolved in the liquid at a rate equal to that of the liquid. In some cases the movement of the diffusant front may take place according to a particular intermediate law

$$x_t = \lambda t + \lambda' t^{1/2} \tag{5.29}$$

which is a combination of Fick diffusion and Type II diffusion [78, p. 63].

A strict mathematical description of the diffusion-convective model is given by

$$\frac{\partial c_i}{\partial t} = -v \operatorname{grad} c_i + D \operatorname{div} \operatorname{grad} c_i \tag{5.30}$$

where v is the rate of movement of the local volume of solution within the polymer, this rate being controlled by the potential set up in the polymer as a consequence of swelling.

The first term in Eq. (5.30) defines the convective component, the second the diffusion component. With $v = 0$, Eq. (5.30) changes into (5.3), while with high values of v we can disregard the diffusion component.

An analytical solution of Eq. (5.30) is possible only with constant D and v.

Peterlin [80] considered a model in which the moving diffusant front was preceded by Fick distribution of the diffusant in the as yet unswollen portion of the polymer, i.e., the value D relates to a portion of a polymer which is in a glassy state.

The diffusion-convective model satisfactorily explains the experimental data, with the exception of diffusion curves which are concave to the time axis [67].

These relaxation and convection models describe practically all anomalous diffusion phenomena (with the exception of those cases where the diffusion process is accompanied by microcracking ("crazing") of the polymer [81]) and, in spite of some physical indefiniteness, remain the only mechanism for the description and interpretation of diffusion processes in glassy polymers.

5.4. SORPTION OF ELECTROLYTES BY POLYMERS

Data on the sorption of electrolytes by polymers are necessary for the calculation of the integral diffusion coefficients when using a variety of methods, and for the interpretation of the kinetic data in the degradation of polymers in aggressive media under heterogeneous conditions.

The solubility of electrolytes in polymers is described thermodynamically as its distribution between the solution and the polymer. In the equilibrium state there is equality of the electrochemical potentials for these two phases [82]:

$$\mu_i^0 + RT \ln a_i + pV_i + Z_i F\psi = \overline{\mu}_i^0 + RT \ln \overline{a}_i + \overline{p}\overline{V}_i + Z_i F\overline{\psi} \qquad (5.31)$$

where μ_i^0 stands for the standard chemical potential of the i-th ion; V_i, the partial molar volume of the i-th ion; and p and ψ, respectively, the hydrostatic pressures and the electrical potentials in the phase. The quantity on the left of the equation relates to the solution phase and the quantity on the right to the polymer phase.

Equation (5.31), chosen on the basis of Matheson and Whewell [83] for creating a theory of sorption of electrolytes by polymers, may be written for an electrolyte CA, which is fully dissociated and obeys the condition of the electrical neutrality both in the polymer and in the solution, as follows:

$$\ln \frac{\overline{a}_C \overline{a}_A}{a_C a_A} = -\frac{\pi \overline{V}_{CA} + F\Delta\psi + \Delta\mu_{CA}^0}{RT} \qquad (5.32)$$

where $\pi = (p - \overline{p})$ is the osmotic pressure; $\Delta\mu_{CA}^0$, the difference between the standard electrochemical potentials; and $\Delta\psi$, the difference between the electrical potentials in the solution and in the polymer.

At equilibrium there is assumed to be equality between the molar volumes of CA in the two phases. Because of the absence of reliable information on the activity of the ions in the polymers, it is assumed, in the majority of cases, that the average activity coefficient of the electrolyte is equal to unity. This, apparently, is correct for low electrolyte concentrations ($<10^{-6}$ mole·cm^{-3}).

Turning from the activities to the concentrations, and using instead the product of the activities of the average electrolyte concentration, Eq. (5.32) may be written in the form

$$K_{\text{distr}} = \frac{\overline{c}_{CA}}{c_{CA}} = e^{-z}$$

$$(5.33)$$

$$z = \frac{\pi \overline{V}_{CA} + F\Delta\psi + \Delta\mu_{CA}^0}{RT}$$

In the absence of an osmotic pressure in the system and a difference between the electrical potentials in the solution and in the polymer Eq. (5.24) (sic) changes into the Nernst equation for the equilibrium distribution of the substance between the two phases.

Let us consider the main rules of behavior in the sorption of electrolytes in hydrophobic and hydrophilic polymers.

Hydrophobic polymers adsorb electrolytes in very small amounts [84]. This results in a lack of experimental data on K_{distr} for the majority of polymer—electrolyte systems, and the emergence of calculation methods for assessing the distribution coefficients. As will be shown below, in hydrophobic polymers electrolytes are predominantly present in undissociated and unhydrated form, i.e., as a first approximation we can consider the distribution of an electrolyte between two nonmixing solvents with dissociation of the electrolyte only in one of them, assuming that the distribution process obeys the Nernst law.

In [84] an attempt is made to calculate K_{distr}:

$$K_{distr} = \exp\left(- \frac{\Delta F^0}{RT}\right) \tag{5.34}$$

where ΔF^0 is the change in Helmholtz energy of the transport of electrolyte from the solution to the polymer.

The transport of an electrolyte from an aqueous solution into a polymer may be presented as consecutive processes of separation of electrolyte from the aqueous solution (a process which is the inverse of what occurs when it goes into solution) and mixing of the electrolyte with the polymer:

$$\Delta F^0_{ext} = -\Delta F^0_{solv} + \Delta F^0_{mix} \tag{5.35}$$

where ΔF^0_{solv} is the change in Helmholtz energy in the formation of an infinitely dilute aqueous solution and ΔF^0_{mix} the change in free energy on mixing of the electrolyte with the polymer.

Exploratory calculations for acids have shown [84] that the major contribution to the value of ΔF^0 is made by the term $(-\Delta F^0_{solv})$, i.e., the distribution constant is to a considerable extent determined by the energy which has to be expended to remove the electrolyte from the aqueous solution. However, differences in the absorption of the electrolytes by the polymers are to a great extent brought about by differences in the values of $(-\Delta F^0_{mix})$, which are directly linked with the sorption, as follows [85]:

$$\Delta F_{mix} = RT + \overline{V}_{electr}(\delta_{electr} - \delta_{pcl}) \tag{5.36}$$

where δ is the solubility parameter, which is equal to the square root of the value for the cohesion energy density; and $\overline{V}$ is the molar volume.

The values of δ for the electrolytes can be calculated as follows:

$$\delta = \left(\frac{L - RT}{\overline{V}} \right)^{1/2}$$

where L is the heat of vaporation of the electrolyte.

The values of δ for the polymers are found by calculation [86]. The following are the calculated values of K_{distr} obtained for a system of polyethylene and solutions of inorganic acids at 25°C [84].

Acid	HNO_3	HCl	H_2SO_4
K_{distr}	10^{-7}	$3 \cdot 10^{-8}$	$3 \cdot 10^{-12}$

For HCl the value of K_{distr} was determined experimentally [84], and practically coincided with that given above.

Thus, electrolytes which have an affinity for water or have high boiling points probably pass from aqueous solutions into hydrophobic polymers only to a very small extent.

In polymers which are below T_g, the sorption of low-molecular substances may become considerably more complex. This is due to the inhomogeneity of the physical structure of such polymers, which becomes apparent in the presence of parts with differing density and, in particular, of voids between the macromolecules, which are treated as a system of micropores.

Vieth [87, 88] proposed a mechanism of "double" sorption, based on the actual solubility of the low-molecular substance and the sorption of this same substance by the surface of the micropores.

Thus, the micropores serve as a "reservoir" for the sorbed substance and increase the effective sorption.

The dependence of the effective sorption of methanol vapors on the maximum volume of micropores for films of F-2 polyarylate at 25°C is as follows [89]:

w_0^{max}, cm^3/g	0.08	0.12	0.18	0.21
σ, $cm^3/(cm^3 \cdot atm)$	0.50	0.90	1.50	1.75

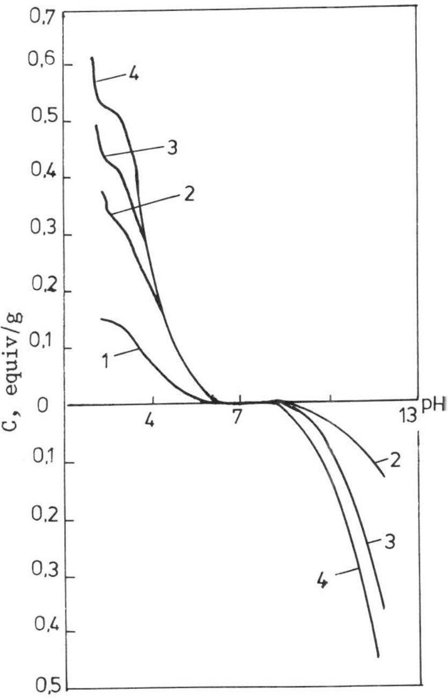

Fig. 5.3. Change in the sorption of aqueous so-
           lutions of hydrochloric acid and of
           sodium hydroxide by polycaproamide
           fibers of varying MM as a function of
           the pH: 1) 5650; 2) 2600; 3) 1700;
           4) 1380 [90].

The value of $w_0^{max} = (1/\rho^*) - (1/\rho_{act})$ (where ρ^* is the density as calculated from the mass and the volume of the film).

Hydrophilic polymers sorb electrolytes in considerable amounts.
Figure 5.3 shows typical isotherms of the sorption of aqueous solutions of HCl and NaOH by polycaproamide fibers of various MMs [90].
All the curves have slight discontinuities at the level of maximum
combination of the terminal groups in the polymer with the acid at
pH 2.3-3.0 and base at pH 12.3-12.4.

Various approaches have been suggested for explaining the resulting data: ion exchange [91], thermodynamic equilibrium [92],
and calculation of the potential of the fibers [93].

However, the proposed equations either do not sufficiently describe the experimental data rigorously, or else do not take into
account the influence of salts present in the polymer on the sorption isotherm. The experimental data can, however, be explained
satisfactorily by means of an equation derived from (5.32)

$$pH = pK_{diss} - \log \frac{1-\alpha}{\alpha} - \frac{0.434}{RT}\left(F\Delta\psi + \pi V_H + \Delta\mu_H^0\right) \qquad (5.37)$$

where K_{diss} and α are, respectively, the dissociation constant and the degree of dissociation in the polymer.

The value of $\Delta\psi$ can be determined from the relationship in [83]

$$\frac{N_0 e \Delta\psi}{RT} = \sin h^{-1}\frac{c_f}{2c} \qquad (5.38)$$

where c_f is the concentration of fixed charges in the polymer and c the concentration of electrolyte in the solution.

With the proviso that c_f/c is large

$$\Delta\psi \simeq \frac{RT}{N_0 e}\ln\frac{c_f}{2c} \qquad (5.39)$$

Thus, Eq. (5.37) may be written as

$$pH = pK_e - \log\frac{1-\alpha}{\alpha} - 0.434\frac{\ln c_f/2c}{N_0 e} \qquad (5.40)$$

where

$$pK_e = pK_{diss} - \frac{0.434}{RT}\left(\pi V_H - \Delta\mu_H^0\right)$$

If the value of pK_e calculated from Eq. (5.40) depends on the ionic strength of the solution, then the osmotic pressure influences the solubility of the electrolytes in the polymer.

If, on the other hand, pK_e depends, at constant ionic strength, on the nature of the sorbed electrolyte, then there is specific interaction of the ionogenic groups of the polymer with the electrolyte. The sorption of acids by various natural and synthetic polyamides is considered in [84]. For hydrophilic polymers which contain a sufficient amount of uncombined water (for instance hydrogels), and in which there is an absence of specific interaction of electrolyte with the polymer matrix, there applies the simple relationship

$$K_{distr} = p\varphi_{H_2O} \qquad (5.41)$$

where p is a coefficient of proportionality and φ_{H_2O} the volume fraction of water in the polymer, and the concentration of electrolyte in the polymer is close to its concentration in the solution [93].

It is shown in [94] that for thin polymer films (<100 Å) the distribution coefficient increases with the diminution in thickness of the layer.

5.5. SPECIAL FEATURES OF THE DIFFUSION OF ELECTROLYTE SOLUTIONS IN POLYMERS

Electrolyte solutions occupy a special place as compared with other diffusants. Let us consider the major features of their diffusion in polymers.

5.5.1. Ions

Electrolyte solutions in a polymer matrix can dissociate into ions which, unlike uncharged particles, move under the action of two forces: the chemical potential gradient of ions of the type in question, and the electric field created by the movement of oppositely charged ions.

In the diffusion of ions Eqs. (5.2) and (5.3) will take the form

$$I = -D_i^* \left[\left(1 + \frac{d \ln y_{i\pm}}{d \ln c_i} \right) \operatorname{grad} c_i + \frac{Z_i F c_i}{RT} \operatorname{grad} \psi \right] \tag{5.42}$$

$$\frac{\partial c_i}{\partial t} = \operatorname{div} \left[D_i^* \left(1 + \frac{d \ln y_{i\pm}}{d \ln c_i} \operatorname{grad} c_i + \frac{Z_i F c_i}{RT} \operatorname{grad} \psi \right） \right] \tag{5.43}$$

These equations, called the Nernst–Planck equations, link the concentration of diffusing ions with the electric potential ψ, the time, and the average molar activity coefficient $y_{i\pm}$. The value of $\operatorname{grad} \psi$ can be determined from the Poisson–Boltzmann equation:

$$\operatorname{div}(\operatorname{grad} \psi) = -4\pi\rho/\varepsilon \tag{5.44}$$

where ρ is the charge density in the system and ε the permittivity of the medium.

Although the ions have differing mobilities, they have practically no independent movement since this should give a considerable electric field. In the great majority of cases there is for electrolyte–polymer systems the condition of electrical neutrality ($\rho = 0$), and this considerably simplifies the solutions of Eqs. (5.42) and (5.43). Solutions for other conditions are considered in [95, p. 205].

Generally, by analogy with the diffusion of electrolytes in solutions [96, p. 331 of Russian translation] the diffusion coefficient of an electrolyte CA in a polymer is

$$D = \left[\alpha\left(D^0_{C^+A^-} + \overline{Z}\Delta n\right) + 2\left(1-\alpha\right)D^0_{CA}\right]\left(1 + \frac{d\ln y_{\pm}}{d\ln c}\right) \qquad (5.45)$$

where α is the degree of dissociation of the electrolyte in the polymer; $D^0_{C^+A^-}$ the limiting value of the diffusion coefficient, which can be expressed by way of the autodiffusion coefficients of the ions $D^*_{C^+}$ and $D^*_{A^-}$

$$D^0_{C^+A^-} = \frac{D^*_{C^+}D^*_{A^-}(Z_{C^+} + Z_{A^-})}{Z_{C^+}D^*_{C^+} + Z_{A^-}D^*_{A^-}} \qquad (5.46)$$

where Δn is a parameter of the electrophoretic effect, which depends on the permittivity of the medium, the temperature, and the concentration, and D^0_{AC} the limiting value of the diffusion coefficient of the undissociated molecule or ion pair.

In electrolyte solutions the electrophoretic effect amounts to no more than a few percent of D^0; it may be expected to be only slight in a system of an electrolyte and a non-ion-exchange polymer.

5.5.2. Concentration Gradient of the Electrolyte at the Interface of the Electrolyte Solution and the Polymer

The concentration gradient of an electrolyte at a phase interface may occur for two reasons: accompanying a rapid diffusion process is a disturbance of the equilibrium statistical distribution of the electrolyte (molecules and ions on the surface of the polymer (concentration effect), and where particles of the electrolyte (molecules and ions) and molecules of the solvent are adsorbed on the surface of the polymer, particularly when the surface is charged (adsorption effect). Both effects, which lead to the formation of a nonmixing layer on the interface, are assessed in Chap. 6.

5.5.3. Strong Interaction of Electrolyte with Polymer

If the polymers contain ionogenic groups, there may be strong interaction of these groups with diffusing ions, leading to a complex character of diffusion. As one example of such interaction we can take the system of aqueous electrolyte solutions with polyamides.

There is at the present time no general quantitative theory
linking any given parameter of a polymer structure with the diffusion coefficients of an electrolyte in it. Study of the diffusion
of electrolytes in various polymers shows that it is desirable to
divide polymers as a whole into two groups, depending on their content of any particular polar solvent. For instance, depending on
their water content, they are divided into polymers in which water
is soluble to a large or to a limited extent (upwards of 1 mass %)
or polymers in which water is poorly soluble (less than 1 mass %).

It must be borne in mind that polymers in the first group, as
a rule, rate as hydrophilic, those in the second group as hydrophobic.

In [99], a method is proposed for assessing the hydrophilicity
of polymers by the amount of adsorbed water on the surface of a dry
specimen held under the conditions of given humidity and temperature. It is suggested that the capacity of the water for being adsorbed depends on the morphology of the polymers; with increase in
the degree of crystallinity there is an increase in the content of
water on the surface of the polymer.

5.6. HYDROPHILIC POLYMERS

This refers to polymers of differing types. We may distinguish
polymers which readily dissolve water [poly(vinyl alcohol), cellulose, etc.], and polymers which dissolve water to a limited extent
(polyamides, various polyesters, etc.). The most detailed study
has been of the diffusion of electrolytes through ion-exchange polymers [95, p. 205; 97; 98, p. 347). Although the diffusion of electrolytes through such polymers is a specific process (diffusion is
accompanied by ion exchange and takes place, as a rule, in porous
materials), it is worthwhile to consider the most significant features.

5.6.1. The State of Electrolytes in Polymers

Numerous attempts have been made to estimate the pH in the ionite phase, but in such cases nonrigorous assumptions have been made.
In [100] the pH value was estimated on the basis of the color change
of an indicator sorbed by an ionite; in [101], on the basis of the
change in the degree of dissociation of amino acids in the adsorbed
state. These data indicate a considerable difference between the
pH in the solution and in a polymer. However, we cannot draw definite
conclusions from this data as to the dissociation constants of the
ionogenic groups. The dissociation constants of electrolytes in
poly(vinyl alcohol) are determined in [102]. If as a first approximation we regard polymers which readily dissolve water as a system
equivalent in properties to mixed solutions, we may expect the law
of mass action to be satisfied and we may use the classical equations

TABLE 5.2. Values of λ_0 and K_{diss} of Phosphoric Acid in Poly(vinyl alcohol) with Varying Water Content [102]

Water content, vol. %	λ_0, cm^2/ ohm·equiv	$K_{diss}\cdot10^3$
100	383	7.51
78	44	7.05 ± 0.6
38	14	7.6 ± 0.5

to determine the dissociation constant in the liquid phase. As a first approximation, K_{diss} can be calculated from the Ostwald equation:

$$K_{diss} = \frac{\alpha^2 c}{1-\alpha} \qquad (5.47)$$

where α is the degree of dissociation, equal to λ/λ_0, and λ the equivalent electrical conductivity.

The equivalent electrical conductivity at infinite dilution, λ_0, can often be calculated on the basis of the Kohlrausch law.

In a number of cases in the determination of K_{diss} it is necessary to take into account the activity coefficient of the ions [103] and the change in the mobility of the ions with increase in ionic strength. Table 5.2 gives the values of K_{diss} and λ_0 of phosphoric acid in poly(vinyl alcohol) (PVA) with differing water content. The calculation was carried out using the Ostwald equation [102].

The constant K_{diss} of phosphoric acid in a thin film of poly(vinyl alcohol) was also determined spectrophotometrically using an indicator (m-nitroaniline) [102]. The values of K_{diss} as found by the two methods were close.

Table 5.2 shows that λ_0 falls considerably with a reduction in the volume fraction of water, while K_{diss} remains practically unchanged. Constant values for the thermodynamic parameters in an ion-exchange resin (water content 60 vol. %) and in water are noted in [110, p. 70].

The absence of a change in K_{diss} of phosphoric acid in PVA with varying water content may be linked either with the presence of micropores in the polymer, an aqueous acid solution being distributed in these pores [105], or else with the formation within the PVA of structures having the same affinity for ions as in aqueous solutions.

The dissociation of sulfuric acid in a polycaproamide (PCA) has been investigated by an electrical conductivity method [106]. On the basis of a number of assumptions the authors conclude that sulfuric acid dissociates mainly as far as the first stage, $pK_{1H_2SO_4}$ being 2.5.

No direct measurements of the hydration of electrolyte ions in polymers of this type have been carried out. It is shown in [107, 108] that the autodiffusion coefficients of ions of alkali metals in sulfonated copolymers of styrene with divinylbenzene rise as follows: Li < Na < K < Rb < Cs, corresponding qualitatively to the radius of the hydrated cations. From the changes in the values for spin-lattice and spin-spin relaxation times as a function of the composition of the polyamide—water system [109] it is concluded that with water content up to 0.5 mass % the water molecules are strongly combined with the amide groups. When the water content is raised from 0.5 to 8.0%, the mobility of the water molecules in the polymer gradually rises and approximates to that for free water. With the addition of sodium chloride the mobility of the water molecules rises; this, in the opinion of the authors, is explained by the disturbing action of the electrolytes on the clusters of water forming in the polymers [109].

5.6.2. Main Theories of Diffusion

Various theories of the diffusion of electrolytes in hydrophilic polymers have been proposed.

The Mackie—Mears Theory [110]. This theory explained the reduction in the diffusion coefficients of electrolytes in hydrophilic polymers as compared with aqueous solutions by an increase in the diffusion route, due to the bending of the macromolecules by the ions. In the absence of interaction between the functional groups of the polymer and the ions this effect is accounted for by the factor

$$f = \left(\frac{1 + \varphi_p}{1 - \varphi_p} \right)^2 \tag{5.48}$$

where φ_p is the volume fraction of polymer.

An attempt has been made to eliminate discrepancy between experimental and theoretical values for the diffusion coefficients by introducing a correction for the change in viscosity as compared with aqueous electrolyte solutions [111]. But the admissibility of such a correction is very doubtful since concepts about viscosity in polymers are ambiguous [112].

The Free Volume Theory [14-16]. This was developed for two-component systems and consequently is applicable to single-component aggressive media. According to this theory the relative change in the autodiffusion coefficient of a low-molecular substance with a change in the parameters of a polymer—solvent system is expressed as follows:

$$\log \frac{D^*}{D^*_{(0)}} = \frac{B}{2.3}\left(\frac{1}{f_{(0)}} - \frac{1}{f}\right) \tag{5.49}$$

where B is a constant of the polymer—solvent system and f the fraction of free volume in the system.

For quantitative calculations we introduce the standard state of the system with respect to the calculation being made; it is characterized by definite values of $D_{(0)}$ and $f_{(0)}$. As the standard state we use either the unmixed polymer or the pure solvent.

The fraction of free volume of the system is an additive function of the fractions of the free volume of the components:

$$f = f_{pol}\,\varphi_{pol} + f_{solv}\,\varphi_{solv} = f_{pol} + \beta\varphi_{solv} \tag{5.50}$$

where f_{pol} and f_{solv} are the fractions of the free volume of the polymer and of the pure solvent, respectively, and φ_{pol} and φ_{solv} the volume fractions of the polymer and solvent in the system; $\beta = f_{solv} - f_{pol}$.

Generally, the free volume of the polymer—solvent system depends on the temperature and composition [113]:

$$\left(\frac{\partial f}{\partial T}\right)\varphi_{solv} = a_{pol} + \Delta a_{solv-pol}\varphi_{solv} \tag{5.51}$$

$$\left(\frac{\partial F}{\partial T}\right)_{T_g} = f_{solv} - f_{pol} = \beta_{(0)} \tag{5.52}$$

where a_{pol} and a_{solv} are the coefficients of thermal expansion of the polymer and solvent, respectively, and $\Delta a_{solv-pol} = a_{solv} - a_{pol}$.

We can now write

$$f = f_{fr} + a_{pol}(T - T_g) + \Delta a_{solv-pol}(T - T_g)\varphi_{solv} + \beta_{(0)}\varphi_{pol} \tag{5.53}$$

where f_{fr} is the fraction of the free volume for amorphous polymers at T_g, equal to approximately 0.025.

On substituting Eq. (5.53) into (5.49), we get

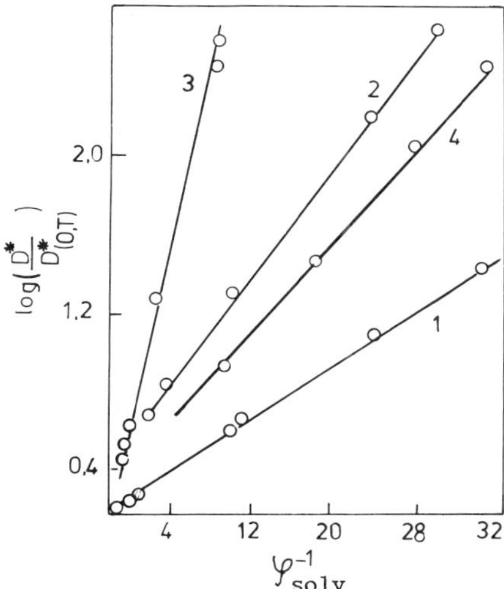

Fig. 5.4.   Dependence of the logarithms of the
autodiffusion coefficients (D^*) on
the volume fraction of solvent, φ solv,
in ethylcellulose at 60°C according
to the coordinates of Eq. (5.55): 1)
dichloroethane; 2) n-propanol; 3) ethanol; 4) methylene chloride [119].

$$\log\frac{D^*}{D^*_{(0)}} = \frac{B}{2.3} \frac{a_{pol}(T-T_g)+[\beta_{(0)}+\Delta a_{solv-pol}(T-T_g)]\varphi_{solv}}{f^2_{fr}+f_{fr}a_{pol}(T-T_g)+f_{fr}[\beta_{(0)}+\Delta a_{solv-pol}(T-T_g)]\varphi_{solv}} \qquad (5.54)$$

which describes the change in the autodiffusion coefficient of the
solvent as a function of the temperature and the concentration in
the system. With T = const, Eq. (5.54) changes into the well-known
Fujita equation [14]

$$\left(\log\frac{D^*}{D^*_{(0,T)}}\right)^{-1} = \frac{2.3f(T)}{B} + \frac{2.3f^2(T)}{B\beta} \frac{1}{\varphi_{solv}} \qquad (5.55)$$

With φsolv = const, Eq. (5.54) takes the following form:

$$\left(\log\frac{D^*}{D^*(\varphi T_g)}\right)^{-1} = \frac{2.3f(\varphi_{solv}T_g)}{B} + \frac{2.3f^2(\varphi_{solv}T_g)}{Ba(\varphi_{solv})} \frac{1}{T-T_g} \qquad (5.56)$$

Typical experimental curves illustrating the above relationships are shown in Fig. 5.4. It is significant that the linear
character of the relationship between $(\log D^*/D^*_{(0,T)})^{-1}$ and φ^{-1}_{solv}
is maintained both above and below the glass transition temperature
of ethylcellulose (about 45°C). This is, on the one hand, evidence
that irrespective of the physical state of the polymer system the

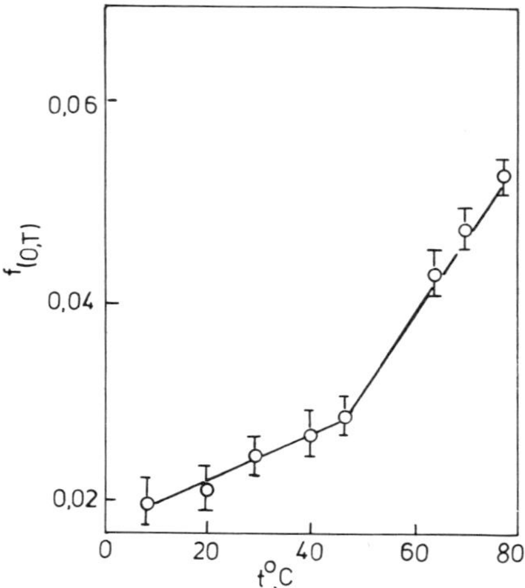

Fig. 5.5.   Dependence of the fraction of free
            volume of ethylcellulose, f(0, T),
            on temperature [113].

change in the fraction of free volume under the influence of a sorbed
solvent follows the same behavior, which as a first approximation can be described by (5.50). On the other hand, this suggests
that the rate of diffusion of the solvent molecules in the polymer
in the glassy state is likewise linked with the fraction of its free
volume. In point of fact, for the realization of the elementary
act of diffusion in the polymer matrix, irrespective of its physical
state, there must be a short-range order regrouping, which is usually effected as a result of the movement of the polymer segments.
Nevertheless, whereas in the high-elastic state the responsibility
for the segment regrouping lies with the thermal movement of the
segments, in the glassy state, under conditions of restricted seg-
mental mobility, this function is apparently fulfilled by local
stresses which are set up in the vicinity of the diffusing molecule.
But even in this case the regrouping of the segments of the molecular chain is determined by the magnitude of the free volume in
the polymer [113].

The fraction of free volume does not depend on the nature of
the molecules of the low-molecular component, and is determined
solely by the temperature (Fig. 5.5). An equation is proposed in
[113] linking the change in the interdiffusion coefficient with the
parameters of the polymer-solvent system:

$$D_V = D_{(*)}(1 - \varphi_{solv} - 2\chi\varphi_{solv}\varphi_{pol})$$
(5.57)

where χ is the Flory-Huggins constant for the system.

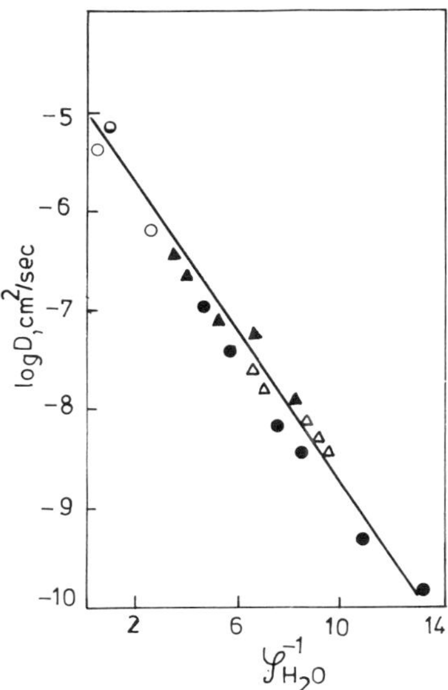

Fig. 5.6.   Dependence of the logarithm of the dif-
             fusion coefficient of sodium chloride
             in copolymers having variable content
             of the initial monomers on the recip-
             rocal of the volume fraction of the
             water in hydroxyethyl methacrylate
             ($\triangle$), and copolymers of methyl meth-
             acrylate with glyceryl methacrylate
             ($\bullet$), hydroxyethyl methacrylate ($\blacktriangle$),
             and hydroxypropyl methacrylate ($\circ$) [114].

     Thus, for single-component aggressive media, within the framework
of the free volume theory, it is possible to make a satisfactory
prediction of the change in the diffusion coefficient as a function
of the volume fraction of the aggressive medium in the polymer, of
the temperature, and of the parameters of interaction of the components of the polymer—solvent system.

     If the aggressive medium is multicomponent (such as solutions
of acids and bases), the application of the free volume theory generally becomes more difficult for such systems. However, there are
examples of application of the free volume theory to certain special
cases. For instance, Yasuda [114], investigating the diffusion of
aqueous solutions of salts in hydrophilic polymers, assumed that
the free volume of the swollen polymer is mainly determined by the
volume fraction of the solvent (water).

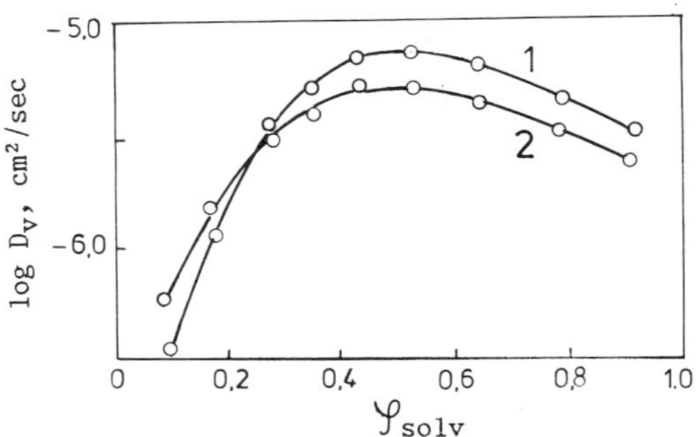

Fig. 5.7.   Dependence of the logarithm of the relative
            diffusion coefficient on the composition of
            the system at 70°C: 1) PVA–H_2O;   2) PVA–0.25
            M aqueous solution of NH_4Cl [115].

Yasuda's equation

$$\log D = \log D_{H_2O} - k\varphi_{H_2O}^{-1} \tag{5.58}$$

is a special case of (5.49). As the standard state in this equation
pure water was used (D_{H_2O} is the diffusion coefficient of the elec-
trolyes in water; k a constant depending on the free volumes of
polymer and water).

Figure 5.6 shows the dependence of the logarithm of the diffusion coefficient of sodium chloride in copolymers having a variable
content of the initial monomers on the reciprocal of the volume
fraction of the water in the polymers.

Indeed, as to be expected, Eq. (5.58) does not hold for low
volume fractions of water, i.e., in systems where the free volumes
of polymer and solvent are commensurate.

In [115] a different approach was made. Assuming that aqueous
solutions of salts are diffusing into a PVA–water–salt system in
a single front without separation into components, the authors reduced the multicomponent system to a two-component system of polymer and aggressive medium. The standard state used was a system
with a volume fraction of the aggressive medium of 0.1. In this
case, Eq. (5.55) can be transformed as follows:

$$\frac{\varphi_{solv} - 0.1}{\log D_V - \log D_{V(\varphi=0.1)}} = A + B\varphi_{solv} \tag{5.59}$$

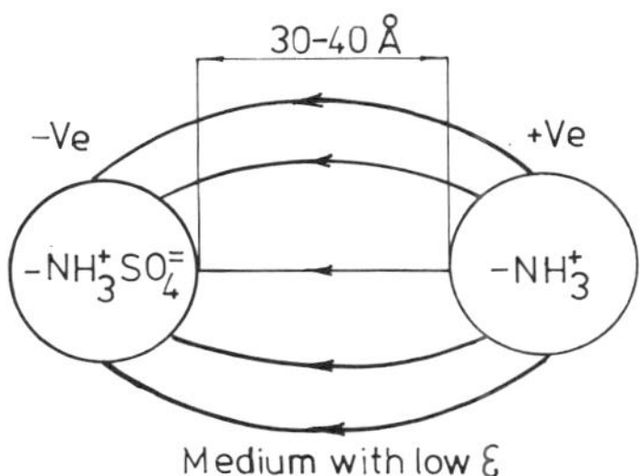

Fig. 5.8.   Schematic representation of terminal
            groups of like charge in poly—hexa-
            methylene adipamide) [118].

where

$$A = \frac{f\text{pol}(f\text{pol} + \beta\,0.1)}{0.43B\beta}; \qquad B = \frac{f\text{pol} + \beta\,0.1}{0.43B}$$

Figure 5.7 shows the change in log D$_V$ vs. the composition of a
system of PVA and a 0.25 M aqueous salt solution, treating the experimental data according to Eq. (5.59).   In the range of concentrations with a high content of aggressive medium the diffusion behavior
of the system PVA—aqueous salt solution is determined mainly by
changes in the structure of the water, which take place under the
influence of the salts dissolved in it, and at such concentrations
it is not possible to express the change in D$_V$ within the framework
of the free volume theory.

Boyd's Theory [116].   This approach represents a further development of prior theories.   Reduction in the mobility of the ions
in ion-exchange polymers as compared with that of the ions in water
is due to three factors:

reduction in the cross-sectional area of diffusion as a consequence of the steric hindrances of the polymer network;

increase in the length of the diffusion route because the ions
have to take an indirect path through the polymer network;

interaction of an ion with the walls of the pores.

The influence of one or the other factor depends on the nature
of the polymer and diffusant.

Theories of Medley [117] and Marshall [118].   These theories
are used to explain the dependence of the diffusion coefficients
on the structure in the case of polymers which dissolve water to
a limited extent.   For instance, in polyamides, as compared with
polymers which readily dissolve water, the dissolved water cannot

form, within the polymer matrix, an aqueous continuum through which
the electrolyte can move. Accordingly, in the polymer matrix the
water is found around polar terminal groups and only to a lesser
extent around the amide bonds.

The diffusion of the ions proceeds by way of activated jumps
from one polar group to another, thus passing through a region with
low dielectric permittivity. For instance, in poly(hexamethylene
diamide) (adipamide, nylon-66) the distance between neighboring
terminal groups of like charge is 30-40 Å (Fig. 5.8).

If we envisage the existence of the polyamides in zwitterion
form, then the active sites in the diffusion of the anions will be
positively charged amino groups, and in the diffusion of the protons they will be carboxylate groups.

5.6.3. Dependence of the Diffusion Coefficients on Concentration

A most characteristic feature in the diffusion of electrolytes
in hydrophilic polymers is the concentration dependence of the diffusion coefficients, which in the general case may be determined by
the following factors:

change in the degree of dissociation of the electrolyte;

nonideal behavior of the polymer-electrolyte system;

swelling of the polymer, leading to a rise in mobility of the
polymer segments;

binding of the electrolyte by ionogenic groups of the polymer.

The influence of the first, second, and fourth factors follows
directly from Eq. (5.45), while the influence of the third factor
is assessed in detail in Sec. 5.4.

Let us consider the influence of each of these factors individually. So as to rule out the second, third, and fourth factors,
we may investigate the diffusion of dilute electrolyte solutions
in polymers which do not contain any ionogenic groups and which
are previously saturated with the solvent.

Concentration Dependence of Diffusion Coefficients Brought
about by Change in the Degree of Dissociation of the Electrolyte.
Most work has been directed to the diffusion of electrolytes in PVA
films. Figure 5.9 shows the concentration dependence of the diffusion coefficients of acids in PVA at 25°C. The diffusion coefficients were determined under steady-state conditions, from the change

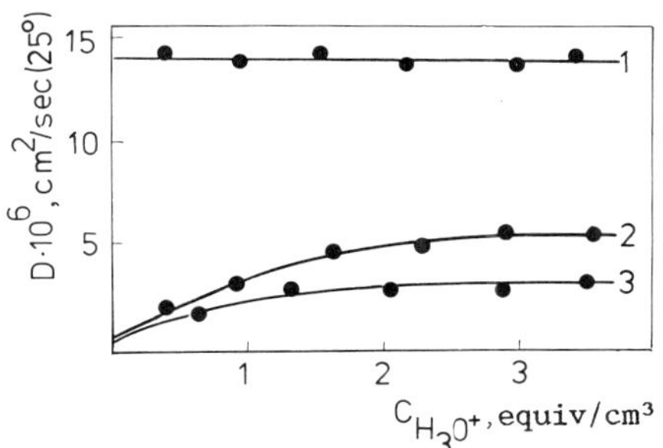

Fig. 5.9. Dependence of the integral diffusion coeffi-
           cients $\bar{D}$ of acids in PVA films at 25°C on
           the concentration [102]:   1) HCl; 2) H_2SO_4;
           3) H_3PO_4.

in concentration of hydrogen ions in the recording chamber of a measuring cell [25].

For an almost completely dissociated electrolyte, e.g., dilute
solutions of HCl, Eq. (5.45) is simplified:

$$\bar{D} \approx D^0_{H^+Cl^-} \tag{5.60}$$

For solutions of H_2SO_4 in a given range there is perceptible second-
stage dissociation of the acid, and hydrogen ions diffuse in the
polymer matrix in the surroundings of the counterions HSO_4^- and
SO_4^{2-}. Calculation of the dissociation leads to the expression

$$\bar{D} = D^0_{H^+ (HSO_4^-)} \left[1 + \frac{2\alpha - 4}{\left(\dfrac{c_{H^+}}{K_{2H_2SO_4}} + 2 \right)^2} \right] \tag{5.61}$$

where $\alpha = D^0_{H^+(SO_4^{2-})} / D^0_{H^+(HSO_4^-)}$; c_{H^+} is the concentration of hydrogen ions
in the polymer; $K_{2H_2SO_4}$ is the second-stage dissociation constant
of the sulfuric acid.

Figure 5.10 shows the graphical solution of Eq. (5.61). In
the case of the weaker acid H_3PO_4 there is dissociation in the first
degree in the concentration range under investigation:

$$\bar{D} = D^0_{H^+ (H_2PO_4^-)} \left[1 + \frac{2\beta}{K_{1H_3PO_4}} c_{H^+} \right] \tag{5.62}$$

where $\beta = D^0_{H_3PO_4} / D^0_{H^+(H_2PO_4^-)}$; $K_{1H_3PO_4}$ is the first-stage dissociation constant of the phosphoric acid.

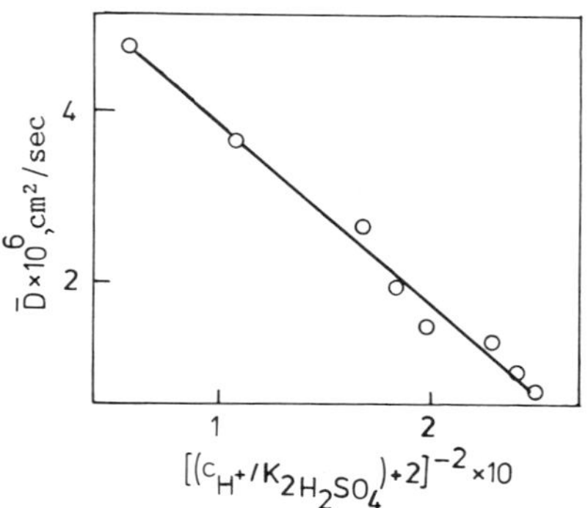

Fig. 5.10.   Graphical solution of Eq. (5.61) for
the diffusion of sulfuric acid in
PVA films at 25°C [102].

With the diffusion of phosphoric acid through PVA a proportion
of the diffusant molecules are combined in the polymer matrix and
do not participate in the transport process.   In Eq. (5.62) all the
parameters relate to free particles of the acid.

The concentration of acid combined with the polymer is found
from the Langmuir equation

$$c_{H^+\,(comb)} = \frac{c_{H^+}}{a + bc_{H^+}} \tag{5.63}$$

where a and b are constants.

From Eqs. (5.62) and (5.63) we get

$$\bar{D} = \frac{D_{H^+(H_2PO_4^-)}\left[1 + 2\beta K_{diss}^{-1}c_{H^+}\right]}{[a\,(a + bc_{H^+})^{-2} + 1]} \tag{5.64}$$

Comparison of the resulting diffusion coefficients of these
acids in PVA with the diffusion coefficients of these same acids
in water shows that they differ less than tenfold (Table 5.3).   It
may be supposed that in a matrix containing up to 78% of water the
influence of the polymer on the diffusing substance is slight since
the particles of the acid are quite strongly screened by the water
from interacting with the polymer and since, in the transport, the
diffusing particles have to take an indirect path through the macromolecules.

The data obtained make it possible to determine the autodiffusion coefficients of individual ions D^*; e.g., by using the Nernst–
Hartley relationship we can write

TABLE 5.3. Diffusion Coefficients of the Ionized Forms of Acids in PVA and in Water ($D \cdot 10^6$, cm^2/sec) [25]

Ionized form of acid	Aqueous solution	PVA	Ionized form of acid	Aqueous solution	PVA
$H^+(Cl^-)$	30.0	14.4	$H^+(SO_4^{-2})$	3.6	1,8
$H^+(HSO_4^-)$	12.0	5.9	$H^+(H_2PO_4^-)$	16.1	1,2

TABLE 5.4. Autodiffusion coefficients of Ions of Acids in PVA films and in Water ($D^* \cdot 10^6$, cm^2/sec)†

Water content, vol. %	Cl^-	HSO_4^-	SO_4^{-2}	H_3PO_4	H^+	$H_2PO_4^-$
100‡	44	14	1,8	7.6	93	4.5
78	14	13	1,0	0,7	11	0,6
38	—	—	—	0,3	3.5	0,4

†The mean error in the determination of D^* in PVA films is ±7%.
‡The values of D^* are taken from [119, 120].

$$D_{H^+(H_2PO_4^-)} = \frac{2RT}{F^2} \frac{\Lambda_H^0 \left(\Lambda_0 - \Lambda_{H^+}^0\right)}{\Lambda_0} \qquad (5.65)$$

where $\Lambda_{H^+}^0$ is the equivalent electrical conductivity of the hydrogen ion.

By solving this equation with respect to $\Lambda_{H^+}^0$ we get a relationship which allows us to determine the autodiffusion coefficient of the hydrogen ion:

$$D_{H^+}^* = \frac{RT}{F^2} \Lambda_{H^+}^0 = \frac{RT}{2F^2} = \Lambda_0 + \sqrt{\Lambda_0^2 - \frac{2D_{H^+(H_2PO_4^-)} \Lambda_0 F^2}{RT}} \qquad (5.66)$$

the value of Λ_0 being determined from the data on the electrical conductivity of the PVA–acid system.

The autodiffusion coefficient of the hydrogen ion is approximately ten times as high as the autodiffusion coefficients of the anions and the undissociated form of the acid (Table 5.4). It is difficult to explain this high mobility of the proton by its small geometrical dimensions, all the more so as in the polymer matrix the proton is in the hydrated state. Apparently the anomalously

high $D^*_{H^+}$ is due to the specific mechanism of movement in the hydrophilic matrix, which is similar to the Grotthus mechanism [121, p.
399 of Russian translation].

Concentration Dependence of the Diffusion Coefficients Brought
about by Nonideal Behavior of the Polymer—Electrolyte System. In
the previous section the diffusion of an acid was regarded as diffusion of an electrolyte with practically ideal thermodynamic behavior.
However, with a reduction in the volume fraction of water we must expect an influence of the polymer molecules on the diffusion of the
electrolyte and thus deviation of the system from ideal thermodynamic behavior. In point of fact, for a polymer with 38% water
content the experimentally determined diffusion coefficients of
phosphoric acid differ from the values calculated from Eq. (5.51).
Let us assume that such an effect is brought about by the nonideal
character of the diffusion system. Then, from the equation

$$D = D^* \left(1 + \frac{d \ln y_\pm}{d \ln c} \right) \tag{5.67}$$

we get an expression for the mean molar activity coefficient of the
ions

$$\log y_\pm = 0.869 \int_0^{c^{0,5}} \frac{\left(\dfrac{D}{D^*} - 1 \right)}{c^{0,5}} \, d\left(c^{0,5}\right) \tag{5.68}$$

Graphical solution of this equation makes it possible to calculate $y_\pm$ [102, 122].

Concentration Dependence of the Diffusion Coefficients Brought
about by Swelling of the Polymer. With the diffusion of aqueous
acid solutions into PVA films there is considerable swelling of the
polymer (up to 78 vol. %) This diffusion cannot be described by
any generalized form of Fick's law with constant boundary conditions
and with diffusion coefficients depending solely on the concentration. Deviations from ordinary behavior may, as already shown in
(5.3), be caused by a change in the boundary conditions with time,
and by dependence of the diffusion coefficient on time brought about
by the relaxation of the polymer molecules in interaction with the
diffusant.

Changes in boundary conditions with time. As the main criterion of a diffusion which obeys Fick's law, we may take rapid attain-
ment of the equilibrium value of the diffusant concentration in an
infinitely slight surface layer of polymer and its remaining constant in subsequent diffusion. At the present time a criterion of
strict assessment of the dimensions of the surface layer of a poly-

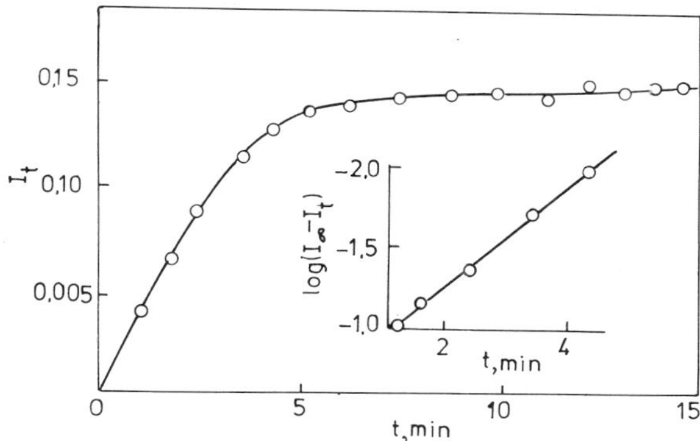

Fig. 5.11.   Change in the absorption intensity of water
             in the surface layer of PVA with the sorption
             of 0.1% HCl into the polymer [on the right,
             the graphical solution of Eq. (5.25)] [123].

mer is not available. Richman and Long [72] obtained, on the basis
of direct experimental measurements of concentration gradients,
proof that in elastomers at temperatures above their T_g equilibrium
surface concentration is established practically instantaneously.
For a PVA which is below its T_g it may be expected that relaxation
processes will be slow and that this will influence the surface concentration of the diffusant. Multiply frustrated total internal
reflection in the IR region has been used to show the change in
the concentration of an aqueous solution in the surface layer of
a PVA film (the thickness of the layer was 1% of the total thickness
of the film) [123]. The choice of this thickness of a surface layer
is an arbitrary one and depends in the first place on the accuracy
required. Since at any particular time there is in this layer a
concentration gradient of the diffusant, we use for calculation the
average concentration $\bar{c}_t$. The change in $\bar{c}_t$ with time is expressed
by Eq. (5.27). Figure 5.11 shows the change in the absorption intensity of diffusant at 1675 cm^{-1} in the surface layer of a PVA film
and the graphical solution of Eq. (5.25). The values of τ, as was
found in [123], depend on the structure of the polymer.

Relaxation of polymer molecules in interaction with the diffusant brings about a change in the diffusion coefficients with time
after a constant diffusant concentration has become established in
the surface layer. The measurement of diffusion coefficients under
nonsteady-state conditions in a multicomponent system often presents
a complex experimental problem.

In investigation of such systems by the sorption method various indicators are generally used (radioactive tracers or dyes)
which make it possible to trace the permeation of one or more com-

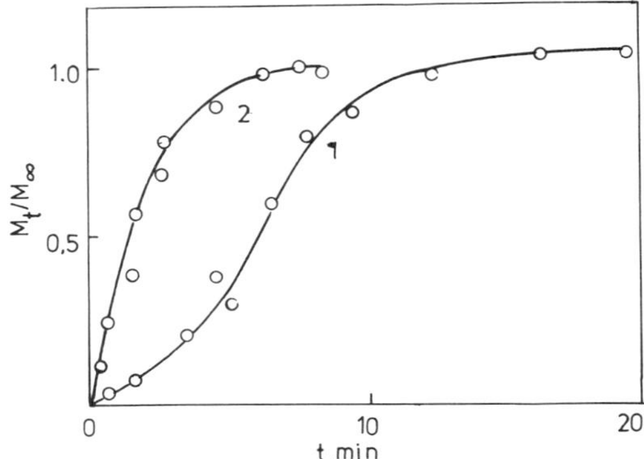

Fig. 5.12.   Kinetics of accumulation of VO^{2+} in dry (1)
and water-saturated (2) PVA films [125].

ponents in the polymer matrix.   Figure 5.12 shows the sorption curves
of the accumulation of vanadyl sulfate in dry and previously water-
saturated films of PVA [125].   For the dry film the sorption curve
has an S-shape with point of inflection (t) at $M_t/M_\infty \approx 0.5$.   At
times longer than t, the relaxation processes in the bulk begin to
show a dominant influence on the diffusion, which leads, for instance,
to an exponential dependence of the diffusion coefficient of the
vanadyl sulfate on its equilibrium concentration in the polymer (c^0):

$$D = D_0 \exp Kc^0 \tag{5.69}$$

where the preexponential term is close to D_{H_2O} in the PVA films.

Thus the inflection point on the absorption curve of the vanadyl
sulfate is governed by the structural transition in the polymer, and
reflects the joint influence, on the diffusion process, of the change
in the surface concentration of the diffusing substance with time,
and of the relaxation processes in the bulk of the polymer.

Unfortunately, a strict expression of the process of diffusion
in polymers accompanied by relaxation of the macromolecules is lacking.   There is an exception in the work of Crank [68], in which the
sorption of a solvent by a polymer is described by means of a diffusion coefficient depending exponentially on time.   The sorption
curve is calculated by means of four, varied, constants which, of
course, makes it difficult to use this approach in a specific situation.

Concentration Dependence of the Diffusion Coefficients Brought
about by Binding of the Electrolyte by Ionogenic Groups of the Polymer.   The concentration dependence of the diffusion coefficients has

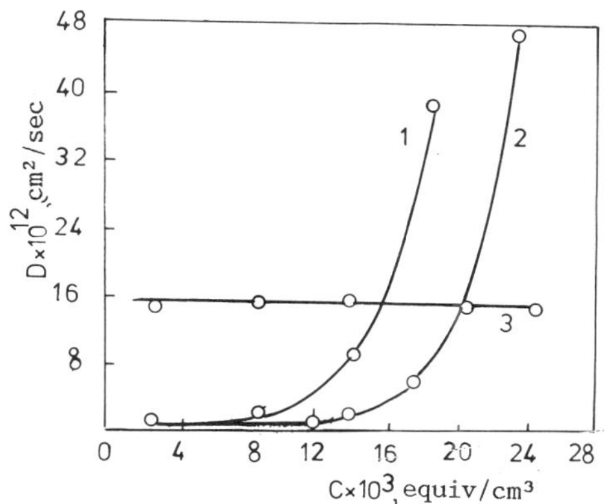

Fig. 5.13. Concentration dependence of $\bar{D}_{HBr}$ (1), $D^*_{H^+}$ (2), and $D^*_{Br^-}$ (3) in poly(hexamethylene adipamide) at 25°C [132].

been studied in the greatest detail in the sorption of acids by polyamides. In [126-130, 137] there is established a considerable increase in the diffusion coefficients with increase in the electrolyte concentration in the solution.

Figure 5.13 shows the change in $\bar{D}_{HBr}$ measured by the desorption method, and in the autodiffusion coefficients of the ions Br^- and H^+ in poly(hexamethylene adipamide) (PHMAA) [132]. $D^*_{Br^-}$ is measured by means of radioactive isotopes; $D^*_{H^+}$ is calculated from Eq. (5.46).

Figure 5.14 shows the corresponding data for a system comprising an aqueous solution of sulfuric acid and PHMAA [132].

Let us consider separately the concentration dependence of the autodiffusion coefficients of protons and anions.

<u>Concentration dependence of the autodiffusion coefficients of protons.</u> By analogy with Eq. (5.45)

$$D^*_{H^+}\, c^0_{H^+} = D^*_{H^+}\,(\text{free})\, c_{H^+}\,(\text{free}) + D^*_{H^+}\,(\text{comb})\, c_{H^+}\,(\text{comb}) \qquad (5.70)$$

where $c^0_{H^+}$ is the total concentration of protons in the polymer;

$$c_{H^+}\,(\text{free}) = \frac{K_{\text{diss}}\, c_{H^+}\,(\text{comb})}{c_{NH_2} - c_{H^+}\,(\text{comb})} \qquad (5.71)$$

where K_{diss} is the dissociation constant of the carboxyl group of the polyamide and c_{NH_2} the concentration of amino groups in the polymer.

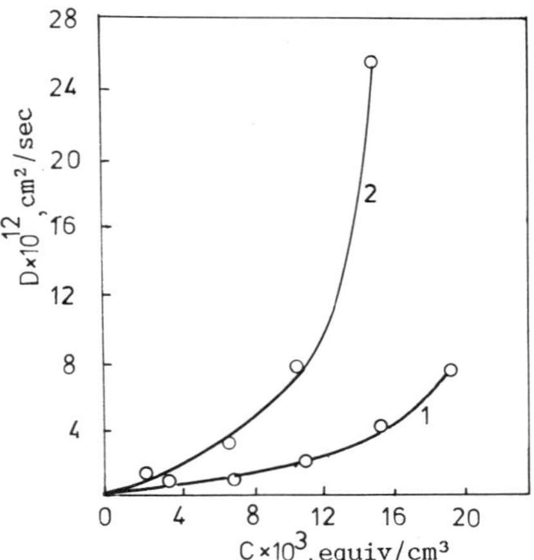

Fig. 5.14. Concentration dependence of $\bar{D}_{H_2SO_4}$ (1)
and $\bar{D}_{H}^{*}{}^{+}$ (2) in poly(hexamethylene
adipamide).

Since K_{diss} is low (in PHMAA about $5 \cdot 10^{-7}$ [128]), it follows
that $c_{H^+(comb)} \gg c_{H^+(free)}$, $c_{H^+(comb)} \approx c_{H}^{0}{}^{+}$, and Eq. (5.70) takes
the form

$$D_{H^+}^{*} = D_{H^+(comb)}^{*} + D_{(free)}^{*} \frac{K_{diss}}{c_{NH^+} - c_{H^+}^{0}} \qquad (5.72)$$

With $c_{H}^{0}{}^{+} \ll c_{NH^+}$ the values of $D_{H}^{*}{}^{+} \approx D_{H^+(comb)}^{*}$ (the minimum value
of $D_{H}^{*}{}^{+}$). With an increase in $c_{H}^{0}{}^{+}$ the value of $D_{H}^{*}{}^{+}$ increases considerably, so that the denominator in the right half of Eq. (5.72)
decreases. The values of $D_{H^+(free)}$ and $D_{H^+(comb)}$ for various polyamides are shown in Table 5.5.

<u>Concentration dependence of the autodiffusion coefficient of</u>
<u>anions.</u> The value of $D_{Br^-}^{*}$ remains constant within the region with-
in which there is combination of the anion with amine terminal groups,
and increases starting with an acid concentration within the polymer
of $3.2 \cdot 10^{-3}$ equiv/g. With an increase in the acid concentration,
the number of active sites for combination (amide groups) increases,
as a result of which the extent of the jump in activation energies
decreases.

Both effects lead to an increase in $D_{Br}^{*}{}^{-}$. It is interesting
to note that in keratin there are 30 times as many active sites for
combination as in PHMAA, and the region between these groups is
more polar [132]; in keratin D_{Br^-} is more than 300 times as much as
in PHMAA (Table 5.5).

TABLE 5.5. Autodiffusion Coefficients of Ions (in $cm^2 \times sec^{-1}$) of Acids and Alcohols in Polyamides at 25°C

	PCA [133]	PHMAA [132]	Keratin [130]		PCA [133]	PHMAA [132]	Keratin [130]
H^+_{free}	—	$3 \cdot 10^{-8}$	$2 \cdot 10^{-6}$	SO_4^{2-}	$1.95 \cdot 10^{-7}$	$2 \cdot 10^{-13}$	$1 \cdot 10^{-11}$
H^+_{comb}	—	$6 \cdot 10^{-3}$	$1,2 \cdot 10^{-10}$	CH_3OH	—	$1 \cdot 10^{-7}$	$1 \cdot 10^{-7}$
Br^-	—	$6 \cdot 10^{-11}$	$2 \cdot 10^{-8}$	C_2H_5OH	—	$6 \cdot 10^{-8}$	$6 \cdot 10^{-8}$
HSO_4^-	$1,1 \cdot 10^{-5}$	$6,8 \cdot 10^{-10}$	$2,2 \cdot 10^{-8}$				

The uncharged molecules, alcohols, diffuse into these polymers with closely similar values of D* [137], which do not differ greatly from $D^*_{Br^-}$ in keratin. Thus, the electrostatic effect plays a big part in the diffusion of acids in PHMAA.

McGregor et al. [127] and Brody [128] explained the increase in $D^*_{A^-}$ in polyamides by a change in the activity coefficient of the anion in the polymer.

Atherton et al. [138] proposed a formula for the calculation of the activity of the anion in a polymer:

$$a_i = \frac{c_i}{c_{NH_2} - z_i c_i} \tag{5.73}$$

In the investigation of the diffusion of aqueous solutions of H_2SO_4, $D^*_{A^-}$ changes more sharply than $D^*_{Br^-}$ with a rise in acid concentration. According to Medley [130], this is due to two factors: the difference in mobility of the sulfate and bisulfate ions, and the combination of the protons with the amino terminal groups

$$D^*_{A^-} = \alpha D^*_{HSO_4^-} + (1 - \alpha) D^*_{SO_4^{2-}} \tag{5.74}$$

where

$$\alpha = \frac{1}{2} \frac{K_{diss}}{K_{2H_2SO_4}} \frac{c^0_{H^+}}{\left(c_{NH_2} - c^0_{H^+}\right)}$$

It is assumed that the ratio $K_{diss}/K_{2H_2SO_4}$ is the same as in aqueous solutions, although the absolute values of these constants may differ in the polymer and in solution. The values of $D^*_{SO_4^{2-}}$ and $D_{HSO_4^-}$ are given in Table 5.5.

5.6.4. Influence of Water on the Diffusion of Electrolytes in Polymers

When considering the diffusion of aqueous electrolyte solutions in polymers which readily dissolve in water, it is necessary to know how strong an influence is exerted on the electrolyte transport by concurrent or independent diffusion of water, and what changes in structure take place in the polymer under the action of such a specific solvent as water.

The diffusion of pure water in polymers is described in detail in a number of surveys [139, p. 259; 140], which consider the kinetic behavior pattern and possible mechanisms of sorption and diffusion of water in polymers. In the present chapter, we are considering only those patterns of water transport which significantly influence the mechanism of diffusion of an electrolyte in polymers.

In the absence of capillary flow, two types of water transport may occur:

by a diffusion mechanism, where the concentration gradient is the driving force;

viscous flow of the water, the driving force of which is the difference in hydrostatic pressure on the two sides of the diaphragm.

The link between the hydraulic (G) and diffusion (P) permeabilities in this case is determined by the dimensionless coefficient ω as given in the equation [141]

$$\omega = \frac{GRT}{P\overline{V}_{H_2O}} \qquad (5.75)$$

where $\overline{V}_{H_2O}$ is the molar volume of the water.

The value of ω is unity under diffusion conditions and increases when we pass to viscous flow conditions [141].

Identification of the type of transport is necessary for correct interpretation of the kinetic results, and also for determining the possibility of interaction between the streams of water and electrolyte in the polymer. For instance, with the diffusion mechanism the probability of such interaction is low. With viscous flow of the solvent through the diaphragm we may expect interaction of the streams of the components of the aqueous electrolyte solutions.

Taking into consideration that the majority of diffusion measurements are carried out in the absence of excess hydrostatic pressure of solvent, it may be supposed that diffusional transport will predominate.

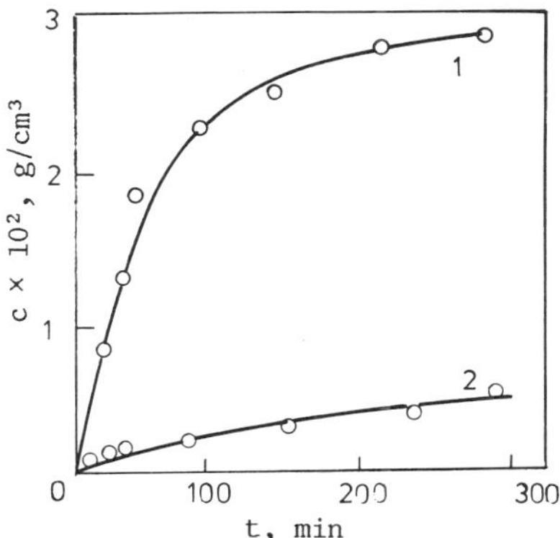

Fig. 5.15. Change in concentration of water (1)
and hydrochloric acid (2) in a film
of ethylcellulose with the diffusion
of vapor of 28% HCl at 35°C [142].

It is known that in gels there is free crystal-like and also
"intermediate" water, these differing in thermal properties. This
great variety of structural forms of water affects the diffusion
of electrolytes in polymers. For instance, Boyd and Soldano [107]
have shown that the autodiffusion coefficient of free water is
higher than the autodiffusion coefficient of ions which are hydrated
with a considerable amount of water, and thus this part of the water
diffuses into the polymer slowly.

5.6.5. Interaction of the Components
of the Diffusing Medium in the Polymer

For the diffusion of electrolytes into hydrophilic polymers
there may, on the one hand, be specific interaction of the components
of the variable medium with the polymer or, on the other hand, interaction between the components of the diffusing medium in the polymer
matrix.

According to the predominance of one or the other type of interaction, the movement of the electrolyte solution in the polymer proceeds on a single front or else with macroscopic separation of the
flows.

The experimental data show that the first type of interaction
is more typical for polymers which dissolve water to a limited
extent.

Figure 5.15 shows the change in the water concentration in a
film of ethylcellulose in the course of diffusion of hydrogen chlor-

ide gas. The water concentration was determined gravimetrically, while the amount of HCl in the polymer was determined by desorption and by a chemical method involving combustion of the polymer [142]. With the diffusion of electrolyte solutions in glassy polymers there is very frequently a sharp boundary between the zone in which the electrolyte is present and the zone where its concentration is unrecorded by the methods being used. The occurrence of a sharp boundary of distribution of diffusant in the polymer is due to the presence of a chemical reaction between the electrolyte and the ionogenic groups, and by an abrupt change in the structure of the polymer at the zone boundary.

This question is considered theoretically in [19]; the penetration of a boundary of constant electrolyte concentration c_x is described by Eq. (5.23), in which $\lambda = [2\overline{D}\left(\frac{c^0}{c_x} - 1\right)]^{1/2}$, if the diffusion of electrolyte and solvent proceeds on a single front, or else $\lambda = (\pi\overline{D})^{1/2}\left(1 - \frac{c_x}{c^0}\right)$, if the electrolyte and solvent diffuse at differing rates in separate flows. For instance, the diffusion of an aqueous solution of sulfuric acid in polycaproamide (PCA) is an example of the first case [19]. The calculated value for $\overline{D}_{25°C}$, $(1.5 \pm 0.2) \times 10^{-9}$ cm^2/sec, is close to that of $\overline{D}_{H_2O}$. In the present case, the reason for the formation of a sharp boundary to the distribution of the diffusant in PCA is the change in the structure of the polymer.

5.7. HYDROPHOBIC POLYMERS

Typical members of this group are polyolefins, polyesters, polyacetals, polysiloxanes, etc.

Investigation of the diffusion of electrolytes in hydrophobic polymers is to a considerable extent related to the use of these polymers as protective coatings, and this has a bearing on the problems to be solved. Accordingly, in the majority of the work recorded the process of diffusion of electrolytes in polymers has been characterized by the depth of penetration of the boundary of a diffusant of given concentration (x_t) or else by the penetration constant (λ); nevertheless, as already indicated, a knowledge of these parameters alone is insufficient for analysis of the processes of diffusion.

For this analysis we need information on the electrolyte concentration in the polymer, on the diffusion and permeability coefficients, and also on the dependence of these values on such factors as the electrolyte concentration in the external medium as well as the temperature. However, there is very little such data. It is this, in particular, in conjunction with the unsatisfactory nature of some of the procedures, which explains such a large number of gaps in our knowledge of the transport of electrolytes in hydrophobic polymers.

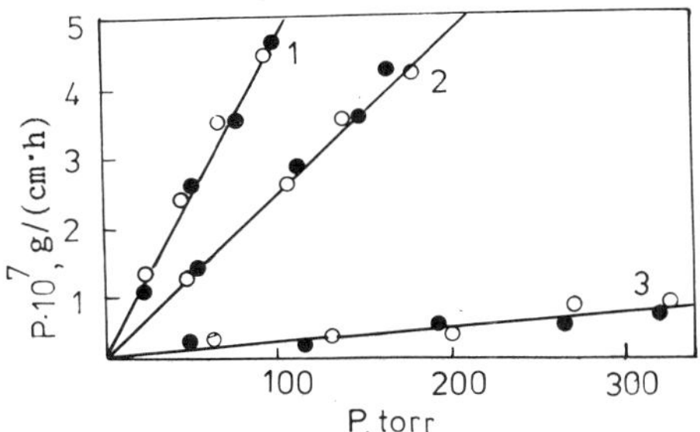

Fig. 5.16.   Permeation of hydrochloric acid through polymer films, from aqueous solution (●), and
from a mixture with air (○) as a function of
the partial pressure of HCl [144]:   1) PE; 2)
PS; 3) PETP.

5.7.1.   The State of the Electrolytes
in the Polymers

In hydrophobic polymers polar groups are practically absent,
and this is responsible for the low solubility of electrolyte solutions in these polymers. The amount of ions in hydrophobic polymers is assessed theoretically in [52]. A calculation using the
Born equation for an ion of radius 2 Å distributed between an aque-
ous medium ($\varepsilon \approx 80$) and the polymer ($\varepsilon \approx 3$) gives $K_{distr} \approx 10^{-20}$,
and this practically excludes any possibility of detecting ions in
the polymer matrix by existing methods. Indeed, in a study of the
electrical resistance of films of a number of hydrophobic polymers
with the diffusion of aqueous electrolyte solutions in them,
D'yachenko and Reka [27, 143] confirmed this conclusion.

A similar conclusion was reached by Shterenzon et al. [144],
who showed that the presence of water in an external phase has no
influence on the rate of transport of HCl through films of polyethylene (PE), polystyrene (PS), and polyethyleneterephthalate (PETP)
(Fig. 5.16). The permeability of the films is determined solely
by the partial pressure of the HCl in the gaseous phase; i.e., the
transport of HCl in hydrophobic polymers is effected in the form of
unhydrated undissociated molecules.

It follows from the calculations of Shterenzon [84] that if
all of the HCl sorbed by PE from concentrated solutions of hydrochloric acid were ionized, then the electrical conductivity of the
polymer would be higher by ten orders than that found experimentally. There is information indicating the possibility of the formation of ion pairs in hydrophobic polymers [145, p. 157 of Russian
translation; 146, 147].

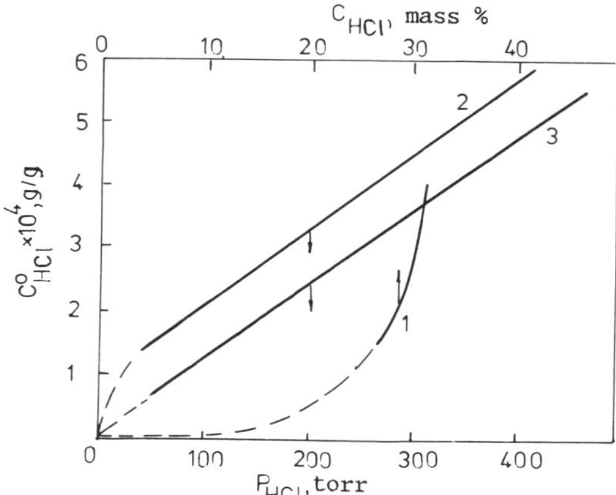

Fig. 5.17. Dependence of the sorption of hydrochloric
acid by polyethylene on the HCl content of
the external medium: 1) polyethylene—HCl
(dependent on the acid concentration); 2)
polyethylene—HCl (dependent on the vapor
pressure of HCl above the solution); 3)
polyethylene—HCl, dry gas (dependent on the
vapor pressure of HCl).

There is described in the literature a case of an increase in
electrical conductivity in hydrophobic polymers on contact with
electrolyte solutions, but this is linked with various indirect effects. For instance, with the interaction of plasticized PVC with
HCl there is a hydrophilization of the polymer, brought about, apparently, by a dehydrochlorination reaction [148]. There is likewise an increase in electrical conductivity for the system of PE
and nitric acid as a result of chemical interaction [148].

According to [149, 150] the activation energy of conduction
considerably exceeds that of the diffusion of small molecules in
polymers, and for polypropylene and PS it reaches 145 and 126 kJ/mole,
respectively. These high values for the activation energy of conduction are determined by the fact that this energy is an overall
value, including both the energy of an ion jump (corresponding to
the activation energy of diffusion of the molecules) and the energy
of ion formation, which is inversely proportional to the dielectric
permittivity of the medium [151]. It is suggested in [150] that
ionic conduction is determined basically by ionization under the
influence of an electric field; the magnitude of the energetic barrier falls and the electrical conductivity rises on account of the
increase in ion concentration [152]. Thus, it may be assumed that
the dissociation of electrolytes in hydrophobic polymers is of low
probability in the absence of an external electrical field.

There is little published information on the solvation and hydration of electrolytes in hydrophobic polymers.

Since water and the electrolyte molecules have high dipole moments, we cannot rule out their interaction in the polymer matrix. Indeed, the following facts support such an interaction: the formation of a new phase in the polymer with the counterdiffusion of electrolyte and water [153] and the difference in the desorption of hydrochloric acid from PE at 50 and 90°C. According to Shterenzon [153], a proportion of the molecules of HCl combine with H_2O, forming hydrates. In the desorption from the polymer at 50°C there are released unhydrated molecules of HCl, while at 90°C there is decomposition of the hydrates with further evolution of HCl.

An attempt to determine the hydration numbers of HCl in PE was made in [154]. Figure 5.17 shows the dependence of the concentration of HCl in PE as a function of the concentration of HCl and as a function of the HCl vapor pressure over the acid. In this same figure we show the dependence of the sorption by PE of HCl from the dry gaseous phase on the partial pressure of the HCl. Figure 5.17 shows that the concentration of HCl in the polymer increases rapidly with an increase in the concentration of hydrochloric acid. Contrary to this, on other coordinates the dependence of the concentration of HCl in the polymer on the vapor pressure over the acid is nearly linear (in the region of relatively high pressures of HCl). In sorption from the dry gaseous phase, Henry's law holds well. A theoretical calculation of the Henry constant for an HCl—PE system using the procedure in [155] gave the value 10^{-8} g/(g·Pa), which is close to the experimental value. In sorption from hydrochloric acid the concentration of HCl in the polymer is higher than in sorption from the dry gaseous phase with identical partial pressures of HCl in the external medium. This may be explained with the assumption that in the polymer containing water molecules there is combination of a proportion of the HCl molecules with the formation of hydrates, and for equilibrium to be established in the polymers an additional amount of HCl passes over from the acid solution. Equation (5.70), the derivation of which is based on the said assumption, satisfactorily describes the experimental data, with the average number of water molecules in the hydrate being close to 5:

$$\log\left(\frac{c^0_{HCl}}{c_{HCl\,(free)}} - 1\right) = \log\left(MK_{eq}K_{distr}\right) + n\log h_{H_2O} \qquad (5.76)$$

where c^0_{HCl} is the total concentration of HCl in the polymer; $c_{HCl(free)}$ the concentration of HCl not combined in hydrates; M the molecular mass of the HCl; K_{eq} the equilibrium constant of the process of hydrate formation; n the number of water molecules in the hydrate; and h_{H_2O} the vapor pressure of the water over the acid solution.

It is shown in [156] that the sorption and the composition of the hydrates of HCl are influenced by the presence of heterogeneous additives in PE. Talc, Al_2O_3, TiO_2, and stabilizers of the phenolic and amine types sharply increase the amount of sorbed hydrochloric acid. In commercial specimens PE hydrates of varying composition ($n = 2-5$) are found. However, in PE which is carefully freed from processing additives there are formed hydrates with $n = 1-2$ and, it would appear, the heterogeneous additives act as nuclei for the association of components of the aggressive medium.

It must be noted that the idea of hydrate formation in an organic phase has found experimental confirmation in a number of studies on extraction [157]. Thus, electrolytes in hydrophobic polymers are practically in an undissociated state but, nevertheless, it appears they can still form hydrates.

5.7.2. Rules of Diffusion Behavior of Electrolytes in Polymers

Let us consider the influence of various factors on the diffusion of electrolyte solutions in hydrophobic polymers.

Influence of the Nature of the Electrolyte. One of the most characteristic features of transport of electrolytes in hydrophobic polymers is the enormous difference in the rates of diffusion of different electrolytes. For instance, the penetration of nitric acid through a film of Ftoroplast 42 (a casting and molding grade of polytetrafluoroethylene) of thickness 0.1 mm may be observed within 20-30 min, whereas in the case of sulfuric acid this requires more than a year [158]. In a number of cases it has proved quite impossible to detect any diffusion of electrolytes in polymers, this being apparently due, on the one hand, to low solubility of the electrolyte and, on the other hand, to inadequate sensitivity of the methods used for investigation.

Let us consider some of the data on the permeability and diffusion of various electrolytes in hydrophobic polymers.

Nitric acid diffuses at an appreciable rate in polyolefins, fluorine-containing plastics, and PVC [159-161]. However, the solubility of nitric acid in PE is very low, as is shown by the absence of a change in mass of polymers in contact with concentrated nitric acid [162, 163]. The retardation time for the diffusion of HNO_3 through a PE film of thickness 160 μm is several weeks [164].

Hydrochloric acid diffuses in all hydrophobic polymers (see Table 5.5).

Sulfuric acid diffuses far more slowly in hydrophobic polymers than do nitric and hydrochloric acids.

The data on the diffusion of sulfuric acid are contradictory. At temperatures of 20-50°C there is absolutely no, or else slight penetration of sulfuric acid from 3-50% solutions through films of low-density PE, high-density PE, polystyrene, PETP, PVC, or Ftorlons (fluorine-containing copolymers generally) [4, 26, 32, 42]. In many cases the penetration is brought about by defects in the films during the experiment [84, 153]. In [165, 166] no penetration of H_2SO_4 into polypropylene was observed when sections taken from specimens immersed in solutions of the acid were stained with acid—base indicators. Nor was penetration of the acid into PETP observed in [167], although on the basis of data on the kinetics of degradation it was possible to confirm that diffusion did in fact take place. Autoradiography has been used to detect the penetration of H_2SO_4 into stretched films of PVC [168].

Phosphoric acid diffuses at a perceptible rate into plasticized PVC [169, 170]. In [171] a gravimetric method was used to observe the increase in mass of specimens of high-density PE in aqueous solutions of phosphoric acid, but a reduction in the change of mass with increase in acid concentration raises doubts on the authors' confirmation of the permeability of PE to H_3PO_4.

Hydrogen fluoride penetrates readily into polymers from solutions of hydrofluoric acid, but in certain cases [e.g., in low-density PE, Pentaplast (poly(3,3-bis(chloromethyl)oxacyclobutane) and Ftoroplast 26 (vinylidene fluoride/hexafluoropropylene copolymer] this is difficult to detect when using the ordinary procedure of staining sections with indicators. This is due to the fact that with staining the desorption of the HF from the polymer is more rapid than the sorption of the indicator. If we first add the indicator to the polymer and then immerse the specimen in the hydrofluoric acid, the penetration of the HF is easily observed [172].

Acetic acid diffuses in various hydrophobic polymers at rapid rates [170-175].

Bases. Inorganic bases practically do not diffuse in hydrophobic polymers [165, 166, 173, 176, 177]. On the other hand, ammonia diffuses readily from aqueous solutions [178, 179].

Salts diffuse very slowly in hydrophobic polymers [170, 180-195]. An exception is NaI, which rapidly penetrates into films of polystyrene, PE, and Ftoroplasts 3 and 4 (PCTFE and PTFE, respectively) [184, 189, 190]. It may, however, be suggested that the diffusing particles are I_2, formed as a result of the reduction of I^-.

It must be noted that proof of the presence of permeability is, as a rule, easier to confirm than proof of the impermeability of the polymer with respect to a particular electrolyte. This is because nondetection of permeability may be a consequence of low sensitivity of the procedure or inadequate time of exposure.

Shterenzon [84, 153, 170] was the first to link the permeability of hydrophobic polymers with the vapor pressure of the electrolytes. He showed that "electrolytes" can be divided into two groups with respect to their capacity for diffusing in hydrophobic polymers: those with high vapor pressure (volatile), which penetrate into hydrophobic polymers with diffusion coefficients close to D_{H_2O}, and those with low vapor pressure (nonvolatile), whose diffusion coefficients in hydrophobic polymers are at least three orders lower than D_{H_2O}.

In [192], on the basis of the study of their transport in polyester resins, electrolytes are divided into "vigorously diffusing" (HCl, HF, and HNO_3) and "weakly diffusing" (H_3PO_4, $NaOH$, and KOH) groups. It is easy to see that this division of electrolytes is equivalent to Shterenzon's classification. Nevertheless, we must stress the arbitrariness of dividing electrolytes into volatile and nonvolatile. For instance, at high temperatures (above 100°C) and high concentrations of solutions of sulfuric acid, when its vapor pressure falls to negligible fractions of a pascal, the penetration of sulfuric acid into hydrophobic polymers can be determined experimentally.

The determination of the diffusion coefficients of electrolytes in hydrophobic polymers is experimentally complex and, therefore, in spite of the numerous studies which have been made of this problem, there is as yet very little correct data.

One typical error in the determination of the diffusion coefficients is the calculation of D from data on the kinetics of sorption by polymers, e.g., [190, 194]. It is evident that where there is simultaneous sorption by the polymer of water and electrolyte from an aqueous solution, we should give particular consideration to the physical meaning of "the diffusion coefficient of an electrolyte solution." Another error in the determination of D arises when it is calculated from the equation of flow through a diaphragm, when the concentration gradient in the diaphragm is replaced by the ratio of the difference in concentrations in the solutions on either side of the diaphragm to its thickness, for instance [184]. The value obtained in this case is the product of D and the distribution coefficient of the electrolyte between the solution and the diaphragm. Finally, sometimes this value is determined from the data on permeability under conditions of counterdiffusion of water and electrolyte in the polymer [172, 195]. Interaction of water and electrolyte leads to delay in the transport of the electrolyte [40, 196, 197] and an increase in the "retardation time," i.e., an uncontrolled lowering of D.

TABLE 5.6. Diffusion Coefficients of Electrolytes in Hydrophobic Polymers

Polymer	Electrolyte	Concn. of electrolyte in aqueous soln., mass, %	Temperature, °C	$D \cdot 10^7$, cm²/sec	Method of determination	References
Polyethylene	HNO_3	98	50	2,9	Sorption	[198]
Low-density polyethylene	HCl	28—36	50	8	Permeability	[154]
Low-density polyethylene	NaCl	1—10	20	$(6—8) \cdot 10^{-4}$	Radioisotope	[171]
Low-density polyethylene	NaI	—	24	$2,2 \cdot 10^{-1}$	"	[190]
Ftoroplast 4	HNO_3	98	50	$2 \cdot 10^{-2}$	Sorption	[198]
Ftoroplast 4	NaI	—	24	$9,6 \cdot 10^{-2}$	Radioisotope	[190]
Ftoroplast 4	NaCl	1—10	20	$(2—3) \cdot 10^{-4}$	"	[171]
Ftoroplast 3	HNO_3	98	50	$7,7 \cdot 10^{-2}$	Sorption	[198]
Ftoroplast 3	NaI	—	24	$9,6 \cdot 10^{-2}$	Radioisotope	[190]
Polystyrene	NaI	—	24	$2,5 \cdot 10^{-1}$	"	[190]

Table 5.6 shows the results of the determination of D using methods which are free of these shortcomings. The following factors attract attention. The diffusion coefficients of NaI are closely similar for all the polymers (around 10^{-8} cm²/sec) and exceed that for NaCl by almost three orders, although the molecular mass of NaI is more than twice that of NaCl. Probably the value of D for NaI is greatly enhanced. As already stated, this is possibly due to reduction of the iodine ion to I_2 [184]. In order of magnitude, the diffusion coefficients of HCl and HNO_3 in PE are close to that of benzene in PE (at zero concentration) and are $1.32 \cdot 10^{-7}$ cm²/sec [17, p. 263].

<u>Influence of the Nature of the Polymer</u>. There is practically no systematic information on the influence of the properties of hydrophobic polymers on their capacity for sorption and transport of electrolytes. Table 5.7 gives data on the sorption of nitric acid in hydrophobic polymers. Although the polymers are not classified by degree of crystallinity, it may be seen that the solubility of the nitric acid in the various polymers varies within quite a wide range. The dependence of the sorption of nitric acid by copolymers of vinylidene fluoride with chlorotrifluoroethylene or hexafluoropropylene on the composition of the copolymers is described by a curve with a maximum which corresponds to a minimum on the curve of the dependence of the tensile strength on the composition. The authors explain the extremal character of the curves of amorphization of polymers containing from 15-20 to 80-85 (mole %) of vinylidene fluoride [200]. It appears that for an approximate assessment of the influence of the chemical nature of an amorphous polymer on its sorption of an electrolyte Eq. (5.34) may be used. With its use, for instance, it is easy to explain the difference in permeability with respect to H_2SO_4 of such polymers as polyolefins ($\delta_{pol} \approx$ 7.9) and Pentaplast ($\delta_{pol} \approx 9.9$). According to Eq. (5.34) we may

TABLE 5.7. Sorption of Nitric Acid by Polymers

Polymer	Extent of sorption, mass %	Reference
Polyethylene	3,1	[198]
Mixture of polyethylene with poly-isobutylene (9:1 to 1:2)	3,3—10	[198]
Copolymer of vinylidene fluoride with tetrafluoromethylene	1,3—7,5	[199]
Polyhexafluoropropylene	~6	[198]
Polyfluorovinylidene	~6	[200]
Floroplast 3	1,1	[158]

TABLE 5.8. Permeability Constants of Various Polymers at 25°C for Nitric Acid, Hydrogen Fluoride, and Water [184]

Polymer	$P^{0} \cdot 10^{10}$, mole/(cm·h·torr)			Polymer	$P^{0} \cdot 10^{10}$, mole/(cm·h·torr)		
	HCl	HNO$_3$	H$_2$O		HCl	HNO$_3$	H$_2$O
Polyethylene	1,3	8,4	3,6	Ftoroplast 42	0,096	4,7	6,5
Copolymer of ethylene with propylene	0,56	4,9	2,4	Ftoroplast 32	0,014	0,4	0,6
				Ftoroplast 26	0.26	18	18
Polypropylene	0,71	3,7	3,5				

expect the sorption of sulfuric acid by Pentaplast to exceed its sorption by PE almost 40-fold and, accordingly, it is not surprising that it is very difficult to detect the penetration of H_2SO_4 into polyolefins, whereas for Pentaplast this has been observed repeatedly [201, 202]. There is very little experimental data on the influence of the nature of polymers on the diffusion coefficients of electrolytes. When we compare the values of D in Table 5.7 we see that nitric acid diffuses in PE considerably more rapidly than in fluorine-containing plastics. This agrees with the conclusion that fluorine-containing polymers are characterized by lower diffusion coefficients, whereas the solubility of gases in them is higher [200]. However, for a nonvolatile electrolyte (NaCl) the diffusion coefficients in PE and the fluorine-containing polymer are of the same order. Possibly this is due to differences in the mechanisms of diffusion of HNO_3 and NaCl. There is considerably more information on the influence of the nature of the polymers on permeability with respect to electrolytes. Table 5.8 gives the permeability constants for nitric acid, hydrogen fluoride, and water.

Thus it is probably possible to classify polymers by their permeability with respect to water and electrolytes, as has been done with respect to gases.

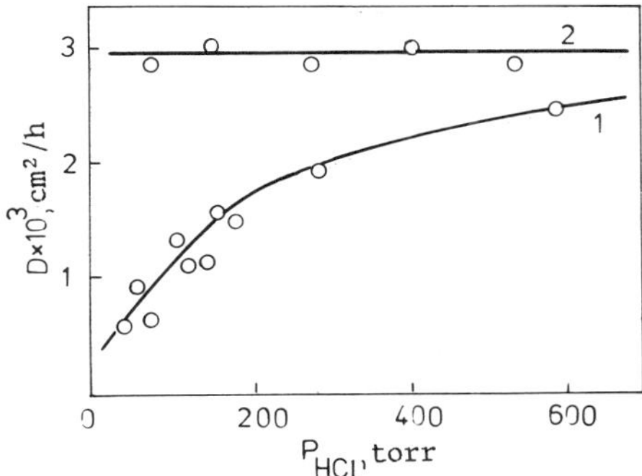

Fig. 5.18.   Dependence of the diffusion coefficient of
hydrochloric acid in PE from an aqueous solution (1) and from a mixture with air (2) on
the partial pressure of HCl [154].

The incorporation of polar substances or polar groups (e.g.,
carboxy, carbonyl, or amine) into hydrophobic polymers quite considerably increases their permeability with respect to electrolytes,
apparently mainly because of increasing the sorption of the pene-
trant [202, 203].

<u>Concentration Dependence of the Diffusion Coefficients</u>. The
question of the influence of the electrolyte concentration in an
external solution on its diffusion coefficient in the polymer is
regarded as most complex and involved. The concentration dependence of the diffusion coefficient of HCl in PE is investigated in
detail in [154]. The integral diffusion coefficients are calculated
according to Eq. (5.7). For the sake of comparison the value of
D for anhydrous HCl was also determined for various partial pressures of it in the external medium. It is found that with an increase in the acid concentration there is a rise in $\overline{D}$, approximating
the value found for anhydrous HCl, which does not depend on the
partial pressure of HCl (Fig. 5.18).

Shterenzon [154] was the first to explain the concentration
dependence of the integral diffusion coefficients of electrolytes,
using the hypothesis of the formation of hydrates of the electrolytes within the polymer matrix.

Volatile electrolytes, e.g., hydrogen chloride, can be present
in a polymer matrix in two forms, unhydrated and hydrated, the diffusion coefficient of the first form being considerably higher. With
reduction in the water content there is a change in the ratio of
the hydrated and unhydrated forms of the hydrogen chloride, which
again leads to an increase in the integral diffusion coefficient,
which in concentrated solutions of hydrochloric acid approximates
to the diffusion coefficient of gaseous hydrogen chloride.

On the basis of Eq. (5.45), we can put

$$\overline{D}c^0_{HCl} = D^0_{HCl\,(free)}\,c_{HCl\,(free)} + D^0_{HCl\,(hydr)}\,c_{HCl\,(hydr)} \qquad (5.77)$$

where $D^0_{HCl(free)}$ and $D^0_{HCl(hydr)}$ are the diffusion coefficients of
the free and hydrated HCl; c^0_{HCl} the solubility of HCl in the polymer, which is equal to the sum of the concentrations of the free
($c_{HCl(free)}$) and hydrated ($c_{HCl(hydr)}$) hydrogen chloride.

It is usually supposed that the hydrates are all of one type
with maximum hydration number n, and $D^0_{HCl(free)}c_{HCl(free)} \gg$
$D^0_{HCl(hydr)}c_{HCl(hydr)}$.

In such a case [150],

$$\overline{D} = \frac{D^0_{HCl\,(free)}}{1 + K_{eq}K^n_{distr}p^n_{H_2O}} \qquad (5.78)$$

The average hydration number for HCl in PE is 4-5 [154]. The
possibility of formation of volatile electrolytes in hydrophobic
polymers is admitted by Dolezel [205], who observed a sharp increase
in the integral diffusion coefficient of HNO_3 in PVC when using a
50% solution of nitric acid. It is at this particular concentration that a maximum appears on the HNO_3–H_2O phase diagram.

The concentration dependence of the diffusion coefficient was
found in [206, 207] for the system PE–nitric acid. It was shown
that D for nitric acid rises from $3.2 \cdot 10^{-10}$ to $9 \cdot 10^{-10}$ cm^2/sec when
the acid concentration in the external medium is altered from 7 to
80%. To determine D, a procedure based on measurement with time
of the electrical conductivity of the acid which is in contact with
the solution was used. Thus, the diffusion of uncharged particles,
i.e., the molecules of the nitric acid, is not recorded in this procedure. It is perhaps this factor which explains the very low values
of D found in this investigation, which are lower by three orders
than the more reliable data given in [198, 204].

In [208], the integral diffusion coefficients of HCl and H_2SO_4
are determined in films of KhSL lac (based on chlorinated PVC) as
a function of the concentration of the acid solutions. It is found
that at 60°C, D_{HCl} changes from $3 \cdot 10^{-9}$ to $7.5 \cdot 10^{-9}$ cm^2/sec with rise
in acid concentration from 25 to 30%, while $D_{H_2SO_4}$ rises from $4 \cdot 10^{-10}$
to $19 \cdot 10^{-10}$ cm^2/sec when the concentration is altered from 50 to
90%. To find D, a procedure was used which consisted of determining
the time τ from the moment of contact of the solution with one side
of the film up to the moment of change in color of an indicator
placed on the other side of the film. Coefficient $\overline{D}$ was calculated
from formula (5.11).

It is evident that τ is the total time necessary both for the diffusion of the electrolyte through the thickness of the film and for the accumulation of acid on the far side of the film in an amount sufficient for the indicator to change color. The last component of the total time is a function of the permeability of the film. As a consequence of this, when using this procedure the concentration dependence of $\overline{D}$ is distorted by the concentration dependence of the permeability. The authors explain the found concentration dependence of $\overline{D}$ as follows: the dimensions of the diffusing pair (hydrated proton + acid anion) are considerably greater than those of an undissociated acid molecule and, accordingly, the latter diffuses at the higher rate. Since with increase in acid concentration there is an increase in the proportion of undissociated molecules, this leads to a rise in $\overline{D}$ with the concentration of the solution. In other words, the authors suggest that the transport of acids through films is effected by particles of differing nature and dimensions.

The concentration dependence of the integral diffusion coefficient has been found for systems comprising an epoxy composition and aqueous solutions of inorganic acids [209]. In the case of sulfuric acid there is a symbatic link between $\overline{D}$ and the water content of the polymer.

A rapid increase in permeability with increase in concentration of electrolyte solutions is noted in [170, 210]. It is shown in [153] that this sharp change may be explained by the proportionality of the permeability to the vapor pressure of the electrolyte above the aqueous solution, while the pressure rises very rapidly with increase in concentration of the solution.

The concentration dependence of the permeability of the films for volatile electrolytes is satisfactorily expressed by the equation

$$P = P^0 p \qquad (5.79)$$

where P^0 is the permeability constant, which does not depend on the electrolyte concentration in the external solution, and p is the vapor pressure of the electrolyte above the aqueous solution [84, 144].

The correctness of Eq. (5.79) has been confirmed in investigations of the permeability of films with respect to HCl, HF, and HNO_3.

Mechanism of Diffusion of Electrolytes in Hydrophobic Polymers. In earlier work the permeability of hydrophobic polymers with respect to electrolytes was explained by their transport by means of a system of "through pores" ("microcracks") [158, 159, 199, 213, 214]. However, these ideas were not enough to explain such facts

TABLE 5.9. Activation Energy of Diffusion of Electrolytes in Polymers

Polymer	Electrolyte	E, kJ/mole	Reference
PETP	H_2SO_4, 53%	50	[165]
Polyvinylchloride (plasticized)	HNO_3, 60%	59	[211]
Polyvinylchloride		14	[211]
Polycarbonate	HNO_3	42	[206]
Ftoroplast 4	HNO_3, 98%	19 at $T > 323\,°C$ 51 at $T < 323\,°C$	Calculated from data in [195]
PE	HNO_3, 98%	18	[188]

as the high selectivity of the permeability of the polymers with respect to differing electrolytes, or the character of the concentration dependence of the integral diffusion coefficients and the high activation energies of the diffusion process (Table 5.9), which are characteristic of activated diffusion.

Analysis of the experimental data on the diffusion, in hydrophobic polymers, of gases and vapor shows that it is helpful to distinguish two cases. In polymers which are above the glass transition temperature there are practically no micropores with radii exceeding 1 nm [215, 216, 89], and transport is effected by an activated diffusion mechanism. At temperatures below T_g there are micropores in the polymers, and the question of the mechanism of diffusion calls for special evidence. For instance, polystyrene with such a microporous structure sorbs alcohols while retaining its rigid structure [215, 216]. Sorption, in the opinion of the authors, is by a mechanism of laminar or volumetric filling of the micropores [217]. In particular, we cannot exclude the possibility of diffusion through a system of through micropores, as happens in microporous zeolites with radii of the micropores 0.6-0.7 nm [218]. For polymers which are below T_g it is now possible to determine, experimentally and by calculation, the total volume of the micropores and their distribution according to diameter [216, 219-221] but, unfortunately, no methods are known for distinguishing between through and closed micropores.

Thus it may be asserted that the transport of electrolytes in hydrophobic polymers takes place in the great majority of cases by an activated diffusion mechanism. Such ideas characterize transport in hydrophobic polymers without taking into account the various kinds of defects which may occur in a particular article (e.g., as a result of the thinness of a film there may be formed "through pores" which are not found in films of the same polymer but are of greater thickness). The presence of transport which is due to microdefects is discussed in [222, 223].

TABLE 5.10. Swelling of Polypropylene in Aqueous Solutions of Various Electrolytes at 90°C [225, 226]

Electrolyte	Concentration, mass %	a_{H_2O}	Degree of swelling, mass %	Concentration, mass %	a_{H_2O}	Degree of swelling, mass %
H_2SO_4	11,0	0,95	1,12	18	0,90	0,85
H_3PO_4	—	—	—	20	0,90	0,85
NaOH	5,5	0,95	1,09	10	0,90	0,83
NaCl	12,0	0,95	1,11	20	0,90	0,83
$CaCl_2$	9,3	0,95	1,11	15	0,90	0,82

The presence of micropores in polymers reduced the effective thickness of the film, but had practically no influence on the value of D [89].

There are a number of points of view with regard to the nature of the diffusing particles. Some writers contend that individual ions, hydrated or unhydrated, diffuse, others that the diffusing particles are undissociated molecules, devoid of hydrate shells.

Deciding on the true picture is made more difficult by the fact that some writers draw no distinction between particles which are present in the polymer and particles by which the transport of the substance is effected. There cannot be particles in the polymer by which transport is not effected, but it is not essential for the transport to be effected by all the types of particles which are present in the polymer. For instance, in [209] it is shown that "the indicator method makes it possible to record only the transport of hydrated ions in polypropylene." It is hard to agree with this assertion because if, in fact, the indicator does record only the ionized forms (which in turn needs to be proved), this merely shows that they are present in the polymer, but it still does not follow that the transport is effected by the hydrated ions, as is asserted in [209]. Taking into consideration the data given above, it may be reckoned that the particles participating in the diffusion process are un-ionized electrolyte molecules, with the unhydrated molecules diffusing far more rapidly than the hydrated ones.

Shterenzon [84, 144] shows that for electrolytes with high vapor pressure the total flow is the additive sum of the individual flows of the components of the diffusant, i.e., the transport of volatile electrolytes conforms to the general behavior of gases and vapor. For instance, for aqueous solutions of volatile electrolytes the total permeability of the polymers is

$$P = P^0_{el}\bar{p}_{el} + P^0_{H_2O}\bar{p}_{H_2O} \qquad (5.80)$$

where P^0 and $\bar{p}$ are the permeability constants and the partial pressures of the components, respectively.

Thus, the transport of each of the components is determined by the partial pressure of the component above the electrolyte solution [84, 224]. The concentration dependence of the integral diffusion coefficient of the electrolyte is described by Eq. (5.79). The diffusion of electrolytes with low vapor pressure in hydrophobic polymers has two characteristic features: an extremely low electrolyte concentration in the polymers ($c^0_{solv} \gg c^0_{electr}$) and independent movement of the flows of solvent and electrolyte in the polymer matrix (with, always, $\bar{D}_{solv} \gg \bar{D}_{electr}$). Accordingly, for hydrophobic polymers the degree of swelling is determined solely by the thermodynamic activity of the solvent (Table 5.10).

Because of the extremely low values of c^0_{electr} and $\bar{D}_{electr}$ the diffusion coefficients of electrolytes with low vapor pressure are available in the literature for only a few polymers (Table 5.6).

It is not clear at present whether the mechanism of transport of electrolytes with low vapor pressures through polymer films is identical with the mechanism of transport of electrolytes with high vapor pressures or whether this transport is described by Eq. (5.79). According to Shterenzon, the differences in the diffusional behavior of electrolytes with high or low vapor pressures are merely quantitative [227].

With suitable choice of conditions (e.g., at 100°C) the permeability of Ftoroplast 26 (vinylidene fluoride/hexafluoropropylene copolymer) depends linearly on the partial pressure of the electrolyte.

It is possible that along with transport by a mechanism of activated diffusion there is transport by a mechanism of slow diffusion, whose contribution to the total flow is not great.

In considering the diffusion of electrolyte solutions in hydrophobic polymers it is of interest to explain how the transport of the solvent proceeds. Polar molecules, for instance water, form aggregates in the matrix of hydrophobic polymers since the cohesion forces between these molecules are greater than the forces of interaction between the diffusing molecules and the polymer. Examples of the formation of aggregates, or clusters, in hydrophobic polymers are given in [228]; the conditions of formation of such aggregates are considered in [228, 229]. The dimensions of these aggregates can be determined from the Zimm–Lundberg relationship [230]:

$$\frac{G}{\bar{V}_{H_2O}} = -\varphi_n \left(\frac{\partial y_{H_2O}}{\partial a_{H_2O}} \right)_{p,T} - 1 \tag{5.81}$$

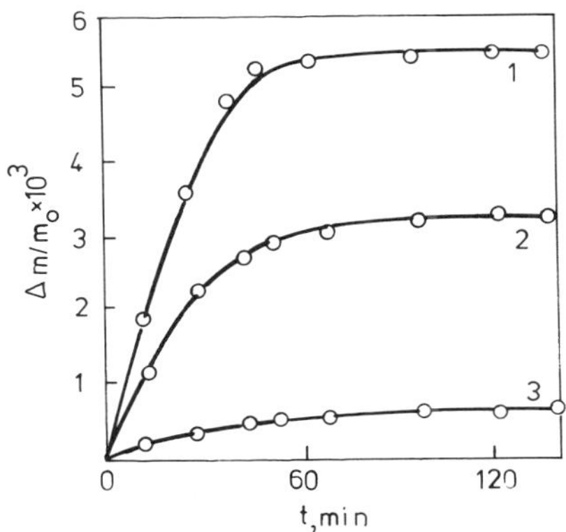

Fig. 5.19. Change in the mass of a PETP film
(thickness 200 ± 5 μm) in the sorption of water and solutions of sulfuric acid at 80°C [167]: 1) water;
2) 25.3% H_2SO_4; 3) 46.3% H_2SO_4.

where G is a parameter characterizing the formation of the aggre-
gates, and $\varphi_{H_2O} G / \bar{V}_{H_2O}$ the excess of the average number of water molecules in relation to their average concentration close to the par-
ticular diffusant molecule.

This value is reduced with a reduction in the volume proportion of water in the polymer.

The aggregation of water molecules in hydrophobic polymers increases with the presence in the polymer matrix of heterogeneous
impurities and various processing additives (stabilizers, catalysts,
fillers, pigments, etc.), and also of polar groups (e.g., ester or
hydroxyl groups) which are formed by oxidation of the polymer.

Thus, heterogeneous substances and polar groups promote the
aggregation of water [157], but there is at present a lack of experimental data on the influence of electrolytes on the extent of
aggregation of the water molecules.

If we suppose that in the polymer matrix aggregates can be
formed only by water molecules (water has a higher aggregating capacity than other solvents), and not by electrolyte molecules, then
with an increase in concentration of the latter φH_2O will fall,
which will lead to a reduction in the amount of aggregates. Let
us suppose that in a particular polymer aggregates of only one type
can form (n = const)

$$K = \frac{c_{H_2O}^n}{c_{(H_2O)_n}} \tag{5.82}$$

TABLE 5.11. Typical Sorption and Diffusion Characteristics of Solutions of Electrolytes in Hydrophilic and Hydrophobic Polymers

Typical sorption or diffusion characteristics	Polymers	
	hydrophilic	hydrophobic
Concentration of sorbed electrolyte	High	Low
Henry's law	Possibly not obeyed	Obeyed
Dissociation into ions	Close to that in water	Practically nonexistent
Diffusion coefficient values	Close to those in water	Low for nonvolatile and high for volatile electrolytes
Main factor deciding rate of diffusion of electrolyte	Dimensions of ions and dissociation of electrolytes	Partial pressure of electrolyte above solution
Influence of sorption of electrolyte on electrical resistant of polymer	Resistivity reduced by several orders of magnitude	Practically no resistivity

Given that $c_{H_2O}^0 > c_{H_2O(free)}$,

$$c_{H_2O(free)} = \left(K c_{H_2O}^0\right)^{1/n}$$

$$\overline{D}_{H_2O}\, c_{H_2O}^0 = D_{H_2O(free)}^0\, c_{H_2O(free)} + D_{H_2O(aggr)}^0\, c_{H_2O(aggr)}$$

(5.83)

If we assume that $D_{H_2O(free)}^0\, c_{H_2O(free)} \gg D_{H_2O(aggr)}^0\, c_{H_2O(aggr)}$ then

$$\overline{D}_{H_2O} = D_{H_2O(free)}^0\, K^{1/n} \left(c_{H_2O}^0\right)^{1/n-1}$$

(5.84)

The concentration of water in the electrolyte solution is conveniently expressed on a scale of molar fractions

$$N_{H_2O} = \frac{K_{distr\,(H_2O)}\, \overline{p}_{H_2O}}{K_{distr(H_2O)}\, \overline{p}_{H_2O} + K_{distr(el)}\, \overline{p}_{el}}$$

For electrolytes with low vapor pressure $K_{distr(H_2O)} p_{H_2O} \gg K_{distr(el)} p_{el}$ and

$$\overline{D}_{H_2O} = D_{H_2O(своб)}^0\, K^{1/n}$$

(5.85)

i.e., the diffusion coefficients determined gravimetrically are practically independent of the concentration of electrolyte in the solution and are equal to the diffusion coefficient of water in these polymers (Fig. 5.19).

For electrolytes with high vapor pressure, Eq. (5.85) is correct in the absence of aggregation of the solvent ($n = 1$).

With high values of n,

$$\bar{D}_{H_2O} \approx D^0_{H_2O\,(free)}\, K^{1/n}\left(1 + \frac{K_{distr\,(el)}\,\bar{p}_{el}}{K_{distr\,(H_2O)}\,\bar{p}_{H_2O}}\right) \qquad (5.86)$$

and $\bar{D}_{H_2O}$ increases sharply with an increase in the electrolyte concentration in the solution.

Thus, in the diffusion of electrolyte solutions in hydrophilic polymers which readily dissolve water the thermodynamic (e.g., K_{diss}) and diffusion parameters (e.g., D^*) are similar to the corresponding parameters in solutions. A reduction in the mobility of molecules and ions of an electrolyte in polymers as compared with that in solution is brought about by a lengthening of the diffusion route and by interaction of the diffusing particles with the polymer. In hydrophilic polymers which dissolve water to a limited extent, the dissolved water does not form a continuous aqueous space in the polymer matrix, and the diffusion of ions proceeds by activated jumps between polar groups; this leads to a considerable reduction in $\bar{D}$ of the electrolyte as compared with the value in solution.

The diffusion of electrolytes in hydrophobic polymers proceeds by a mechanism analogous to the transport of gases and vapor. Accordingly, for electrolytes with high vapor pressure c^0_{electr} and D^0_{electr} are closely similar to the corresponding parameters for water in these polymers. Electrolytes with low vapor pressure are characterized by extremely low values of c^0_{electr} and $\bar{D}_{electr}$. Typical characteristics of sorption and diffusion of electrolytes in hydrophilic and hydrophobic polymers are shown in Table 5.11.

REFERENCES

1. R. Haase, Thermodynamics of Irreversible Processes, Addison-Wesley, Massachusetts (1968).
2. E. Lightfoot, Transport Phenomena and Living Systems, Wiley, New York (1974).
3. J. Crank, The Mathematics of Diffusion, Oxford University Press, London (1970).
4. R. M. Barrer, Diffusion in and through Solids, Cambridge University Press, Cambridge (1941).

5. W. Jost, Diffusion in Solids, Liquids, and Gases, Academic Press, New York (1960).
6. S. A. Reitlinger, Permeability of Polymeric Materials, Khimiya, Moscow (1974).
7. J. S. Darken, Trans. Am. Inst. Min. Met. Pet. Eng., 175, 184 (1948).
8. P. C. Carman, J. Phys. Chem., 72, 1707 (1968).
9. R. N. Howard, J. Macromol. Sci., C, 4 (2), 191 (1970).
10. I. P. Terletskii, Statistical Physics, Vyssh. Shkola, Moscow (1966).
11. S. Glasstone, K. Laidler, and H. Eyring, Theory of Rate Processes, McGraw-Hill, New York (1941).
12. R. B. Parlin and H. Eyring, Ion Transport across Membranes, Academic Press, New York (1954).
13. R. M. Barrer, Trans. Faraday Soc., 38, No. 8, 322 (1942); 39, No. 2, 48 (1943).
14. H. Fujita, A. Kishimoto, and K. Matsumoto, Trans. Faraday Soc., 56, 424 (1960).
15. A. E. Chalykh and R. M. Vasenin, Vysokomol. Soedin., 7, No. 4, 586 (1965); 8, No. 11, 1908 (1966).
16. M. L. Williams, R. F. Zandel, and J. D. Ferry, J. Am. Chem. Soc., 77, No. 17, 3701 (1955).
17. K. Rodgers, in: Physics and Chemistry of the Organic Solid State, Wiley, New York (1963-1967).
18. H. L. Frisch, J. Phys. Chem., 61, 93 (1957).
19. A. L. Iordanskii, Candidate's Dissertation, Inst. Khim. Fiz. Akad. Nauk SSSR, Moscow (1975).
20. J. H. Petropolous and P. P. Rousis, J. Chem. Phys., 47, No. 4, 1491 (1967).
21. E. M. Grozhan and Yu. S. Zuev, Kauch. Rezina, No. 9, 10 (1966); translated in Sov. Rubber Technol., No. 9, 13 (1966).
22. A. V. Dzene and A. E. Kreitus, in: Diffusion Phenomena in Polymers, Riga (1977), p. 252.
23. P. M. Terfizzi and A. E. Zenehau, Mod. Plast., 30, 140 (1953).
24. T. N. Nikolaeva, N. S. Kudryavtseva, and L. V. Zakharova, Plast. Massy, No. 5, 45 (1964); translated in Soviet Plast., No. 5, 46 (1965).
25. A. L. Iordanskii et al., Vysokomol. Soedin., A, 14, No. 4, 801 (1972).
26. C. Ysusi, Tetsudo Gijutsu Kenkyu Shiryo, 18, 292 (1961).
27. O. P. D'yachenko, B. A. Reka, and R. V. Isaev, Lakokras. Mater. Ikh Primen., No. 3, 40 (1965).
28. A. L. Geass and J. Smith, Isot. Radiat. Technol., 8, 60 (1970).
29. M. Svoboda, D. Kychunka, and B. Knapek, Farbe Lack, 77, 11 (1971).
30. W. W. Kittelberger and A. C. Elin, Ind. Eng. Chem., 44, 326 (1952).
31. A. B. Pakshver and I. V. Bykova, Kolloidn. Zh., 16, 381 (1954).
32. I. L. Rozenfel'd, F. I. Rubinshtein, and S. V. Yakubovich, Lakokras. Mater. Ikh Primen., No. 2, 58 (1962).

33. R. V. Rao, M. Yaseen, and I. S. Aggerwal, Paint Manuf., $\underline{35}$, 49 (1965).
34. H. H. Jellinek and B. Blan, Polym. Prepr., $\underline{11}$, 1363 (1970).
35. A. O. Jakubovic, G. J. Hills, and J. A. Kitchener, J. Chem. Phys., $\underline{55}$, No. 4, 263 (1958).
36. A. Despic and G. J. Hills, Trans. Faraday Soc., $\underline{53}$, 1262 (1957).
37. A. E. Lagos and J. A. Kitchener, Trans. Faraday Soc., $\underline{56}$, No. 8, 1245 (1960).
38. M. A. Laufer, Biophys. J., $\underline{1}$, 3 (1961).
39. H. J. Spencer and J. M. Ibrahim, J. Polym. Sci., A-2, $\underline{6}$, 2067 (1968).
40. Yu. A. Lobanov and A. L. Shterenzon, Lakokras. Mater. Ikh Primen., No. 5, 42 (1966).
41. A. U. Wilson, Philos. Mag., $\underline{39}$, 48 (1948).
42. A. L. Shterenzon, S. A. Reitlinger, and L. P. Topina, Zavod. Lab., $\underline{34}$, 892 (1968).
43. B. Dolezel and V. Chytry, Plast. Hmoty Kauc., $\underline{2}$, 289 (1965).
44. T. Gamabe et al., Bull. Soc. Sea Water Sci. Jpn., $\underline{25}$, No. 4, 259 (1972).
45. K. Umerawa and T. Gamabe, J. Inst. Ind. Sci. Univ. Tokyo, $\underline{23}$, 457 (1971).
46. A. L. Iordanskii et al., Vysokomol. Soedin., A, $\underline{16}$, No. 4, 849 (1974).
47. J. E. Tanner and E. O. Steyskal, J. Chem. Phys., $\underline{49}$, 1768 (1974).
48. O. F. Bezrukov et al., in: Nuclear Magnetic Resonance, Vol. 4, Leningrad State University, Leningrad (1971), p. 90.
49. V. S. Markin et al., Vysokomol. Soedin., A, $\underline{12}$, No. 10, 2174 (1970).
50. J. F. May and J. Vallet, Rev. Gen. Electr., $\underline{81}$, 255 (1972).
51. M. Petit et al., J. Polym. Sci., C, No. 30, 195 (1970).
52. V. S. Markin and Yu. A. Chizmadzhev, Induced Ion Transport, Nauka, Moscow (1974).
53. N. P. Grusin, V. D. Grebenyuk, and M. V. Levnitskaya, Electrochemistry of Ionites, Nauka, Novosibirsk (1972).
54. C. W. Saltonstall and W. M. King, Desalination, $\underline{4}$, 309 (1968).
55. E. Emerich and J. Zdenor, Chem. Listy, $\underline{50}$, 1869 (1956).
56. B. I. Boltaks, V. Ya. Vodyanoi, and I. A. Fedorovich, Biofizika, $\underline{16}$, 824 (1971).
57. B. I. Laskovin, I. M. Smirnova, and I. I. Glazkova, Elektrokhimiya, $\underline{10}$, 805 (1974).
58. V. V. Kryuchenkov, L. A. Gubareva, and V. S. Musinova, Zh. Fiz. Khim., $\underline{46}$, No. 4, 936 (1972).
59. A. D. Nesterenko, Fundamentals of Calculation of Compensating Electrical Measurement Circuits, Izd. Akad. Nauk Ukr. SSR, Kiev (1960).
60. V. N. Manin and A. I. Gromov, Mekh. Polim., No. 4, 744 (1967).
61. J. H. Deterding, D. W. Singleton, and R. W. Wilson, Trav. Cent. Rech. Etud. Oceanogr., $\underline{6}$, Nos. 1-4, 275 (1965).
62. B. Dolezel, Corrosion of Plastics and Rubbers, Khimiya, Moscow (1964).

63. B. Dolezel, Chem. Ing. Tech., $\underline{40}$, 846 (1968).
64. A. L. Shterenzon and I. B. Gemusova, Summaries of Papers, "Diffusion Phenomena in Polymers," Moscow (1974), p. 76.
65. A. D. Aliev, in: Diffusion Phenomena in Polymers, Riga (1977), p. 7.
66. H. B. Hoffenberg and H. L. Frisch, Polym. Prepr., $\underline{10}$, 999 (1969).
67. J. H. Petropolous and P. P. Rousis, in: Permeability of Plastic Film and Coat by Gases, Vapors, and Liquids, Pergamon Press, New York (1974), p. 205.
68. J. Crank, J. Polym. Sci., $\underline{11}$, 151 (1953).
69. F. A. Long and D. Richman, J. Am. Chem. Soc., $\underline{82}$, No. 3, 513 (1960).
70. T. Alfrey, E. F. Gurnu, and W. G. Loyd, J. Polym. Sci., C, No. 12, 249 (1966).
71. R. M. Vasenin, Vysokomol. Soedin., $\underline{6}$, No. 4, 624 (1964).
72. D. Richman and F. A. Long, J. Am. Chem. Soc., $\underline{82}$, No. 3, 509 (1960).
73. A. Kishimoto and T. Kitahara, J. Polym. Sci., A-1, $\underline{5}$, No. 8, 2147 (1967).
74. C. E. Rogers, V. Y. Stannett, and M. Szwarc, J. Phys. Chem., $\underline{63}$, No. 9, 1406 (1959).
75. H. Yasuda and V. Y. Stannett, J. Polym. Sci., $\underline{57}$, 907 (1962).
76. H. B. Hoffenberg et al., J. Polym. Sci., C, No. 28, 243 (1969).
77. J. D. Wellons and V. Y. Stannett, J. Polym. Sci., A-1, $\underline{4}$, 593 (1966).
78. A. S. Michaels, H. J. Bixler, and H. B. Hoffenberg, J. Appl. Polym. Sci., $\underline{12}$, 991 (1968).
79. T. K. Kwei and T. Wang, in: Permeability of Plastic Film and Coat by Gases, Vapors, and Liquids, Pergamon Press, New York (1974), p. 37.
80. A. Peterlin, Makromol. Chem., $\underline{124}$, 136 (1969).
81. C. H. M. Jacques, H. B. Hoffenberg, and V. Y. Stannett, J. Appl. Polym. Sci., $\underline{18}$, 223 (1974).
82. N. Lakshminarayanaiah, Transport Phenomena in Membranes, Academic Press, New York (1969).
83. A. R. Matheson and C. S. Whewell, J. Appl. Polym. Sci., $\underline{8}$, 2029 (1964).
84. A. L. Shterenzon, Candidate's Dissertation, Ural'sk. Gos. Univ., Sverdlovsk (1969).
85. A. A. Askadskii, Usp. Khim., $\underline{46}$, No. 6, 1122 (1977).
86. J. H. Hildebrand, Solubility of Nonelectrolytes, Reinhold, New York (1936).
87. W. R. Vieth and K. Y. Sladek, J. Colloid Sci., $\underline{20}$, 1014 (1965).
88. W. R. Vieth, C. S. Fraugoulis, and J. A. Rionda, J. Colloid Interface Sci., $\underline{22}$, 454 (1966).
89. S. A. Reitlinger, M. V. Tsilipotkina, and A. A. Tager, Vysokomol. Soedin., A, $\underline{19}$, No. 7, 1495 (1977).
90. A. R. Matheson, C. S. Whewell, and P. E. Williams, J. Appl. Polym. Sci., $\underline{8}$, 2009 (1964).

91. A. B. Pakshver and V. A. Myagkov, Zh. Prikl. Khim., 29, 845, 1329 (1956).

92. R. H. Peters, J. Soc. Dyers Colour., 65, 63 (1949).

93. G. A. Gilbert and J. Rideal, Proc. R. Soc. London, A, 182, 335 (1944).

94. J. E. Anderson and H. W. Sackson, J. Phys. Chem., 78, No. 22, 2259 (1974).

95. Yu. A. Kokotov and V. A. Pasechnik, Equilibrium and Kinetics of Ion Exchange, Khimiya, Leningrad (1970).

96. R. A. Robinson and R. H. Stokes, Electrolyte Solutions, Butterworth (1959).

97. R. Griessbach, Theory and Practice of Ion Exchange [Russian translation], Izd. Inostr. Lit., Moscow (1963).

98. F. Hellferich, Ion Exchange, McGraw-Hill, New York (1962).

99. D. Paul, V. Gröbe, K. H. Göebel, H. Buschatz, and H. H. Schwarz, Acta Polym., 3, No. 2, 117-129 (1979).

100. M. Honda, Jpn. Anal., 1, 22 (1952).

101. J. Feitelson, Biochim. Biophys. Acta, 66, No. 2, 229 (1963).

102. A. L. Iordanskii et al., Vysokomol. Soedin., A, 16, No. 10, 2248 (1974).

103. M. A. Peterson and H. P. Gregor, J. Electrochem. Soc., 106, No. 12, 1051 (1969).

104. P. S. N's and E. M. Savitskaya, in: Physical Chemistry of Solutions, Khimiya, Moscow (1972).

105. T. I. Rozhanskaya et al., Zh. Prikl. Khim., 43, No. 9, 2034 (1970).

106. G. Hodouto, J. Polym. Sci., C, No. 30, 205 (1970).

107. B. A. Soldano and G. E. Boyd, J. Am. Chem. Soc., 75, No. 24, 6099, 6107 (1953).

108. G. E. Boyd and B. A. Soldano, J. Am. Chem. Soc., 75, No. 24, 6091 (1953).

109. E. Golling, Z. Angew. Phys., 14, 717 (1962).

110. I. S. Mackie and P. Mears, Proc. R. Soc. London, A, 232, 498 (1955).

111. A. Despic and G. I. Hills, Trans. Faraday Soc., 53, 1262 (1957).

112. P. H. Elworthy, A. T. Florence, and A. Rahman, J. Phys. Chem., 76, No. 12, 1763 (1972).

113. M. I. Artsis et al., Vysokomol. Soedin., A, 15, No. 1, 63 (1973).

114. H. Yasuda et al., Makromol. Chem., 125, 177 (1968).

115. E. A. Korytova, Candidate's Dissertation, Moscow Chemical-Technology Institute of Light Industries (1976).

116. G. E. Boyd, J. Phys. Chem., 78, 737 (1974).

117. J. A. Medley, Trans. Faraday Soc., 53, 1380 (1957).

118. J. Marshall, J. Polym. Sci., A-1, 6, No. 7, 1913 (1968).

119. O. W. Edwards and E. O. Huffman, J. Phys. Chem., 63, No. 11, 1830 (1959).

120. Handbook of Chemistry and Physics, Cleveland (1955), p. 2026.

121. I. Klotz, Horizons in Biochemistry [Russian translation], Mir, Moscow (1964).
122. A. L. Iordanskii et al., Chem. Prum., No. 25/50, 149 (1975).
123. A. L. Iordanskii et al., Dokl. Akad. Nauk SSSR, 219, No. 1, 316 (1974).
124. A. L. Iordanskii et al., Summaries of Papers, "Diffusion Phenomena in Polymers," Moscow (1974), p. 70.
125. A. L. Iordanskii et al., Vysokomol. Soedin., A, 16, No. 4, 849 (1974).
126. M. Hayashi, Bull. Chem. Soc. Jpn., 33, No. 9, 1184 (1960).
127. R. McGregor, R. H. Peters, and J. H. Petropolous, Trans. Faraday Soc., 58, 1054 (1962).
128. H. Brody, Text. Res. J., 35, 844 (1965).
129. M. L. Wright, Trans. Faraday Soc., 49, No. 1, 95 (1953).
130. J. A. Medley, Trans. Faraday Soc., 53, 1380 (1957); 60, 1010 (1964).
131. J. Marshall and J. A. Medley, Nature, 210, 729 (1966).
132. J. Marshall, J. Polym. Sci., A-1, 6, No. 7, 1913 (1968).
133. T. Iijima, T. Obara, and S. Ikeda, J. Chem. Soc. Jpn., Ind. Chem. Sect., 74, No. 9, 1874 (1971).
134. V. A. Myagkov and A. B. Pakshver, Kolloidn. Zh., 17, 120 (1955).
135. M. B. Rodell, L. Juess, and G. Aktian, J. Pharm. Sci., 53, No. 8, 873 (1964).
136. A. Kapadia et al., J. Pharm. Sci., 53, No. 7, 720 (1964).
137. J. A. Medley, Proc. 3rd Wool Text. Res. Conf., 3, 117 (1965).
138. E. Atherton, R. H. Peters, and D. A. Downey, Text. Res. J., 25, 977 (1955).
139. J. A. Barrie, in: Diffusion in Polymers, J. Crank and G. S. Park (eds.), Academic Press, New York (1968), p. 259.
140. M. M. Mikhailov, Moisture Absorption of Organic Dielectrics, Gosenergoizdat, Moscow—Leningrad (1960).
141. H. Yasuda and A. Peterlin, J. Appl. Polym. Sci., 17, No. 2, 433 (1973).
142. M. I. Artsis et al., Vysokomol. Soedin., A, 17, 128 (1975).
143. O. R. D'yachenko, B. A. Reka, and B. Ya. Kanterov, Lakokras. Mater. Ikh Primen., No. 2, 63 (1967).
144. A. L. Shterenzon, S. A. Reitlinger, and L. P. Topina, Lakokras. Mater. Ikh Primen., No. 1, 32 (1969).
145. G. Zundel, Hydration and Intermolecular Interaction, Academic Press, New York (1969).
146. A. V. Tobolsky and P. F. Zgons, Macromolecules, 1, No. 6, 515 (1968).
147. A. Eisenberg, Macromolecules, 3, No. 2, 147 (1970).
148. G. D. Shishov, V. A. Murov, and S. P. Rozanov, in: Tr. Mosk. Inst. Khim. Mashinostr., 67, 130 (1975).
149. D. Sinor, Khim. Tekhnol. Polim., No. 12, 22 (1966).
150. R. A. Foss and W. Dannliauser, J. Appl. Polym. Sci., 7, No. 3, 1015 (1963).
151. T. Miyamoto and K. Shibayama, J. Appl. Phys., 44, No. 12, 5372 (1973).

152. B. T. Sazhin, N. G. Podosenova, and V. S. Skyrikina, Plast. Massy, No. 6, 34 (1970).
153. A. L. Shterenzon, S. A. Reitlinger, and L. P. Topina, in: Proc. 3rd Intern. Conf. on Corrosion of Metals, Vol. 3, Mir, Moscow (1968), p. 130.
154. A. L. Shterenzon, S. A. Reitlinger, and L. P. Topina, Vysokomol. Soedin., A, 11, No. 4, 887 (1969).
155. A. S. Michaels and H. J. Bixlor, J. Polym. Sci., 50, No. 154, 393 (1961).
156. A. M. Gransbergs et al., in: Diffusion Phenomena in Polymers, Riga (1977), p. 265.
157. E. Högfeldt, in: Ion Exchange [Russian translation], Mir, Moscow (1968), p. 512.
158. D. D. Chegodaev, Z. K. Naumova, and I. S. Dunaevskaya, Fluorine-Containing Plastics, Goskhimizdat, Leningrad (1960).
159. F. H. Garner, S. R. M. Ellis, and J. C. Gill, J. Appl. Chem., 6, No. 9, 407 (1956).
160. D. D. Chegodaev, Leningr. Promst., No. 2, 84 (1958).
161. T. N. Nikolaeva, N. S. Kudryavtseva, and L. V. Zakharova, Plast. Massy, No. 5, 18 (1967).
162. C. A. R. Wilson, Plastics, 30, No. 331, 86 (1965).
163. L. L. Vol-Rabinovich and I. V. Marakhovskii, Plast. Massy, No. 5, 18 (1967).
164. V. G. Plisov and Yu. V. Zelenev, Khim. Promst., No. 12, 33 (1975).
165. É. A. Vishnevskaya and A. A. Shevchenko, Summaries of Papers, 27th Conf. Mosc. Inst. Chem. Eng., Sec. 3 (1967), p. 69.
166. V. A. Kuz'min, I. N. Palin, and G. F. Severov, Khim. Neft. Mashinostr., No. 4, 28 (1973).
167. T. E. Rudakova et al., Vysokomol. Soedin., A, 16, No. 16, 1356 (1974).
168. M. Koga and H. Kawaguichi, J. Chem. Eng. Jpn., 33, No. 11, 1062, 1065 (1969); 34, No. 1, 127.
169. W. A. Murow and K. H. Baumann, Plaste Kautsch., 21, No. 12, 922 (1974).
170. G. Rüprich, W. A. Murow, and K. H. Baumann, Plaste Kautsch., 22, No. 6, 487 (1975).
171. L. Peeva, M. Natov, and E. Ivanova, in: Collective works of Sci. Res. Inst. for Plastics and Polym. Process. (Hungary), No. 2 (1972), p. 101.
172. A. L. Shterenzon, I. B. Belousova, and Yu. E. Lobanov, in: Properties, Processing, and Use of Pentaplasts, Khimiya, Leningrad (1975), p. 96.
173. A. L. Shterenzon, Yu. E. Lobanov, and S. F. Konovalova, Vysokomol. Soedin., 6, No. 9, 1668 (1964).
174. V. N. Kestel'man et al., in: Synthesis and Physical Chemistry of Polymers, Vol. 11, Naukova Dumka, Kiev (1973), p. 57.
175. A. Ya. Metrya, M. M. Kalnin', and V. P. Karlivan, in: Modification of Polymer Materials, Zinatne, Riga (1972), No. 3, p. 115.
176. N. N. Pavlov et al., Vysokomol. Soedin., A, 18, No. 7, 1591 (1976).

177. V. N. Padeiskii, Lakokras. Mater. Ikh Primen., No. 3, 37 (1963).
178. A. I. Artem'ev, in: Sb. Nauchn. Tr. Tsentr. Aptechn. Nauchno-Issled. Inst., No. 5, 22 (1964).
179. H. T. Hoffman, J. Polym. Sci., A-2, $\underline{8}$, No. 12, 2177 (1970).
180. A. L. Glass and J. Smith, J. Paint Technol., $\underline{38}$, $\underline{495}$, 203 (1966).
181. I. E. Krichevskii, in: Sb. Dokl. Gidrotekh., Vses. Nauchno-Issled. Inst. Gidrotekh. im. Vedeneeva, Moscow—Leningrad (1962), Issue 4, p. 131.
182. B. I. Borisov and S. K. Noskov, Zashch. Met., $\underline{7}$, No. 1, 88 (1971).
183. M. B. Kilcullen, J. Oil Colour Chem. Assoc., $\underline{58}$, No. 1, 365 (1975).
184. M. Naumann, W. Blime, and D. Schneider, Radioanal. J. Chem., $\underline{30}$, No. 2, 489 (1976).
185. B. A. Reka, N. I. Obshchetko, and O. P. D'yachenko, Lakokras. Mater. Ikh Primen., No. 6, 34 (1973).
186. M. V. Gorelik, N. I. Shilov, and I. V. Fodiman, Khim. Promst. za Rubezhom., No. 9, 31 (1965).
187. British Patent No. 876045.
188. L. V. Sukhareva, in: Toxicology of High-Molecular Materials and Chemical Starting Materials for Their Synthesis, Khimiya, Moscow—Leningrad (1966), p. 47.
189. E. G. Smith and D. N. Robbian, Polymer, $\underline{15}$, No. 11, 713 (1974).
190. N. S. Tikhomirova and V. N. Kotrelev, Primen. Radioakt. Izotopov i Yadernykh Izluchenii, Nos. 17-63, 194 (1963).
191. N. S. Tikhimirova and V. I. Kotrelev, in: Plastics, Khimiya, Moscow (1970), p. 268.
192. B. Alt, Kunststoffe, $\underline{62}$, No. 12, 809 (1972).
193. B. I. Borisov and N. A. Meshchanskii, Plast. Massy, No. 3, 61 (1966); translated in Soviet Plast.
194. S. Kh. Shliner and V. I. Govorukhina, Rybn. Khoz., No. 1, 70 (1969).
195. G. Menges and W. Schneider, Kautsch. Gummi, Kunstst., $\underline{25}$, No. 5, 213 (1972).
196. V. Khityi, in: Trudy GNIIZM im. Akimova, Issue 3, 276 (1960).
197. A. E. Kreitus et al., in: Modification of Polymeric Materials, Zinatne, Riga, Issue 5, 16 (1975).
198. N. S. Tikhomirova, K. I. Zernova, and V. N. Kotrelev, Plast. Massy, No. 12, 40 (1962); translated in Soviet Plast., No. 12 (1962).
199. D. D. Chegodaev and N. A. Bugorkova, Zh. Fiz. Khim., $\underline{33}$, No. 2, 262 (1959).
200. N. S. Gilinskaya et al., Vysokomol. Soedin., B, $\underline{11}$, No. 3, 215 (1969).
201. R. K. Pavlova et al., Tekh. Zashch. Korr., No. 2 (81), 11 (1973).
202. M. F. Kaizer and S. I. Sukov, in: Equipment in the Chemical Industry, Its Service, Repair, and Corrosion Protection, Abstracts, NIITEKhIM, Moscow, Issue 5, 18 (1975).

203. R. A. Pasternak, M. V. Christensen, and J. Heller, Macro-
      molecules, $\underline{3}$, No. 3, 366 (1970).
204. A. N. Radaev, A. G. Parshin, and V. A. Murov, in: Tr. Mosk.
      Inst. Khim. Mashinostr., Issue 67, 121 (1975).
205. B. A. Dolezel et al., Plaste Kautsch., $\underline{14}$, No. 9, 699 (1967).
206. O. F. Shlenskii, N. T. Khovanskaya, and V. V. Lavrent'ev,
      Plast. Massy, No. 5, 52 (1966); translated in Soviet Plast.
207. N. I. Khovanskaya, Plast. Massy, No. 7, 72 (1966); translated
      in Soviet Plast.
208. O. R. D'yachenko and B. A. Reka, Lakokras. Mater. Ikh Primen.,
      No. 3, 44 (1966).
209. A. A. Shevchenko, Candidate's Dissertation, Inst. Neftekhim.
      Gazov. Promst., Moscow (1972).
210. T. N. Nikolaeva and N. S. Kudryavtseva, Plast. Massy, No. 7,
      41 (1962); translated in Soviet Plast., No. 7 (1962).
211. H. Kowaguchi, R. Fukuoka, and M. Koga, Proc. 18th Jpn. Congr.
      Mater. Res., Kyoto (1975), p. 160.
212. L. V. Glukhov et al., Plast. Massy, No. 2, 66 (1976).
213. N. D. Tomashov, A. F. Luneva, and K. I. Gedgovd, in: Tr.
      Inst. Fiz. Khim. Akad. Nauk SSSR, No. 8, 264 (1960).
214. I. L. Rozenfel'd, F. I. Rubinshtein, and S. V. Yakubovich,
      Lakokras. Mater. Ikh Primen., No. 2, 58 (1962).
215. A. A. Tager, M. V. Tsilipotkina, and D. A. Reshet'ko, Vysok-
      omol. Soedin., A, $\underline{17}$, No. 11, 2566 (1975).
216. M. V. Tsilipotkina et al., Vysokomol. Soedin., A, $\underline{18}$, No. 11,
      874 (1976).
217. M. M. Dubinin, in: Main Problems in the Theory of Physical Ad-
      sorption, Nauka, Moscow (1970), p. 251.
218. M. M. Dubinin, in: Adsorption and Porosity, Nauka, Moscow
      (1976), p. 105.
219. O. V. Nechaeva, M. V. Tsilipotkina, and A. A. Tager, in:
      Adsorption and Porosity, Nauka, Moscow (1976), p. 224.
220. A. A. Tager, A. A. Askadskii, and M. V. Tsilipotkina,
      Vysokomol. Soedin., A, $\underline{17}$, No. 6, 1346 (1975).
221. W. E. Statton, in: Newer Methods of Polymer Characterization,
      Ke. Bacon (ed.), Wiley-Interscience, New York (1964).
222. L. S. Koretskaya and E. G. Il'ina, Izv. Akad. Nauk BSSR, Ser.
      Fiz.-Tekhn. Nauk, No. 3, 122 (1973).
223. E. G. Il'ina et al., Dokl. Akad. Nauk BSSR, $\underline{18}$, No. 12, 1084
      (1974).
224. A. L. Shterenzon et al., Dep. 251/14, Ural'sk. NI Khim. Inst.,
      Sverdlovsk (1974).
225. A. A. Shevchenko, E. A. Skuratova, and I. Ya. Klimov, Fiz.-
      Khim. Mekh. Mater., $\underline{9}$, No. 6, 106 (1973).
226. E. A. Skuratova, Candidate's Dissertation, Moscow Institute
      of Chemical Engineering, Moscow (1969).
227. A. L. Shterenzon, T. V. Kazantseva, and V. V. Chasova, in:
      Diffusion Phenomena in Polymers, Riga (1977), p. 350.
228. U. Starkweather, Polym. Prepr., $\underline{14}$, No. 2, 376 (1976).
229. R. M. Barrer and J. A. Barrie, J. Polym. Sci., $\underline{28}$, 377 (1958).

230. B. H. Zimm and J. L. Lundberg, J. Phys. Chem., $\underline{60}$, No. 4, 425
      (1956).

Chapter 6

DEGRADATION OF POLYMERS
IN
AGGRESSIVE MEDIA

The degradation of polymers in aggressive media is a complex physicochemical process including adsorption, diffusion, and the dissociation of chemically unstable bonds. The course of degradation has a number of special features, which are linked both with the specific structure of the polymeric materials and with the specific kinetics of reactions in solids.

Let us consider, in order, the main features of the degradation of polymer articles in aggressive media.

6.1. FUNDAMENTAL KINETIC EQUATIONS

In the degradation of polymers in aggressive media there are the following fundamental processes:

1) adsorption of the aggressive medium on the surface of the polymer article;

2) diffusion of the aggressive medium within the polymer article;

3) chemical reaction of the aggressive medium with chemically unstable groups of the polymer;

4) diffusion of the degradation products to the surface of the polymer article;

5) desorption of the degradation products from the surface of the polymer article.

179

A mathematical consideration of these processes is very difficult, and, accordingly, it is usually assumed that one or at the most two stages are slow in comparison with the others, and consequently limit the course of the whole degradation process. The first and fifth stages usually take place more rapidly than the second, third, and fourth.

The diffusion of the degradation products to the surface of a polymer article usually plays no significant part in the assessment of the change in the service properties of polymer articles in the course of degradation, but it may restrict, for instance, the process of biodegradation of a polymer within a living organism. In the course of chemical degradation the polymer and the aggressive medium (liquid or gas) are in separate phases. Chemical reaction between the aggressive medium and chemically unstable groups in the polymer may take place either at the phase interface or within the polymer phase (components of the aggressive medium which are dissolved in the polymer take part in the reaction).

Regarding the two phases as a single closed system, the rate of the chemical reaction w may be expressed as dn/Vdt, where n is the number of groups which have decomposed at time t, and V is the volume of the polymer.

In the first case $w \sim S/V$, where S is the surface area of the polymer article and, in the second case, this relationship is more complex. Let us consider this in more detail. Assuming that the law of mass action is observed, the volume of the polymer remains practically unchanged during the course of degradation and the polymer is isotropic in properties; then we can write the equation for the rate of decomposition of the chemically unstable groups as follows:

$$w = \frac{dc_n}{dt} = k\,(c_n^0 - c_n)\,c_{\text{cat}}c_{\text{solv}} \tag{6.1}$$

where c_n^0 is the initial concentration of chemically unstable groups in the polymer; c_n the concentration of decomposed groups; c_{cat} the concentration of catalyst in the polymer; c_{solv} the concentration of solvent in the polymer; k the rate constant of decomposition of the unstable groups.

The concentration of catalyst in the polymer, e.g., acid or base, can be found from the equation

$$\frac{dc_{\text{cat}}}{\partial t} = D_{\text{cat}}\nabla^2 c_{\text{cat}} - \sum_i c_{\text{cat}}c_i k_i \tag{6.2}$$

where ∇ is the Laplace operator; c_i the concentration of functional groups in the polymer which are capable of taking part in a complex-formation or substitution reaction; k_i the rate constant of the complex-formation or substitution reaction of the catalyst with functional groups of the polymer.

The second term on the right-hand side of Eq. (6.2) takes into account the possibility of such reactions as protonation, interaction with a hydroxide ion, and so forth.

If the solvent is involved in the decomposition of the chemically unstable groups in the polymer, for instance the removal of water in hydrolysis reactions, then the solvent concentration may be found from the equation

$$\frac{\partial c_{solv}}{\partial t} = D_{solv} \nabla^2 c_{solv} - k(c_n^0 - c_n) c_{cat} c_{solv} \tag{6.3}$$

In writing this equation it is necessary to justify a number of assumptions.

First, the polymer—aggressive medium system is, as a rule, very much diluted in relation to the aggressive medium, i.e., it may be assumed that D_{cat} and D_{solv} do not depend on the concentration of the corresponding components in the polymer. If the aggressive medium is soluble in the polymer to any considerable extent, it becomes necessary to use an equation which takes into account the relationship between D and the diffusant concentration.

Second, the value of k is constant. This is correct when the aggressive medium has low solubility in the polymer. In general, the influence of the aggressive medium on k can be reduced, as a first approximation, to the influence of the solvent on the rate of the chemical reaction, which can be predicted on the basis of existing theories [1, p. 143].

Third, the decomposition reaction of chemically unstable groups is practically irreversible. This condition is realized when degradation takes place under some particular reaction conditions to quite low degrees of dissociation (<0.05). In this case the concentration of chemically unstable groups changes only slightly and $c_n^0 - c_n \approx c_n^0$.

Thus, determining the rate of decomposition of chemically unstable groups in a polymer under the action of aggressive media means the joint solution of Eqs. (6.1)-(6.3).

Usually the following boundary conditions are assumed:

the concentration of substances diffusing to the surface of a polymer article is a constant value (in practice this condition is fulfilled with a rapid flow over the article of a stream of solution with constant concentration of the aggressive medium);

the concentration of substances diffusing to the surface of
the polymer article is a function of time, and the dependence of the concentration on time is expressed by one of the adsorption equations;

the concentration of substances diffusing to the surface of
the polymer article is determined by some particular law of mass transfer,
the simplest being

$$j = \beta \, (c_{\text{surf}} - c_{\text{p}})$$

where j is the flow of substance diffusing; β the mass transfer coefficient; c_{surf} and c_{p} are, respectively, the concentration of the
substance on the surface and within the volume of the solution of
aggressive medium.

The majority of polymer articles may, to a first approximation,
be regarded as simple geometrical bodies.

As a model of polymer films and coatings we can take a parallelepiped, as a model of a filament a cylinder, and so forth.

We shall now consider the solution of Eqs. (6.1)-(6.3) for a
parallelepiped and a cylinder.

6.2. MACROKINETICS OF CHEMICAL DEGRADATION

The process of degradation may take place in various regions,
depending on the ratio of the rates of the diffusion process and
of the chemical reaction.

The rate of diffusion of the aggressive medium is commensurate
with the rate of the chemical reaction, and degradation takes place
in a particular reaction zone whose size increases with time and
ultimately reaches the limits of the dimensions of the polymer article, i.e., the reaction takes place under internal diffusion-kinetic conditions.

The rate of diffusion of the aggressive medium considerably
exceeds that of the chemical reaction. After the dissolution of
the aggressive medium in the polymer is finished, degradation takes
place over the total volume of the polymer, i.e., under internal
kinetic conditions.

The rate of diffusion of the aggressive medium is considerably
lower than that of the chemical reaction. In this case the degradation takes place in a particular thin reaction surface layer or, as
it is usually put, from the surface of the polymer article, i.e.,
under external diffusion-kinetic conditions.

6.2.1. Internal Diffusion-Kinetic Conditions

The joint solution of Eqs. (6.1)-(6.3), even with the above assumptions, is a difficult mathematical problem and is possible only by using a computer. In considering the diffusion of aggressive media in polymers, it has been established that the diffusion of the catalyst (acid or base) and of the solvent takes place at an identical rate on a single front and that the diffusion of the solvent is at a considerably higher rate than that of the catalyst.

The first variant occurs, as a rule, in the diffusion of aggressive media in hydrophilic polymers and in the diffusion of acids and bases with high vapor pressure in hydrophobic polymers.

The second variant occurs in the diffusion of acids and bases with low vapor pressure in hydrophobic polymers.

These patterns of behavior simplify the solution of the problem, since, in the first case, the diffusion of the catalyst and solvent can be characterized by a single diffusion coefficient $D_{cat} = D_{solv}$. In the second case, $D_{solv} \gg D_{cat}$, and it can be reckoned that the concentration of solvent in the reaction zone of the polymer article becomes constant after a certain time, and equal to its solubility c^0_{solv}, i.e.,

$$\partial c_{solv}/\partial t = 0 \quad \text{when} \quad c_{solv} = c^0_{solv}$$

Thus, the problem is simplified and is reduced to the solving of Eqs. (6.1) and (6.2), of which, taking into account the relationship $c^0_n - c_n \approx c^0_n$, the second takes the form

$$\frac{dc_n}{dt} = k_{eff} c_{cat} \tag{6.4}$$

where

$$k_{eff} = k c^0_n c^0_{solv}$$

In Eq. (6.2), for simplicity, we restrict ourselves to one term and make $c = c_i$.

Depending on whether or not there takes place combination of the catalyst with functional groups of the polymer, we may obtain two equations for the change in the number of macromolecular scissions in the degradation.

$\underline{K_p \to \infty}$, i.e., the functional groups present in the polymer
formed in the course of degradation react with the catalyst prac-

<u>tically irreversibly</u>. Bearing in mind that these reactions (e.g., protonation or complex-formation) proceed far more rapidly than degradation reactions, Eq. (6.2) takes the form

$$\frac{\partial c_{cat}}{\partial t} = D_{cat}\nabla^2 c_{cat} - k_{eff}c_{cat}$$

(6.5)

The solution of this equation was first achieved by Danckwerts [2], on the basis of the solution of the analogous problem of thermal conductivity considered by Carslaw and Jaeger [3]. The solution of Eq. (6.5) for a parallelepiped one of whose dimensions (ℓ) is much less than the other two with the boundary condition $c_{cat} = c_{cat}^0$ with x = 0 and ℓ with t → 0 and the initial condition $c_{cat} = 0$ with t = 0 and $0 \leqq x \leqq \ell$ takes the form

$$c_{cat}(x, t) = c_{cat}^0\left\{1 - \frac{4}{\pi}\sum_{m=0}^{\infty}\frac{\sin b_m x}{(2m+1)(b_m^2 D_{cat}+1)}[k_{eff} + b_m^2 D_{cat}\exp - (b_m^2 D_{cat} + k_{eff})t]\right\}$$

(6.6)

where $b_m = \pi(2m+1)/\ell$; c_{cat}^0 the solubility of the catalyst in the polymer.

Substituting Eq. (6.6) into (6.4), and carrying out double integration from 0 to t and from 0 to $\ell/2$, we get

$$c_n = k_{eff}c_{cat}^0 t\left\{1 - \frac{8}{\pi^2}\sum_{m=0}^{\infty}\frac{k_{eff}(b_m^2 D_{cat}+k_{eff}) + b_m^2 D_{cat}[1 - \exp - (b_m^2 D_{cat} + k_{eff})t]}{(2m+1)b_m(b_m^2 D_{cat} + k_{eff})^2\,\ell t}\right\}$$

(6.7)

Examination of Eq. (6.7) shows that if $b_m^2 D_{cat} > k_{eff}$, it is sufficient to limit oneself to the first term of the series. A detailed consideration of this case will be given below.

Solution of Eq. (6.5) for a cylinder of radius r much less than the length ℓ with the boundary conditions $c_{cat} = c_{cat}^0$ with x = r with t > 0 and the initial condition $c_{cat} = 0$ with t = 0 and x ≤ r, takes the form

$$c_{cat}(x, t) = c_{cat}^0\left\{1 - \frac{2}{r}\sum_{n=1}^{\infty}\frac{\left[k_{eff} + D_{cat}\frac{\mu_n^2}{r^2}\exp - \left(D_{cat}\frac{\mu_n^2}{r^2} + k_{eff}\right)t\right]J_0\left(\frac{x}{r}\mu_n\right)}{\left(D_{cat}\frac{\mu_n^2}{r^2} + k_{eff}\right)\frac{\mu_n}{r}J_1(\mu_n)}\right\}$$

(6.8)

where J_0 and J_1 are Bessel functions of the first kind of zero and first order, respectively; and μ_n the roots of the Bessel function.

An expression for c_n is obtained from Eqs. (6.4) and (6.8) after integration from 0 to t and from 0 to r:

$$c_n = k_{eff} c_{cat}^0 \sum_{n=1}^{\infty} \frac{D_{cat}}{r^2 z} \left[t - \frac{1 - \exp{(-zt)}}{z} \right] \qquad (6.9)$$

where

$$z = \frac{D_{cat}}{r^2} + k_{eff}$$

For the initial time of degradation, if $\exp{(-zt)} > 0.9$, we can use for calculation the approximated equation

$$c_n \approx k_{eff} c_{cat}^0 \left[\frac{8}{3\pi^{1/2}} \frac{D_{cat}^{1/2}}{r} t^{3/2} - \frac{D_{cat} t^2}{2r^2} - \frac{2}{15\pi^{1/2}} \frac{D_{cat}^{1/2}}{r} \left(\frac{D_{cat}}{r^2} + k_{eff} \right) t^{5/2} \right] \qquad (6.10)$$

If $\exp{(-zt)} < 0.1$, then Eq. (6.9) takes the form

$$c_n \approx k_{eff} c_{cat}^0 \sum_{n=1}^{\infty} \frac{D_{cat}}{z r^2} \left(t - \frac{1}{z} \right) \qquad (6.11)$$

If $D_{cat} \mu_n^2 / r^2 > k_{eff}$, then, as in the case considered above, it is sufficient to restrict ourselves to the first term of the series, which simplifies Eqs. (6.9) and (6.11).

$\underline{K_p \to 0}$, i.e., $\underline{\text{the functional groups in the polymers do not re-}}$ $\underline{\text{act appreciably with the catalyst.}}$ Equation (6.2) takes the form of the Fick diffusion equation, the solution of which, for the parallelepiped (film) and cylinder (filament) mentioned above, is given in Chap. 5.

On substituting the solution into the Fick equation for the parallelepiped in (6.4) and integrating within the same limits as in Eq. (6.9), we get

$$c_n = k_{eff} c_{cat}^0 t \left\{ 1 - \frac{8}{\pi^2} \sum_{m=0}^{\infty} [1 - \exp{(2m+1)^2 y}] \frac{1}{(2m+1)^4 y} \right\} \qquad (6.12)$$

where $y = \pi^2 D_{cat} t / \ell^2$.

With $y < 1$, this case is realized in the initial period of degradation for films of any particular thickness

$$c_n = k_{eff} c_{cat}^0 t \frac{8}{\pi^2} \varphi(y) \qquad (6.13)$$

where

$$\varphi(y) = \sum_{m=0}^{\infty} \frac{\exp - (2m+1)^2 y - 1 + (2m+1)^2 y}{(2m+1)^4 y}$$

By using a computer there is established a simple relationship $\varphi(y) = 0.589y^{1/2}$, i.e.,

$$c_n = \frac{4}{\pi^{1/2}} \, k_{\text{eff}} c_{\text{cat}}^0 \, D_{\text{cat}}^{1/2} \, l^{-1} \, t^{3/2} \tag{6.14}$$

Equation (6.14) is conveniently expressed as follows:

$$n = \frac{8}{\pi^{1/2}} \, k_{\text{eff}} c_{\text{cat}}^0 \, D_{\text{cat}}^{1/2} \, st^{3/2} \tag{6.15}$$

Thus, in the initial period of degradation the number of groups which have decomposed depends on the rate constant of the chemical reaction, on the diffusion coefficient, on the solubility of the catalyst, on the surface area of the film, and on time.

For a cylinder, after the corresponding operations, we get

$$c_n = k_{\text{eff}} \, c_{\text{cat}}^0 \left\{ t - \sum_{n=1}^{\infty} \frac{4r^2}{\mu_n^4 D_{\text{cat}}} \left[1 - \exp - \left(\frac{D_{\text{cat}}}{r^2} \, \mu_n^2 t \right) \right] \right\} \tag{6.16}$$

In solving this equation it is sufficient to restrict oneself to the first term of the series.

For the initial period of degradation with $\left(-\frac{D_{\text{cat}}}{r^2} \mu_1^2 t \right) < 0.1$ it is possible to use the equation

$$c_n \approx k_{\text{eff}} c_{\text{cat}}^0 \left[\frac{8}{3\pi^{1/2}} \frac{D_{\text{cat}}^{1/2}}{r} \mu_1^2 t^{3/2} - \frac{D_{\text{cat}} t^2}{2r^2} - \frac{2}{15\pi^{1/2}} \left(\frac{D_{\text{cat}}}{r^2} \right)^{3/2} t^{5/2} \right] \tag{6.17}$$

If $\left(-\frac{D_{\text{cat}}}{r^2} \mu_1^2 t \right) \leqslant 0.9$, we get the linear equation

$$c_n \approx k_{\text{eff}} c_{\text{cat}}^0 \left(t - \frac{4r^2}{\mu_1^4 D_{\text{cat}}} \right) \tag{6.18}$$

where the intercept on the time axis

$$t_0 \approx 4r^2 / \mu_1^4 D_{\text{cat}} \tag{6.19}$$

while the slope is equal to the product $k_{\text{eff}} c_{\text{cat}}^0$.

An approximate solution of Eq. (6.2) for a chemically reversible reaction (K_{eq} does not approach infinity) is considered in [4].

6.2.2. Internal Kinetic Conditions

The kinetic equation for the parallelepiped may be obtained from Eq. (6.12) with $y \gg 1$, and for the cylinder, from Eq. (6.16) with $\left(-\frac{D_{cat}}{r^2}\mu_1^2 t\right) \longrightarrow 1$.

In both cases,

$$c_n = k_{eff} c_{cat}^0 t \tag{6.20}$$

Thus, at fairly low degrees of dissociation the concentration of groups which have decomposed is proportional to the catalyst concentration under saturation conditions.

6.2.3. External Diffusion-Kinetic Conditions

The process of degradation under these conditions takes place in a particular thin reaction surface layer, the dimensions of which usually cannot be determined because of the lack of values for k_{eff} and D_{cat}. It is usually assumed that this layer is infinitely small and that degradation takes place practically from the surface of the polymer article. In this case,

$$\frac{dn}{dt} = k_{surf} c_{n\,(surf)} c_{cat\,(surf)} c_{solv\,(surf)} S \tag{6.21}$$

where "surf" indicates that the concentration relates to the polymer—aggressive medium interface.

In solving this equation it is assumed that the polymer is orientated, and that the macromolecules are arranged parallel to the larger part of the interface (e.g., for a filament the curved surface will be the greatest). The concentration of active sites on the surface $c_{n(surf)}$ for the process in question depends on the type of decomposition of the macromolecules, their packing, the cohesive energy, and the solubility of the decomposition products (monomers and oligomers) in the surrounding medium.

With this type of degradation there is established within a very short time on the surface of the polymer article a steady-state concentration of active sites $c_{n(surf)}^0 z^{-1}$ (where z is a constant depending on the factors indicated above). For instance, if the decomposition of the macromolecules is at a terminal bond, and the monomers which are formed are soluble in the aggressive medium, then $c_n(surf) \approx$

$c_n^0(\text{surf})/\bar{P}_{n_0}$ and k_{surf} is the rate constant of depolymerization. If the decomposition of the macromolecules is by a random process with subsequent rapid single-act depolymerization, and the degradation products are soluble, then $c_n(\text{surf}) \approx c_n^0(\text{surf})$ and $k_{(\text{surf})}$ is the rate constant for decomposition by a random process.

Let us consider how the linear dimensions and mass of polymer articles of differing geometrical forms will change with time.

Let us express the degree of dissociation by the number of chemically unstable groups which have decomposed to the maximum possible number of scissions (n_∞)

$$\alpha = n/n_\infty \tag{6.22}$$

We can determine the value of n_∞ from the formula

$$n_\infty = \frac{m_0 A}{M}\left[1 - \frac{1}{(\bar{P}_{n_\theta} - 1)}\right] \tag{6.23}$$

where M is the MM (molecular mass) of the monomer unit.

For quite large $\bar{P}_{n_0}$ in formula (6.23) we can restrict ourselves to the first term. If the degradation products are soluble in the surrounding medium, α may be expressed by the mass of the polymer

$$\alpha = \left(1 - \frac{m}{m_0}\right) \tag{6.24}$$

In such a case,

$$\frac{d\alpha}{dt} = \frac{1}{n_\infty}\frac{dn}{dt} = -\frac{1}{m_0}\frac{dm}{dt} \tag{6.25}$$

Substituting (6.21) into (6.25), we get

$$-\frac{dm}{dt} = \frac{M}{A}\,k_{(\text{surf})}\,\frac{c_n^0{}_{(\text{surf})}}{z}\,c_{\text{cat (surf)}}\,c_{\text{solv (surf)}}\,s \tag{6.26}$$

For a film of thickness ℓ,

$$m = sl\rho$$

The thickness of the specimen will change as follows:

$$l = l_0 - k_{(\text{surf})}\,\frac{M}{A}\,c_n^0{}_{(\text{surf})}\,\frac{1}{z\rho}\,c_{\text{cat (surf)}}\,c_{\text{solv (surf)}}\,t \tag{6.27}$$

The time for complete decomposition of the polymer τ is

$$\tau = \frac{l_0 \rho}{k_{(surf)} \dfrac{M}{A} c_n^0{}_{(surf)} \dfrac{1}{z} c_{cat\ (surf)} c_{solv\ (surf)}} \qquad (6.28)$$

Since the surface area of the polymer article remains practically unchanged right up to high degrees of dissociation, we get from Eq. (6.28)

$$m = m_0 - k_{(surf)} \frac{M}{A} c_n^0{}_{(surf)} \frac{1}{z} s c_{cat\ (surf)} c_{solv\ (surf)} t \qquad (6.29)$$

Equations (6.27) and (6.28) can be expressed in dimensionless parameters

$$\frac{m}{m_0} = 1 - \frac{t}{\tau} \qquad (6.30)$$

$$\frac{l}{l_0} = 1 - \frac{t}{\tau} \qquad (6.31)$$

For a filament of radius r we find from Eq. (6.24)

$$\alpha = 1 - \frac{r^2}{r_0^2} \qquad (6.32)$$

and

$$\frac{d\alpha}{dt} = \frac{1}{n_\infty} \frac{dn}{dt} = - \frac{2r\,dr}{r_0^2\,dt} \qquad (6.33)$$

We obtain via the corresponding transforms

$$r = r_0 - k_{(surf)} \frac{M}{A} c_n^0{}_{(surf)} \frac{1}{z\rho} c_{cat\ (surf)} c_{solv\ (surf)} t \qquad (6.34)$$

From Eq. (6.34) we determine the change in mass of the polymer with time, using the relationship between the surface area of the side and the volume of the cylinder:

$$m^{1/2} = m_0^{1/2} - k_{(surf)} \frac{M}{A} c_n^0{}_{(surf)} \frac{(\pi l)^{1/2}}{(z\rho)^{1/2}} c_{cat\ (surf)} c_{solv\ (surf)} t \qquad (6.35)$$

Using dimensionless parameters, Eqs. (6.34) and (6.35) take the following form:

$$\frac{r}{r_0} = 1 - \frac{t}{\tau} \qquad (6.36)$$

$$\frac{m}{m_0} = \left(1 - \frac{t}{\tau}\right)^2 \qquad (6.37)$$

where

$$\tau = \cfrac{r_0 \rho}{k_{(surf)}\,\dfrac{M}{A}\,c_{n\,(surf)}\,\dfrac{1}{z}\,c_{cat\,(surf)}\,c_{solv\,(surf)}}$$
(6.38)

With degradation of this type it is very important to know the state of the catalyst solutions on the surface of the polymer, i.e., what the values of $c_{cat}(surf)$ and $c_{solv}(surf)$ are.

When an electrolyte solution is in contact with a solid polymer, the electrolyte may be in solution or in the polymer, or on the phase interface (solution–polymer). The state of the electrolyte in the solution and in the polymer is assessed in Chaps. 1 and 5. The concentration and activity gradients which may occur at the solution–polymer interface are brought about by the occurrence of degradation, including the diffusion of the electrolyte and of the reaction products and the chemical reaction, which disturbs the equilibrium of statistical distribution of the electrolyte (molecules and ions) at the surface of the polymer specimen (a concentration effect), and also by the fact that the electrolyte and reaction products can be adsorbed on the surface of the polymer specimen (an adsorption effect).

The concentration effect was investigated in detail mostly by McGregor and Peters [5], who established a link between the thickness of the surface layer δ and the diffusion parameters of the substances in solution and in the polymer. They showed that at high rates of transport of the solution $\delta \to 0$.

Thus, by using the transport of the solution, it is possible to reduce the concentration effect to a minimum.

The adsorption effect can be assessed from the Gibbs equation [6]:

$$G = -\frac{a}{RT}\left(\frac{\partial \sigma}{\partial a}\right)_T$$
(6.39)

where G is the relative adsorption and σ the surface tension at the polymer–solution interface.

The value of σ is found from the Young equation:

$$\cos\theta = \frac{\sigma_{pg} - \sigma}{\sigma_{sg}}$$
(6.40)

where σ_{pg} and σ_{sg} are, respectively, the surface tensions at the polymer–gas and solution–gas interfaces, and θ the wetting angle.

Aqueous electrolyte solutions are surface-inactive and the value of G is negative. Calculations for hydrophobic polymers show that G is low and that, in the majority of cases, it can be assumed that the thermodynamic parameters of the catalyst and solvent will be closely similar in solution and at the polymer—solution interface.

The physicochemical basis of the biodegradation of synthetic polymers and of the resorption of polymer implants is given by Livshits [6a]. The equations which describe the external diffusion—kinetic conditions are considered.

6.3. SPECIAL FEATURES OF THE DEGRADATION
OF POLYMER ARTICLES OF INHOMOGENEOUS STRUCTURE

In polymer articles there is practically always anisotropy of structural features and properties throughout the volume (the so-called structural differential) [7].

In single-phase systems the following are examples of such differentials:

glasslike and high-elastic portions (copolymers of methyl methacrylate and methyl acrylate);

heavily and lightly crosslinked regions (vulcanizates and elastomers);

orientated and nonorientated regions (crystallizing polymers);

regions with differing MMDs (fractionation under shear flow);

regions within single-phase amorphous polymers with differing degrees of ordering (association).

In hetero-phase systems there occur the following major structural differentials:

amorphous and crystalline regions (crystallizing polymers);

regions with differing degrees of microinhomogeneity (stereoregular polymers);

regions with differing three-dimensional structure (cis-trans isomerism) or chemical structure.

In addition, in polymeric materials there are varied structural differentials created by additions of organic and inorganic substances, such as stabilizers, plasticizers, blowing agents, fillers (glass fibers in glass reinforced plastics and carbon black in rubber), pigments, etc.

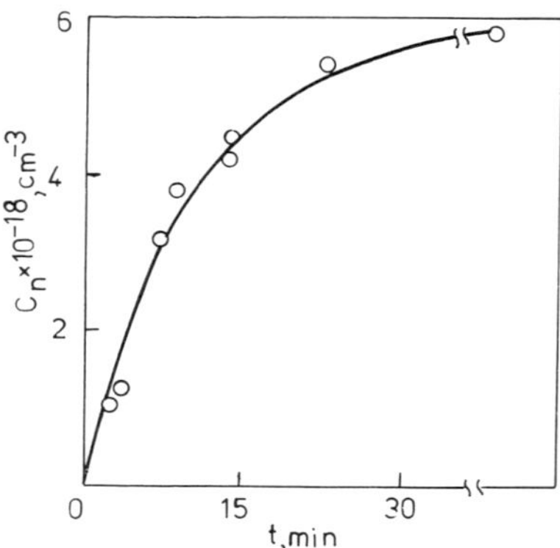

Fig. 6.1. Kinetics of accumulation of carboxyl
groups formed in the hydrolytic degradation of a film of PK-4 polycaproamide at 60°C in water [11].

These structure differentials bring about anisotropy of chemical and diffusion properties in polymers articles. At the present time
there is a lack of precise quantitative data to show the influence
of structural factors, on the one hand, on the diffusion of aggressive media, and, on the other hand, on the reactivity of the functional groups in the polymers. This is discussed more fully in [8,
9].

Generally, there are, in a polymer, regions with differing reactivity (differing k values) and different capacity for sorbing and
transporting aggressive media (differing c^0 and D).

Thus, the overall rate of the degradation process is

$$w_\Sigma = \sum_i [k\,(c_n^0 - c_n)\,c_{cat}\,c_{solv}]_i\,v_i \qquad (6.41)$$

where v_i is the relative volume of each "region."

Equation (6.41) shows that degradation can occur at differing
rates in different parts of a polymer article, i.e., there is particular
"stepwise" chemical degradation in solid polymers.

This effect was first described in [10] for radical reactions
in the solid phase. Under isothermal conditions there is an apparent overstepping of the reaction rate at a certain degree of dissociation of the total number of chemically unstable groups, while
with the increase of the temperature the reaction begins afresh and
finishes at a higher degree of dissociation.

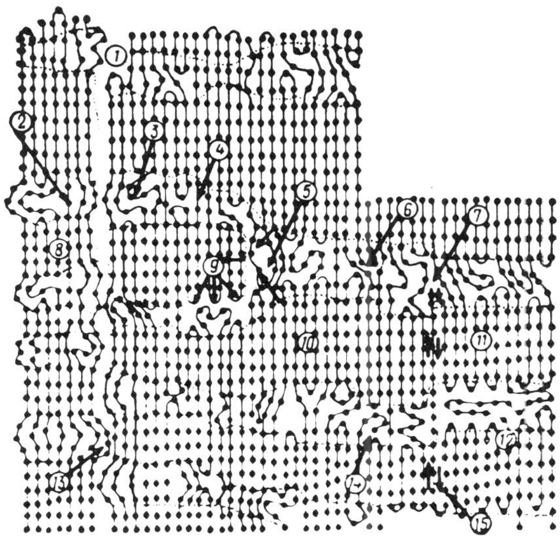

Fig. 6.2. Faulty structure of macromolecular crystals [12]: 1) voids; 2) the Statton model; 3) amorphous regions; 4) straight chains; 5) crystallites of block material; 6) a long fold (as proposed by Flory); 7) a migrating fold; 8) individual fibrils (cold drawing); 9) a short fold (as proposed by Keller); 10) paracrystalline laminar lattice; 11) cluster fibrils (hot drawing); 12) monocrystals; 13) four-point pattern; 14) chain end; 15) region of slipping.

Figure 6.1 shows the kinetic curve for the accumulation of carboxyl groups which are formed during the hydrolytic degradation of a film of polycaproamide at 60°C in water. Under these conditions there decomposes less than 1% of all the amide groups [11]. At a higher temperature (>150°C) there is, to a perceptible degree, a decomposition of most of the remaining amide groups within the same time.

Let us consider the principal causes of stepwise chemical degradation.

6.3.1. Accessibility of Chemically Unstable Groups in Polymers

Generally, there may be, in regions of the polymer with differing structural differentials, differing concentrations of the components of the aggressive medium (c_{cat} and c_{solv}) around the chemically unstable groups, which leads to differing rates of chemical degradation. Moreover, it may be that the rate constants of decomposition of the chemically unstable groups in these regions will differ. An example of how k differs considerably within a single region is considered below.

Let us consider crystallizing polymers. One possible model of a partially crystalline oriented polymer is shown in Fig. 6.2, which presents different types of ordering and disordering of segments of the macromolecule within the volume [12].

In the amorphous regions the microvoids are larger and, consequently, there may be set up a considerably higher average concentration of the components of the aggressive medium than in the ordered regions.

All the functional groups in the polymer, including the chemically unstable groups, may be divided conventionally into "accessible" and "inaccessible" in relation to the first functional groups, the concentration of the components of the aggressive medium around these being much higher than around the other functional groups.

Figure 6.2 shows that as "accessible" groups we can assume not only functional groups in the amorphous regions but also groups on the surface of the lamellae.

Quantitatively the "accessibility" of the functional groups in the polymer is expressed as follows:

$$d = \frac{n}{n_0} \qquad\qquad (6.42)$$

where n_0 and n are, respectively, the total amount of the functional groups and the amount of accessible groups.

Accessibility is always defined in relation to a particular reagent. For instance, within one and the same region the functional groups of a polymer may react with water molecules and not react with certain acid molecules. It is thus natural to expect that d depends not only on the nature of the reagent but also on the external conditions.

Let us consider some examples of determination of the accessibility of the functional groups of the polymer.

Our investigation [13] relates the influence of the structure of cellulose on the accessibility of glucose groups to acid-catalyzed degradation.

In [14] there is a detailed study of the structure of the accessible regions in cellulose. There was added to the cellulose a

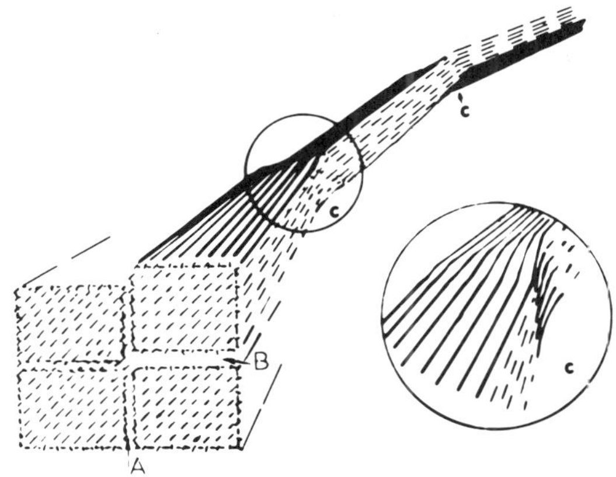

Fig. 6.3. Schematic representation of the surfaces of crystalline and nonordered segments of cellulose fibrils (see text [15]).

TABLE 6.1. First-Order Rate Constants of Formation of Various Substituted D-glucopyranose Residues in Acid-Catalyzed Hydrolysis of Cellulose at 25°C in 2.5 N HCl

Glucopyranose residues	Nonordered segments	Crystalline segments	Glucopyranose residues	Nonordered segments	Crystalline segments
	$k \cdot 10^3$, sec^{-1}	$k \cdot 10^6$, sec^{-1}		$k \cdot 10^3$, sec^{-1}	$k \cdot 10^6$, sec^{-1}
Unsubstituted	1.1	6.5	3-O-Substituted	1.9	38.6
2-O-Substituted	4.8	15.8	6-O-Substituted	3.9	16.7

quite small amount of diethylaminoethyl substituent, which caused practically no change in the structure of the polymer (as shown by x-ray diffraction analysis) and which it was easy to determine quantitatively in soluble and insoluble products of acid-catalyzed hydrolysis by means of GLC. This method made it possible to determine the presence of O-diethylaminoethyl groups in the 2, 3, and 6 positions of the D-glucopyranose residues.

The hydrolysis reaction takes place on the surface of the fibrils, the authors distinguishing two types of accessible surfaces, which they call crystalline and nonordered segments.

Table 6.1 shows the effective first-order rate constants of the formation of various substituted D-glucopyranose residues during the course of acid-catalyzed hydrolysis of cellulose at 25°C in 2.5 N HCl.

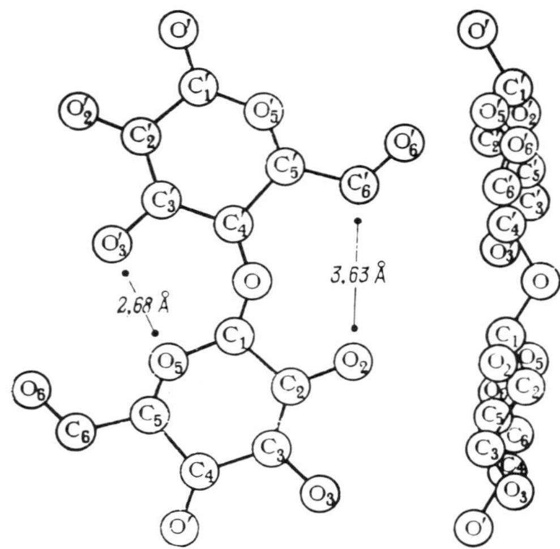

Fig. 6.4. Conformation of a cellobiose segment of a crystalline cellulose chain (see [15]).

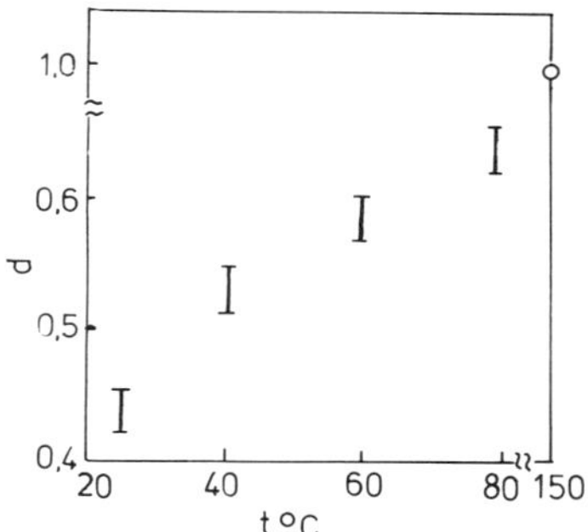

Fig. 6.5. Accessibility of amide groups d in PK-4 polycaproamide as a function of temperature.

This difference in the reactivity of the glycoside group is brought about, according to Miller [15], by the higher accessibility of the groups for protonation in the nonordered segments and greater ease of transition from the chair conformation to the half-chair conformation (the limiting stage of the hydrolysis reaction).

The distribution of accessible surface between nonordered and crystalline segments has been determined from data on the distribution of 2-O-substituted D-glucopyranose residues in these regions.

This ratio is 0.26:0.74 in the initial period of reaction; it changes in the course of hydrolysis to 0.36:0.64 because of recrystallization of the cellulose.

Surfaces of crystalline and nonordered segments of fibrils are shown diagrammatically in Fig. 6.3. In the crystalline regions (surfaces A and B) the segments are oriented and the hydroxyl groups are in regularly repeating positions, with the hydroxyl group in position 3 linked by a strong hydrogen bond to the oxygen in the ring of the adjacent D-glucopyranose residue (Fig. 6.4). In the nonordered regions (surface C) there are concentrated the main part of the defects of the crystalline structure, which are caused by twisting, flexure of the fibrils, and internal stresses. The hydrogen bonds $O_3'H—O_5$ in these regions are the least strong, and this increases the rate constant of formation of 3-O-substituted glucopyranose residues.

For polymers with labile hydrogen atoms one convenient method of determining the accessibility of the functional groups is the reaction of H—D exchange since in the course of this reaction the concentration of functional groups remains unchanged.

In [6], using the change in absorption intensity at 2480 cm^{-1} (νND) (IR spectroscopy), the accessibility of the amide bonds in a film of polycaproamide in relation to D_2O was determined. Figure 6.5 shows how d changes with temperature. With increase of the temperature from 25 to 80°C the accessibility increases and becomes greater than the proportion of the amorphous phase (the degree of crystallinity was determined by x-ray diffraction analysis). The result may be explained by deuteration of the amide groups in the faulty structures on the surface of the crystalline formations.

At 150°C, all the amide groups, including those in the ordered regions, are accessible to D_2O [17].

6.3.2. Reactivity of Chemically Unstable Groups in Polymer Articles

Generally, chemically unstable groups in different structural regions must have differing reactivity.

In amorphous polymers in which the accessibility of the functional groups is, as a rule, practically identical, the differences in the values of k are not great for the different regions of the polymer, and we may therefore use the parameter of the mean rate of the degradation process, which is equal to the ratio of the overall rate of the process to the total volume of the polymer article:

$$\overline{w} = \frac{\sum\limits_{i} [k\,(c_n^0 - c_n)\,c_{\mathrm{cat}}\,c_{\mathrm{solv}}]_i V_i}{\sum\limits_{i} V_i} \tag{6.43}$$

In crystalline polymers there may be, even among the accessible groups, unstable groups with differing reactivity. For instance, in cellulose there are two types of glucoside bonds, differing in reactivity toward acid-catalyzed degradation: highly reactive glucoside bonds in the tips of the folds and those in the linear portions of the folds [18].

6.4. PHYSICAL MODEL OF THE PROCESS OF DEGRADATION OF POLYMER ARTICLES

The investigation of the kinetics of degradation of polymer articles in aggressive media involves great difficulties since the reaction takes place in the condensed phase and the reagents (the components of the aggressive medium and the chemically unstable groups of the macromolecules) are in close contact with the surrounding fragments of the macromolecules and the low-molecular particles.

Although at present there is a lack of precise molecular theories of electrolyte solutions in solid polymers, in some cases there are a number of semiempirical models which make it possible, at the molecular level, to consider approximately the kinetics of the degradation reactions of chemically unstable groups in polymers.

Let us consider the simplest degradation reaction; the interaction of chemically unstable groups of a polymer R with a catalyst cat. With the collision of R and cat in the condensed phase these particles will collide with one another for a more or less prolonged time. The average lifetime of such a pair can be determined approximately from the theory of random processes [19, p. 423 of Russian translation]:

$$\overline{\tau} \approx \frac{2r^2}{6D_{\mathrm{cat}}} \exp\left(-\frac{E_S}{kT}\right) \tag{6.44}$$

where r is the distance between R and cat in the couple, and E_S is the energy necessary to remove cat from R by a distance 1.7r in medium S.

For $E_S = 0$, we can estimate the limiting value of $\overline{\tau}$ in solution and within the matrix of a solid polymer for typical values of D_{cat} in these media, assuming $r = (4 \pm 1)$ Å:

	D_{Cat}, cm^2/sec	$\bar{\tau}$, sec^{-1}
Solution	10^{-5}	10^{-11}
Polymer	10^{-8}	10^{-8}

From the data it may be seen that within the matrix of the solid polymer the average lifetime of the couple R and cat is quite long, and in the couple there is established equilibrium for all degrees of freedom, i.e., these couples can be regarded as kinetically independent particles. The specific kinetics of bimolecular reactions in a matrix of a solid polymer are considered in [10, 20]. In the event of there being only two stages, microdiffusion and kinetic (the catalyst and chemically unstable groups in the polymer article being distributed uniformly), chemical interaction between cat and R begins only when they form a couple in microvoids of volume V*.

As a result of any external activation cat may move relative to R at rate k_t, but if in couples it may react with constant k_p.

Under steady-state conditions (the rates of formation and of destruction of couples being equal) the kinetic equation for the rate takes the form

$$\frac{-dc_m}{dt} = kc_m c_{cat} \tag{6.45}$$

where

$$k = \frac{k_t k_p}{k_t + k_p}\, V^*$$

For $k_t \gg k_p$ the kinetic stage is the limiting one and

$$k = k_p V^* \tag{6.46}$$

For $k_t \ll k_p$, the microdiffusion stage (the frequency of approach) is the limiting one

$$k = k_t V^* \tag{6.47}$$

The movement of cat and R in the polymer matrix is effected as a cooperative process. Accordingly the kinetics of the degradation reactions prove to be linked with the kinetics of the molecular movements, with the physical properties, and with the structure of the polymer.

The character of this relationship is such that the processes of structural relaxation of the macromolecules within the matrix of a solid polymer are hindered, and that within the time for an elementary process of the chemical reaction it is not always possible to bring about the most favorable state of the transition complex, i.e., the elementary reactions take place over a higher potential barrier than the same reactions in liquids.

Such ideas were first put forth by Kooyman and Strang [21]. There are now a number of investigations confirming this postulate.

In [22] a correlation is found between the rate constant of the bimolecular reaction k (the interaction of 2,4,6-tert-butylphenol with hydrogen peroxide groups in polyolefins) and the frequency of rotation ν of the stable nitrosyl radical; this frequency characterizes the mobility of the macromolecules in polymers with differing solvent content. For the reaction of a phenoxy radical with a hydrogen peroxide group it is necessary for the reagents to assume the necessary configuration and to have sufficient energy. The movement of the reagents takes place within the microvoids, and the greater the "stiffness" of the walls of the microvoids the higher the energy which has to be expended for a chemical reaction to take place.

The rate constant of the reaction within the matrix of a solid polymer may be presented as follows:

$$k = p_{\mathrm{prob}} p \exp\left(- \frac{E_{\mathrm{prob}} + E}{RT} \right) \tag{6.48}$$

where p and E are, respectively, the steric factor and the activation energy in the gaseous phase, and $p_{\mathrm{prob}} \exp(-E_{\mathrm{prob}}/RT)$ the probability of rotation of the reagents in the condensed phase for creation of the activated complex configuration.

With increase of the solvent concentration in the polymer there is an increase in the size of the microvoids due to the solvent molecules and a reduction in the "stiffness" of their walls. These two effects lead to a considerable rise in the rate constant and reduction in the activation energy of the chemical reaction. With a solvent content in the polymer around 40 mass %, ν and k cease to depend on the solvent content in the matrix of the polymer and, hence, become equal to the corresponding figures in pure solvent (the number of solvent molecules in the microvoids is sufficient to screen the reagents from the influence of the walls of the microvoids).

It must be noted that in the reactions of chemically unstable groups with aggressive media ion pairs and ions can participate as reagents; where such reactions occur the kinetic parameters may be significantly influenced by the microdielectric permittivity and the specific interactions in the microvoids.

TABLE 6.2. Features of the Occurrence of Chemical Degradation

Features	Internal diffusion-kinetic region	Internal kinetic region	External diffusion-kinetic region
Molecular mass of the polymer (film, thread, section from surface of polymer)	Changes	Changes	Remains practically unchanged
Dependence of the rate of degradation on: the surface of the specimen	Is dependent	Is independent	Is dependent
the volume of the specimen	Is practically independent	Is dependent	Is independent
the structure differentials	Is dependent, change in accessibility, c_{cat} and c_{solv}	Is dependent, change in accessibility, c_{cat}^0 and c_{solv}	Is practically independent
concentration (or activity) of the catalyst and solvent	Depends on c_{cat} and c_{solv}	Depends on c_{cat}^0 and c_{solv}	Depends on h_x and α_{solv} in the solution

There are in polymers, at temperatures below T_g, micropores of various sizes which, as already indicated in Chap. 5, influence the sorption and permeability of the components of the aggressive medium, and this considerably raises the rate of degradation on the surface of the micropores.

All these factors can influence the kinetic parameters in differing ways, making the dependence of k on the solvent content of the polymers often more complex than in the example given above.

Thus, degradation of polymer articles in aggressive media can take place in the following regions:

an internal diffusion-kinetic region (degradation takes place in a reaction zone whose size increases with time and ultimately attains the dimensions of the polymer article;

an internal kinetic region (degradation takes place at an identical rate throughout the volume of the polymer article);

an external diffusion-kinetic region (degradation takes place in a thin surface layer of constant dimension, which at the limit can be a monolayer of the polymer).

Table 6.2 shows the characteristic features of the occurrence of chemical degradation of polymers in each of the three regions.

Where these processes take place in the internal diffusion-kinetic and internal kinetic regions, it is necessary, when determining the kinetic parameters, to take into account the differences in accessibility and in reactivity of the chemically unstable groups in the polymer articles.

REFERENCES

1. S. G. Entelis and R. P. Tiger, Kinetics of Reactions in the Liquid Phase, Khimiya, Moscow (1973).
2. P. V. Danckwerts, Trans. Faraday Soc., 46, 300 (1950).
3. H. S. Carslow and J. C. Jaeger, Conduction of Heat in Solids, Oxford University Press, London (1947).
4. J. Crank, The Mathematics of Diffusion, Oxford University Press, London (1956).
5. R. McGregor and R. H. Peters, J. Soc. Dyers Colour., 81, 393 (1965).
6. D. A. Fridriksberg, Course in Colloid Chemistry, Khimiya, Leningrad (1974).
6a. V. S. Livshits, Vysokomol. Soedin., A, 14, No. 2, 394-402 (1972).
7. M. Shen and M. Bever, J. Mater. Sci., 7, 741 (1972).
8. S. W. Lososki and W. H. Cobbs, J. Polym. Sci., 36, 21 (1959).

9. G. Farrow, D. A. S. Ravens, and J. M. Ward, Polymer, $\underline{3}$, 17 (1962).
10. Ya. S. Lebedev, Kinet. Katal., $\underline{8}$, No. 2, 245 (1967).
11. V. A. Bershtein and L. M. Egorova, Vysokomol. Soedin., A, $\underline{19}$, No. 6, 1260 (1977).
12. R. Hosemann, J. Appl. Phys., $\underline{34}$, No. 1, 25 (1963).
13. S. A. Haworth et al., Carbohydr. Res., $\underline{10}$, 1 (1969).
14. S.P. Rowland and E. J. Roberts, J. Polym. Sci., A-1, $\underline{10}$, 2447 (1972).
15. J. N. Be Miller, Adv. Carbohydr. Chem., $\underline{22}$, 25 (1967).
16. V. S. Markin, Candidate's Dissertation, Inst. Khim. Fiz. Akad. Nauk SSSR, Moscow (1974).
17. P. Schmidt and B. Schneider, Collect. Czech. Chem. Commun., $\underline{31}$, 1896 (1966).
18. M. Chang, T. C. Pound, and R. St. J. Manley, J. Polym. Sci., Polym. Phys. Ed., $\underline{11}$, 399 (1973).
19. S. Benson, Foundations of Chemical Kinetics, McGraw-Hill, New York (1960).
20. V. S. Pudov and A. L. Buchachenko, Usp. Khim., $\underline{39}$, No. 1, 130 (1970).
21. E. Kooyman and A. Strang, Recl. Trav. Chim., $\underline{72}$, 329 (1953).
22. A. P. Griva and E. T. Denisov, Dokl. Akad. Nauk SSSR, $\underline{219}$, No. 3, 640 (1974).

Chapter 7

DEGRADATION

OF

HETERO-CHAIN POLYMERS

At the present time it is the chemical degradation of hetero-chain polymers which has been studied the most completely on a quantitative basis. We shall now consider the approaches to this investigation and the principal results obtained by various authors.

7.1. POLYAMIDES

Polyamides decompose in water and acidic and basic solutions.

The degradation of polycaproamide (PCA) in water was investigated by Rogovin et al. [1]. According to them, at 25°C the rate constant of decomposition of the amide groups in polyamide fibers under heterogeneous conditions in water (k_{H_2O}) is $(7 \pm 1) \cdot 10^{-11}$ min^{-1}. Nevertheless, it has been shown [2] that in PCA (PK-4 film) there are about 1% of reactive amide groups which decompose readily under the action of water. These authors [2] determined the number of scissions from the accumulation of carboxyl terminal groups using differential IR spectroscopy. The results could be expressed by a first-order equation. In the temperature range 60-130°C the degradation process is expressed by the equation

$$k^{*}_{H_2O}(min)^{-1} \approx 1{,}6 \cdot 10^7 \exp\left(-\frac{23\,000}{RT}\right) \qquad (7.1)$$

Apparently the high reactivity of a proportion of the amide groups is brought about by mechanical activation due to residual tensile stresses on individual chains which occur as a result of drawing followed by rapid cooling.

Comparison of k_{H_2O} and $k^{*}_{H_2O}$ as determined from Eq. (7.1) shows that the mechanically activated amide bonds in the PCA differ by more than three orders from the other amide bonds in reactivity.

The degradation of PCA has been studied mainly in acidic media. In concentrated acid solutions PCA dissolves, and degradation has been investigated under homogeneous conditions in solution (Appendix 7.1).

The kinetic parameters for PCA and its oligomers are practically identical whether in H_2SO_4 [3] or in HCl [4, 5]. An exception is the cyclic dimer of PCA, whose lower reactivity is brought about by the high conformational stability of the ring and by the electrostatic interaction of the closely spaced amide groups [3].

These data, and also the linear change in the inverse value of $\bar{M}_n$ as a function of the time of degradation for the initial stages of dissociation, according to Eq. (4.3), show that the degradation of PCA in acid solutions follows a random process [6-9]. The change in k_{eff} (in min^{-1}) of the degradation of PCA in aqueous solutions of H_2SO_4, as a function of the thermodynamic parameters of the medium, is expressed by the equation

$$k_{eff} = 2.5 \cdot 10^9 \exp\left(-\frac{23\,000}{RT}\right) \frac{c_{H_S^+}}{1 + h_A/5.0} \tag{7.2}$$

where $h_A > 5 k_{eff}$ decreases with increase in the acid concentration; at 108°C, in 92% H_2SO_4, degradation does not occur to a perceptible degree over a period of 7 h [10].

The kinetics of degradation of PCA fibers under heterogeneous conditions in 0.05-0.2 N H_2SO_4 at 96 and 101°C are investigated in [11]. The high value of the diffusion coefficient of sulfuric acid in PCA (see Chap. 5) brings about degradation under internal kinetic conditions. With an increase in the degree of orientation of the fibers the rate of degradation decreases, apparently as a consequence of the change in the accessibility of the amide groups with respect to the acid and water.

The influence of the accessibility of the amide groups on the kinetics of degradation of PCA films in HCl vapor is investigated in detail in [12].

The kinetics and mechanism of acid-catalyzed degradation of polyamido acids are described in [13].

The degradation of polyacrylamides under homogeneous conditions in basic solutions was discussed in Chap. 3.

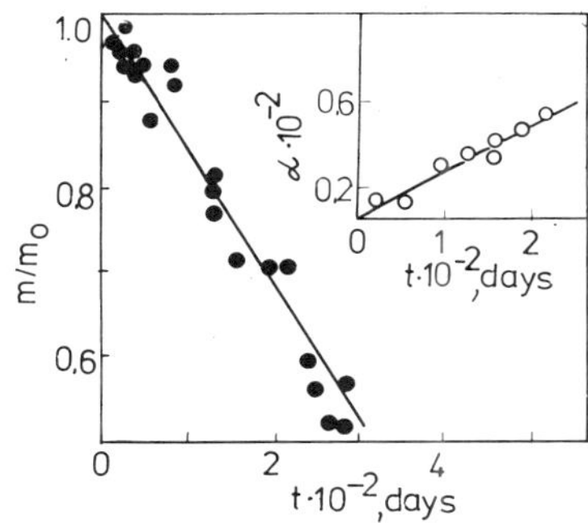

Fig. 7.1.   Change in the mass of a filament and
            in the degree of dissociation of amide
            groups in PCA as a function of the
            time of implantation in the subcutane-
            ous tissue of rabbits [14].

The macrokinetics of degradation of fibers of PCA and polydodecanamide were investigated in the subcutaneous tissue of rabbits
by Daurova et al. [14].   In the organism, polyamides degrade both
from the surface [the reduction in mass of the implants is expressed
by Eq. (6.30)] and throughout the whole volume, under internal kinetic conditions (Fig. 7.1).   The values of $k_{rand} \cdot c_{cat}$ at 37°C for
amide groups in the amorphous regions of the polymer, for PCA and
polydodecanamide, are, respectively, $(2.7 \pm 0.3) \times 10^{-5}$ and $(5 \pm 1) \times$
10^{-5} day^{-1}.

7.2. POLYESTERS

Most investigations have been on the chemical degradation of polyethyleneterephthalate (PETP), which decomposes in water, acids, and
bases.

The degradation of PETP in water is studied in [15].   Water
diffuses in PETP at 25°C with a diffusion coefficient around 10^{-9}
cm^2/sec.   The dependence of D_{H_2O} on temperature for films of PETP
with degree of crystallinity approximately 40% is expressed as follows [15]:

$$D_{H_2O} = (4.5 \pm 0.5) \cdot 10^{-2} \exp\left(-\frac{10\,000}{RT}\right)$$

The following shows the change in solubility of water as a function of temperature in PETP films of thickness 100 ± 5 μm of similar
degree of crystallinity:

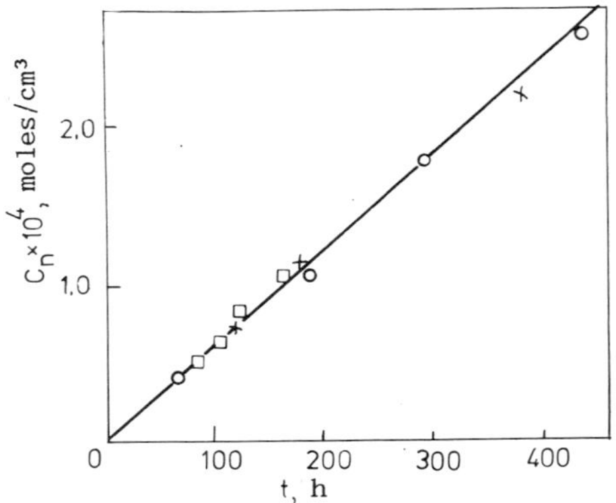

Fig. 7.2. Dependence of the number of scissions
           on the duration of degradation of
           PETP of varying linear density in 5 N
           HCl at 70°C [21]: ○) 0.22 tex; □)
           0.333 tex; +) 0.667 tex.

Temperature, °C	25	37	60	80	100
Solubility, g per 100 g of polymer	0.22	0.29	0.53	0.64	0.71

In [16] there is described the degradation of PETP films of
thickness 125 μm in saturated water vapor at 60-175°C. This takes
place under internal kinetic conditions, and is satisfactorily expressed by Eq. (6.20). It must be noted that Golike and Zasoski [16]
mistakenly suggested that the degradation process takes place under
diffusion-kinetic conditions and is limited by the diffusion of
water into the polymer. Their conclusions are criticized in [17].
If we take the activation energy of the degradation of PETP in water
as 109-113 kJ/mole [18, 19], then k_{H_2O} at 25°C is approximately
10^{-10} min^{-1}, i.e., it can be reckoned that at room temperature PETP
is practically stable in water; over 10 years around 10^{-3} of all
the ester groups decompose.

At high temperatures (>80°C) the degradation of PETP film and
fibers may occur at a perceptible rate in an autocatalytic fashion
[19]. The carboxyl groups which are formed at the terminus of the
polymer fragments and molecules of terephthalic acid may act as
active catalysts of the degradation. The autocatalysis constant
in the degradation of PETP in aqueous solution at 100°C is (7.5 ±
1.5)·10^{-3} cm^3/(mole·min) [20].

The degradation of PETP in acid solutions is investigated in
detail in [15, 21]. In the presence of HCl solutions (1-7 M), the

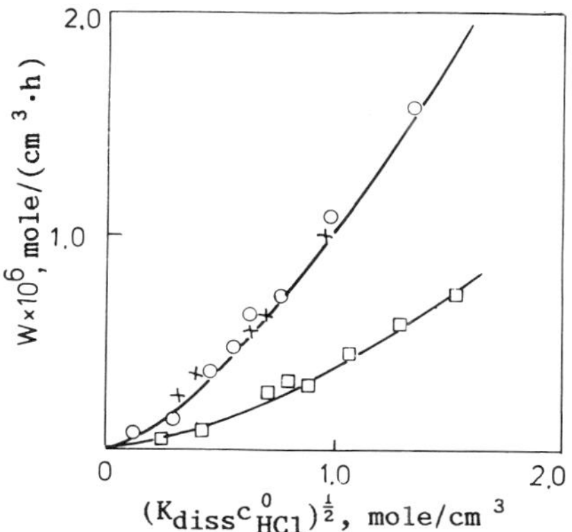

Fig. 7.3. Dependence of the degradation rate of
           PETP filaments of varying structure
           on the concentration of hydrogen
           chloride at 70°C: +) amorphous, non-
           oriented; o) nonoriented, with degree
           of crystallinity 48%; □) oriented,
           with degree of crystallinity 30% [21].

degradation of PETP fibers takes place under internal kinetic conditions [21]. This is confirmed by the absence of the dependence of
the degradation rate on the diameter of the filament (Fig. 7.2).
The influence of the concentration of HCl on the degradation rate
is shown in Fig. 7.3.

If we assume that the HCl in the polymer is a weak acid (the
ε value of the polymer saturated with water is approximately 4) [21],
then the proton concentration in the PETP $(c_{H^+}^0)$ will be

$$c_{H^+}^0 = (K_{diss}c_{HCl}^0)^{1/2} = (K_{diss}K_{distr}c_{HCl})^{1/2} \qquad (7.3)$$

where K_{diss} is the dissociation constant of HCl in the polymer, and
K_{distr} is the distribution constant of HCl between the polymer and
the external phase, as determined from the sorption isotherms.

Figure 7.3 shows the correlation between the degradation rate
and the square root of the concentration of HCl in the solution
(c_{HCl}).

The relationship obtained by Ravens should be regarded with
caution since K_{diss} can change as a function of the concentration
of water in the amorphous portion of the polymer and of the structure differentials of the polymer (the degrees of crystallinity and
orientation).

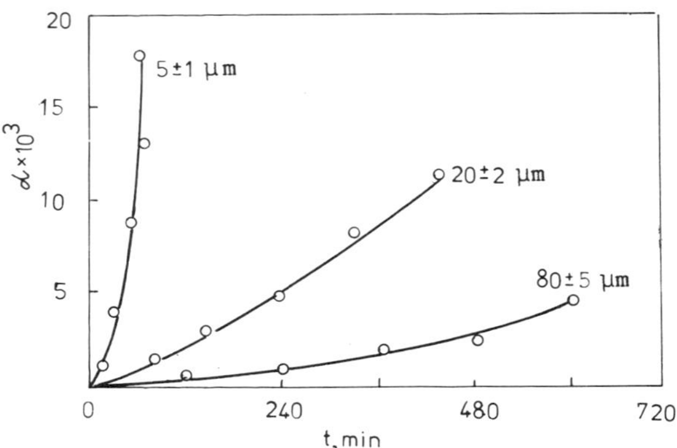

Fig. 7.4. Dependence of the degree of dissociation of
          ester groups on the duration of degradation
          of PETP films of various thicknesses in 53%
          H_2SO_4 at 116°C [15].

The activation energy of the degradation process was 92 kJ/mole.

Investigation of the degradation of PETP films in aqueous solutions of sulfuric acid shows [15]:

with contact of the PETP films with aqueous solutions of sulfuric acid the diffusion of water in the polymer takes place far more rapidly than the diffusion of sulfuric acid;

the diffusion coefficient of water in PETP does not depend on the concentration of the sulfuric acid, and changes with the temperature according to Eq. (7.2);

the solubility of water in the PETP decreases with increase in the concentration of sulfuric acid.

Attempts to determine the diffusion coefficient of sulfuric acid in PETP were unsuccessful because of the very low solubility of the sulfuric acid.

Figure 7.4 shows the typical dependence of the degree of dissociation of the ester groups on time, for the degradation of PETP films of various thicknesses at 116°C in 53% H_2SO_4.

The considerable change in the molecular mass (MM) of PETP specimens during the course of degradation is evidence that degradation takes place in a particular reaction zone which changes with time. Generally we cannot exclude the additional occurrence of the degradation process from the surface of the specimen since the ester groups on the surface make contact with a larger number of acid molecules than do the ester groups within the PETP.

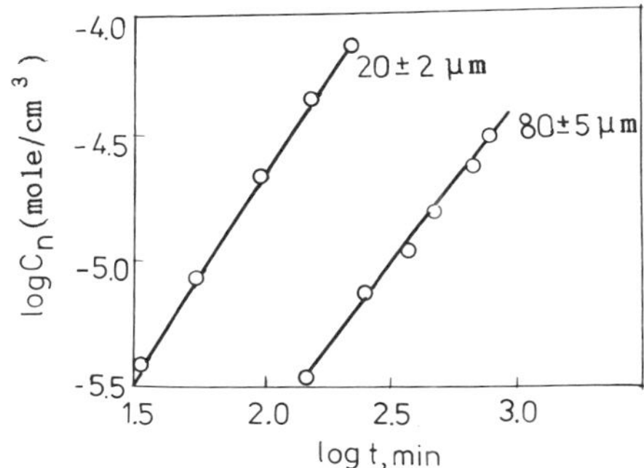

Fig. 7.5. Dependence of log C_n on log t for the
degradation of PETP films of various
thicknesses in 53% H_2SO_4 at 116°C [15].

The practically identical slope of the α vs. t relationship
with prolonged degradation of PETP films of thicknesses 5 and 20
μm indicates that degradation of PETP films in acid takes place
mainly in the reaction zone and to a far lower degree from the surface. Sulfuric acid is practically not removed during the course
of degradation since neither the ester groups in the PETP nor the
degradation products are protonated to a perceptible extent in the
sulfuric acid solutions in question [22, p. 195]. During degradation, according to x-ray structural analysis data, the structure
of the PETP films remains practically unchanged. In handling the
experimental results, Eqs. (6.14) and (6.20) were used. For a PETP
film 5 μm thick there is, after 30 min exposure, a linear dependence
of α on t (Fig. 7.4). From Eq. (6.20) there can be found the product

$$k_{eff}c^0_{H_2SO_4} = (1.0 \pm 0.1) \cdot 10^{-8} \, sec^{-1} \, equiv \cdot cm^{-3} \tag{7.4}$$

Using this value, it is possible to determine from Eq. (6.14) the
value of $D_{H_2SO_4}$ (Fig. 7.5). For PETP films of various thicknesses
we get differing values for the diffusion coefficients [for a film
of thickness 20 μm: $(4.0 \pm 0.5) \cdot 10^{-12}$ cm²/sec, and for a film of
thickness 80 μm: $(2.0 \pm 0.5) \cdot 10^{-12}$ cm²/sec]. This may bring about
differences in the structure of films of differing thicknesses.

X-ray investigations carried out in this work show that the
films differ in their crystalline structure at long-range order and
relative content of paracrystalline phase. In addition, differing
values of $D_{H_2SO_4}$ may be caused by the presence of degradation from
the surface of the specimen. The average values of the diffusion
coefficients of sulfuric acid in PETP film at various temperatures
are as follows:

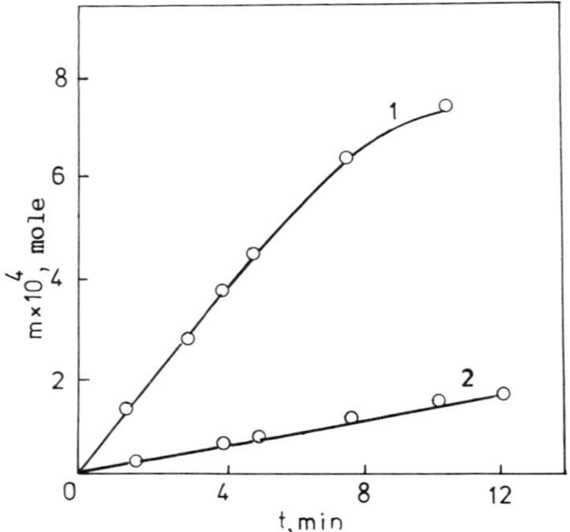

Fig. 7.6. Kinetics of the accumulation of terephthalic acid in the degradation of
a PETP film of surface area 4 cm² in
10 cm³ of a potassium hydroxide solution at 108°C [25]: 1) 49% KOH; 2) 22%
KOH.

Temperature, °C	90	116	126
$D \cdot 10^{12}$, cm²/sec	1.5 ± 0.5	3.0 ± 1.5	7 ± 2

The activation energy of the diffusion of sulfuric acid in PETP
film is 50 ± 8 kJ/mole. In the work in question, it was not found
possible to determine k_{eff} and $c_{H_2SO_4}^0$ separately. Table 7.1 shows
the products of these values for differing temperatures and concentrations of sulfuric acid in the solution.

Thus, the degradation of PETP films in aqueous solutions of
sulfuric acid is described by the equation

$$c_n = k_{eff} c_{H_2SO_4}^0 t \left\{ 1 - \frac{8}{\pi^2} \sum_{m=1}^{\infty} \left[1 - \right. \right.$$

$$\left. \exp(2m-1)^2 \frac{\pi^2 t}{l^2} 2.9 \cdot 10^{-7} \exp\left(-\frac{12\,000}{RT}\right) \right] \frac{4l^2}{(2m-1)^4 \pi^2 t \cdot 2.9 \cdot 10^{-7} \exp\left(-\frac{12\,000}{RT}\right)} \right\} \quad (7.5)$$

The degradation of PETP in basic solutions is investigated in
detail in [23, 25]. The following results were obtained:

a zero order of reaction with respect to the polymer (Fig. 7.6)
under the condition of removal of the degradation products, which
can accumulate on the surface and retard the degradation process;

TABLE 7.1. Values of the Product of $k_{eff} \times c_{H_2SO_4}^0 \cdot 10^8$ ($sec^{-1} \cdot equiv \cdot cm^{-3}$) for Various Temperatures and Concentrations of Sulfuric Acid

Concentration of	Temperature, °C		
H_2SO_4 , mass %	90	116	126
53.0	0.3	1.7	4
60.0	0.7	3.7	8
70.0	1.6	8.0	17

a constant MM of the PETP film right up to the termination of the process;

the absence, in the multiply frustrated total internal reflectance spectra, in the IR region, of the PETP films, of absorption bands of the carboxyl groups which are formed by the decomposition of the ester groups.

All this indicates that the degradation process takes place within a thin reaction layer which remains unchanged with time, i.e., practically from the surface of the specimen. It may be supposed that the PETP fragments which form will pass into solution and decompose under homogeneous conditions. However, oligomers of PETP ($\bar{M}_n$ from 500 to 800) do not dissolve in water, and it may be assumed that the degradation of PETP in aqueous basic solutions practically does not take place under external kinetic conditions.

During degradation, a rough surface is formed. Changes in the surface of PETP films can be assessed as follows. The microrelief obtained on a profilogram can be represented to a first approximation as an aggregate of cones, the curved surface of which, s_S, is determined from the formula

$$s_S = \pi \bar{r} \sqrt{Ra^2 + \bar{r}^2} \qquad (7.6)$$

where $\bar{r}$ and Ra are, respectively, the mean radius of the base and the height of the cone.

From a comparison of profilograms obtained at various time intervals it was apparent that there is practically no change either in the number of cones on the surface or in the value of $\bar{r}$. Then the relative change in the surface area is determined as follows:

$$\frac{s_S}{s_0} = \sqrt{\left(\frac{Ra}{\bar{r}}\right)^2 + 1} \qquad (7.7)$$

where s_0 is the area of the base of the cone.

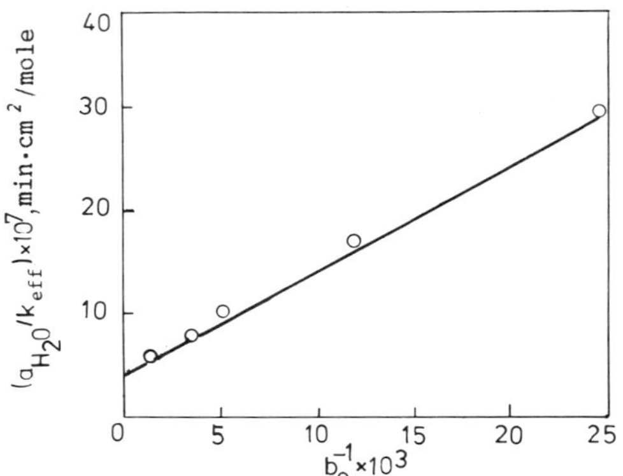

Fig. 7.7. Experimental data for the degradation
           of PETP films at 25°C in aqueous so-
           lutions of KOH plotted on the coordi-
           nates of Eq. (7.9).

Thus, from Eq. (6.22) it is possible to calculate the effective
rate constant

$$k_{eff} = k_{(surf)}c_{cat(surf)}c_{solv(surf)}$$

For a comparison of k_{eff} with the thermodynamic parameters of the
medium it is necessary to determine $c_{cat(surf)}$ and $c_{solv(surf)}$. In
[25] the concentration effect (cf. Chap. 6) was reduced to a minimum
by agitating the solution, and the adsorption effect was estimated
from Eq. (6.29).

It has been shown that instead of $c_{cat(surf)}$ and $c_{solv(surf)}$ it
is possible to use the volumetric thermodynamic parameters of the
medium.

As a consequence of the high reactivity of compounds with ester
groups their hydrolysis has been investigated under homogeneous conditions only in dilute basic solutions [26, 27]. On the basis of
the data it is not possible to draw a clear conclusion as to the
mechanism of hydrolysis of compounds with ester groups. Under heterogeneous conditions the hydrolytic degradation of PETP takes place
quite slowly, and this makes it possible to obtain kinetic data in
concentrated basic solutions and on this basis to solve the question
of the mechanism of the hydrolysis of esters.

Generally, the hydrolysis of esters can, by analogy with amides,
take place by the following three mechanisms:

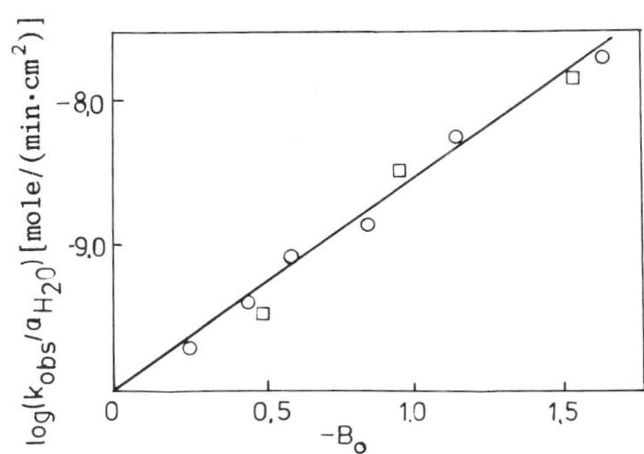

Fig. 7.8.   The dependence of log k_{eff}/a_{H_2O} on B_0 for the degradation of PETP filaments in aqueous solutions of KOH ($\bigcirc$), NaOH ($\square$), and LiOH ($\bullet$) at 25°C.

To these mechanisms there correspond the equations (not taking into account the activity coefficients)

$$k_{eff} = \frac{k_{act_1}/b_0}{1 + K_{eq}} \qquad (7.8)$$

$$k_{eff} = \frac{k_{act_2} a_{H_2O}}{1 + K_{eq}} \qquad (7.9)$$

$$k_{eff} = \frac{k_{act_3}}{1 + K_{eq}/b_0} \qquad (7.10)$$

Equations (7.8) and (7.10) are kinetically indistinguishable; Eq. (7.9) includes the thermodynamic activity of the water, which makes it possible to identify easily the mechanism of hydrolysis in concentrated basic solutions, in which the thermodynamic activity of the water changes considerably. Figure 7.7 shows the experimental data plotted on the coordinates of Eq. (7.9) for the degradation of PETP in aqueous solutions of KOH. The values of K_{act} and K_{eq} at 25°C are $(4.0 \pm 0.5) \cdot 10^{-8}$ mole/(min·cm^2) and $(4.0 \pm 0.5) \cdot 10^2$ mole × (min·cm^2)$^{-1}$, respectively [25].

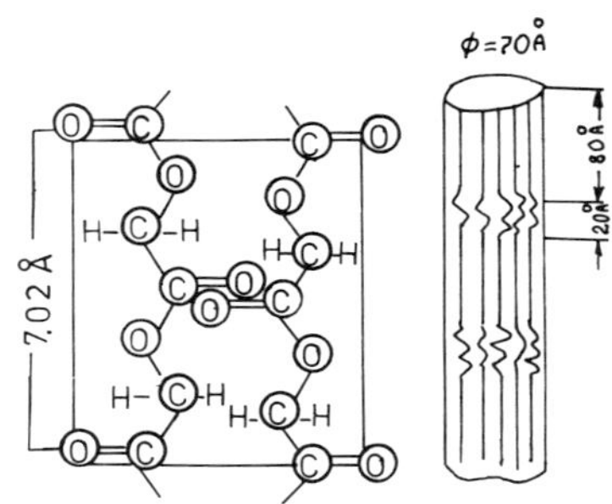

Fig. 7.9. Structure of the elementary unit of
crystalline polyglycolid (Dexon)
(Exxon Corp.) and diagram of the
microfibrils of PG fibers, or crystallites in the form of an ellipsoid.
d = a + z; d) major axis; a) crystallite; z) amorphous intermediate.
d ≅ 100 Å; a = 80; z = 20 Å.

Thus, the hydrolysis of the ester groups in PETP is by a mechanism analogous to that of the hydrolysis of amides in basic solutions, rapid equilibrium attachment of a hydroxide ion to the carbonyl group of the ester group with subsequent slow attack of the
ionized form by a water molecule.

According to the mechanism put forth, the nature of the base
should not influence the kinetic parameters. This is shown in Fig.
7.8, which presents a relationship, plotted on the coordinates of
Eq. (7.11), which is a special case of Eq. (7.9) [24]:

$$\log \frac{k_{eff}}{a_{H_2O}} + B_0 = \log \frac{k_{act}}{K_{eq}} \tag{7.11}$$

The following are the equations of the change in mass of film
and fibers of PETP with time, for degradation in aqueous basic solutions:

$$m_n = m_n^0 - \frac{6,4 \cdot 10^6 \exp\left(-\dfrac{16\,500}{RT}\right) a_{H_2O}\, st}{1 + \dfrac{4 \cdot 10^2}{b_0}} \tag{7.12}$$

$$m_n = m_n^0 \left[1 - \frac{6,4 \cdot 10^6 \exp\left(-\dfrac{16\,500}{RT}\right) a_{H_2O}\, t}{\left(1 + \dfrac{4 \cdot 10^2}{b_0}\right) \bar{r}_0 \rho} \right]^2 \tag{7.13}$$

where $\bar{m}_n$ is given in g, t in min, s in cm^2, and $\bar{r}_0$ in cm.

In connection with the problem of creating materials which dissipate rapidly in the living organism, a detailed study has been made of the degradation of highly crystalline polyglycolid (PG) [14]. The polymer filaments consisted of fibrils separated by amorphous portions (Fig. 7.9).

The polyglycolid molecule, $\sim O\!-\!\overset{\overset{\displaystyle H}{|}}{\underset{\underset{\displaystyle H}{|}}{C}}\!-\!\overset{\overset{\displaystyle O}{\|}}{C}\!-\!O\!-\!\overset{\overset{\displaystyle H}{|}}{\underset{\underset{\displaystyle H}{|}}{C}}\!-\!\overset{\overset{\displaystyle O}{\|}}{C}\!-\!O\!-\!\overset{\overset{\displaystyle H}{|}}{\underset{\underset{\displaystyle H}{|}}{C}}\!-\!\overset{\overset{\displaystyle O}{\|}}{C}\!-\!O\!\sim$, has a flat zigzag shape [27a, 27b]. Because of the planar structure of the macromolecule a very high packing density is achieved, with the close approach of the ester groups of neighboring chains. The degree of crystallinity of Dexon fibers, according to small-angle x-ray diffraction data, is 80-90% [27a, b]. The crystallites, which have the form of an ellipsoid (elliptical prism) with axes measuring 70 and 80 Å, are linked to each other by entangled sections of chains, "straight-through" chains, forming amorphous portions, the proportion of which, according to x-ray structural analysis, is 10-20 mass % [27a, b].

The degradation of PG in water, unlike the degradation of PETP, takes place at a considerable rate. Figure 7.10 shows a typical kinetic curve of the change in relative mass of PG fibers, Fig. 7.10a the typical dependence of the tensile stress of an elementary filament of Dexon fibers on the time of implantation [27c]. On the initial section of the kinetic curve the mass of the fiber changes only slightly, and the diameter of the elementary filament remains practically unchanged. After a certain time (about 30 days), there is a considerable increase in the rate of the process, and the filaments, while not changing in diameter, begin to separate into small sections.

The mechanical strength of these fibers begins to fall sharply after 10 days (see Fig. 7.10a) and after 20-25 days falls to zero [27c]. Moreover, the relative deformation of the elementary filament falls, and fracture becomes more brittle (Fig. 7.10b) in nature.

Taking into account that the sorbed water is localized mainly in the amorphous regions, it is concluded that degradation takes place both throughout the volume of the amorphous regions and from the surface of the crystallites [14].

Figure 7.10c shows electron micrographs of the initial fiber (1) and of specimens which are in the organism for 13 (2), 25 (3), and 28 (4) days. It may be noted that the initial fiber has a smooth surface. As the time of implantation lengthens there takes place, under the action of the medium of the organism, a breakdown of the morphology of the surface, which is preceded by the appearance of

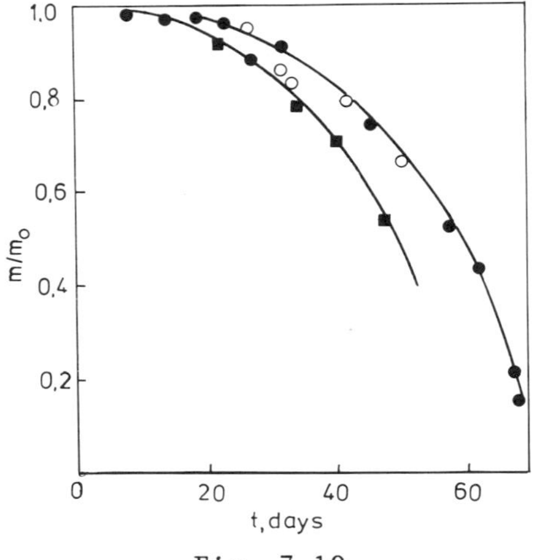

Fig. 7.10

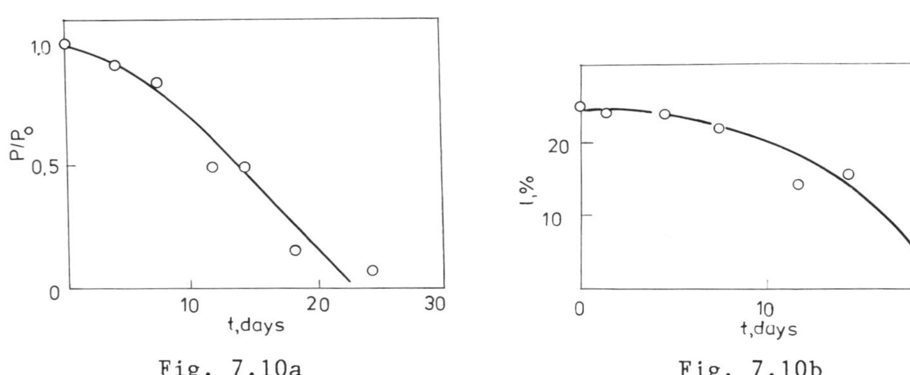

Fig. 7.10a   Fig. 7.10b

Fig. 7.10.   Kinetics of the change in relative mass of PG fibers at
             37°C in water (●) (initial pH 6.5) and in buffer solutions
             with pH 4.05 (○) and 2.01 (■) [27c].

Fig. 7.10a. Dependence of the tensile strength of an elementary fil-
            ament of Dexon (PG) fibers on the time of implantation.

Fig. 7.10b. Dependence of the relative deformation of an elementary
            filament of Dexon fibers on the time of implantation.

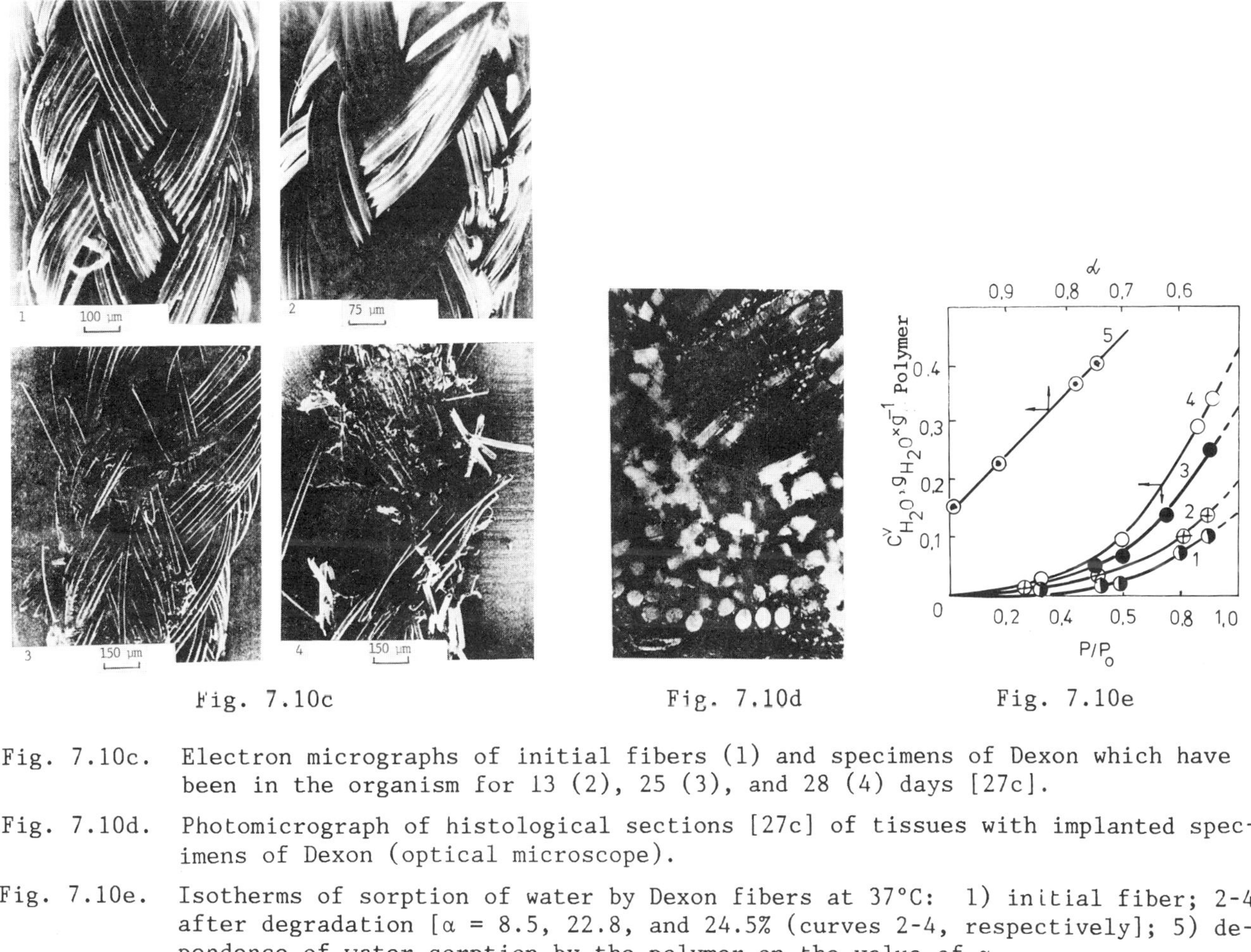

Fig. 7.10c

Fig. 7.10d

Fig. 7.10e

Fig. 7.10c. Electron micrographs of initial fibers (1) and specimens of Dexon which have been in the organism for 13 (2), 25 (3), and 28 (4) days [27c].

Fig. 7.10d. Photomicrograph of histological sections [27c] of tissues with implanted specimens of Dexon (optical microscope).

Fig. 7.10e. Isotherms of sorption of water by Dexon fibers at 37°C: 1) initial fiber; 2-4) after degradation [α = 8.5, 22.8, and 24.5% (curves 2-4, respectively]; 5) dependence of water sorption by the polymer on the value of α.

individual random transverse defects which then lead to lamination
of the elementary filaments into small fragments (20-100 μm). At
the same time, the diameter of the filaments remains practically
unchanged in all cases. Similar behavior occurs in the investigation of histological sections of tissues with implanted specimens
of Dexon, using an optical microscope (Fig. 7.10d).

The amorphous regions contain fragments of polymer chains with,
in general, two ester groups. Since the only soluble decomposition
product is a monomer (hydroxyacetic acid), the degree of dissociation by mass is

$$\alpha = \frac{m}{m_0} = 1 - 0.2\left[1 - \exp\left(-k_{rand}t\right)\bar{c}^{v}_{H_2O}\right]^2 \tag{7.14}$$

where m and m_0 are, respectively, the present and initial mass of
the polymer; $\bar{c}^{v}_{H_2O}$ the average concentration of water in the amorphous portion of the polymer; 0.2 the volume proportion of amorphous
regions in PG.

The experimental data are satisfactorily explained by Eq.
(7.14) right up to a degree of dissociation of 0.9, but with lower
values of α the contribution of degradation from the surface of
the crystallites becomes considerable (Fig. 7.11).

It is interesting to note that with α = 0.9 the PG filaments
lose practically all their strength because of considerable decomposition of the fragments of those macromolecules which contain
crystallites. This particular factor explains the separation of
the filaments into individual sections.

The degradation of the crystallites proceeds mainly from their
curved surface since, apparently, water dissolves, only to a very
slight degree, in the crystallites, and the ester groups which are
accessible to the water are located on the curved surface of the
crystallites.

To a first approximation the crystallites may be represented
in the form of elliptical prisms, the mass change of which is determined from the equation

$$\frac{m}{m_0} = 0,8\left(1 - \frac{k^{S-V}_{eff}t\bar{c}^{S-V}_{H_2O}}{N\bar{b}_0\rho}\right)^2 \tag{7.15}$$

where k^{S-V}_{eff} is the effective rate constant of degradation of the
ester groups on the surface of the crystallites, which depends on
the type of decomposition of the fragments of the PG macromolecules
on the surface of the crystallite and on the adhesion and solubility
of the decomposition products in the surrounding medium; $\bar{c}^{S-V}_{H_2O}$ the
average concentration of water on the curved surface of the crystal-

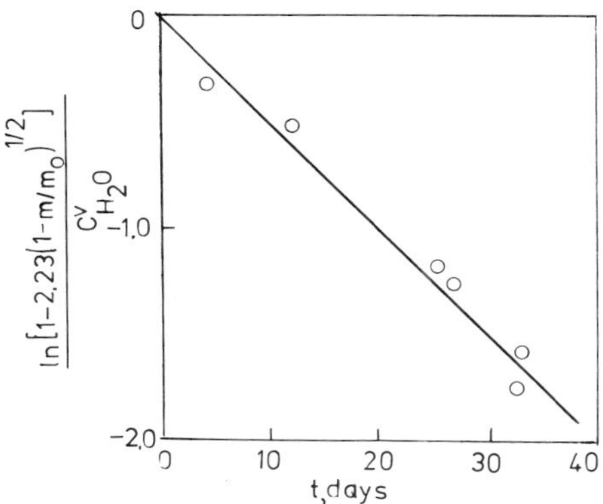

Fig. 7.11. Treatment of the experimental data
            for the degradation of PG fibers at
            37°C in water on the coordinates of
            Eq. (7.14).

lites; $\bar{b}_0$ the average radius of the crystallites, which is equal
to half the shorter axis of the elliptical prism; and N the number
of crystallites in the cross section.

With joint solution of Eqs. (7.14) and (7.15) we get a general
expression for the change in mass of the PG filaments as a function
of time of degradation:

$$\frac{m}{m_0} = 0.2 - 0.2\left[1 - \exp(-K_{\text{rand}}t)\,\bar{c}_{H_2O}^V\right]^2 + 0.8\left(1 - \frac{k_{\text{эф}}^{S-V}\,t\bar{c}_{H_2O}^{S-V}}{N\bar{b}_0\rho}\right)^2 \qquad (7.16)$$

In order to use Eq. (7.16) it is necessary to determine the depen-
dence of $\bar{c}_{H_2O}^V$ and $\bar{c}_{H_2O}^{V-S}$ on the degree of dissociation of the polymer (Fig. 7.10e) [27d-h].

It has been found experimentally that

$$\bar{c}_{H_2O} = \bar{c}_{H_2O}^0 + p\,(1 - \alpha) \qquad (7.17)$$

where $\bar{c}_{H_2O}$ is given in g/g (of polymer) and p is a coefficient of
proportionality.

Equation (7.17) means that with degradation of the polymer,
whether in the amorphous regions or from the surface of the crystallites, the microvoids which are being formed are filled with water.
Moreover, as already noted, the diameter of the elementary filaments
does not change perceptibly.

The concentration $\bar{c}_{H_2O}^{V-S}$ was calculated with the assumption
that it changes in direct proportion to the degree of dissociation
of the polymer in the crystalline regions since, at the same time,

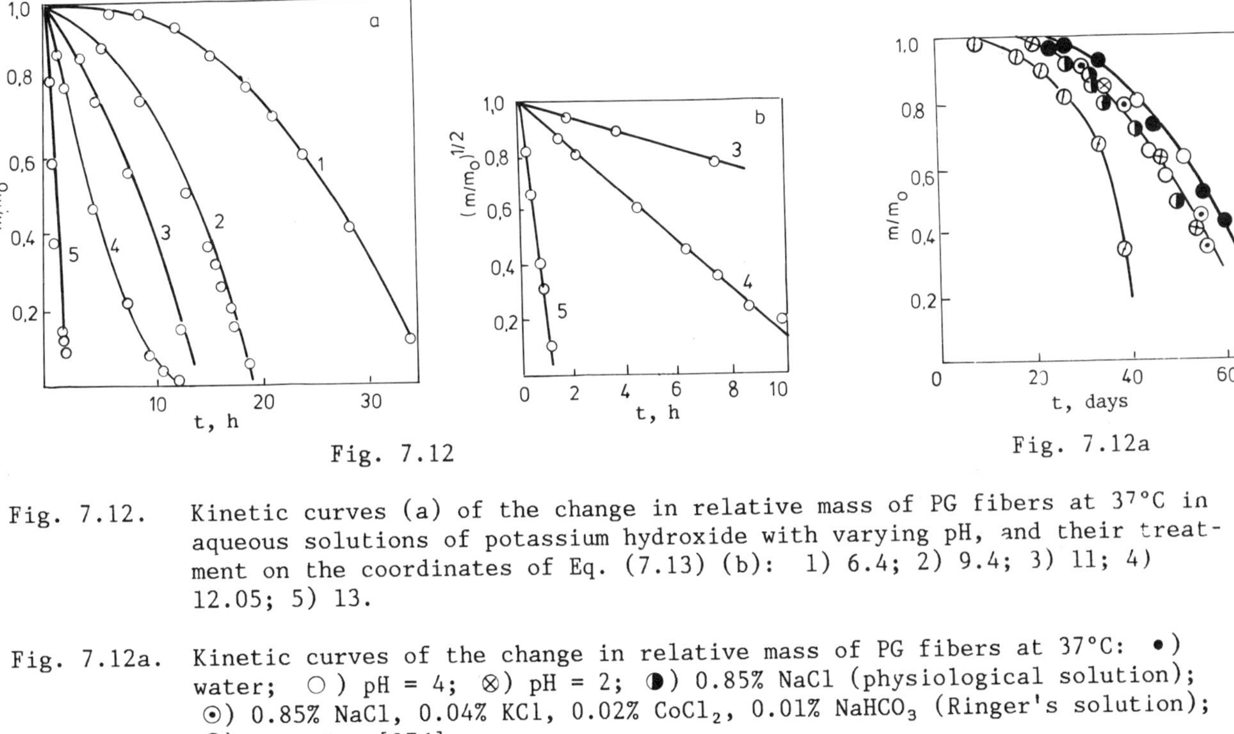

Fig. 7.12

Fig. 7.12a

Fig. 7.12. Kinetic curves (a) of the change in relative mass of PG fibers at 37°C in aqueous solutions of potassium hydroxide with varying pH, and their treatment on the coordinates of Eq. (7.13) (b): 1) 6.4; 2) 9.4; 3) 11; 4) 12.05; 5) 13.

Fig. 7.12a. Kinetic curves of the change in relative mass of PG fibers at 37°C: •) water; ○) pH = 4; ⊗) pH = 2; ◐) 0.85% NaCl (physiological solution); ⊙) 0.85% NaCl, 0.04% KCl, 0.02% $CoCl_2$, 0.01% $NaHCO_3$ (Ringer's solution); ⊕) organism [27d].

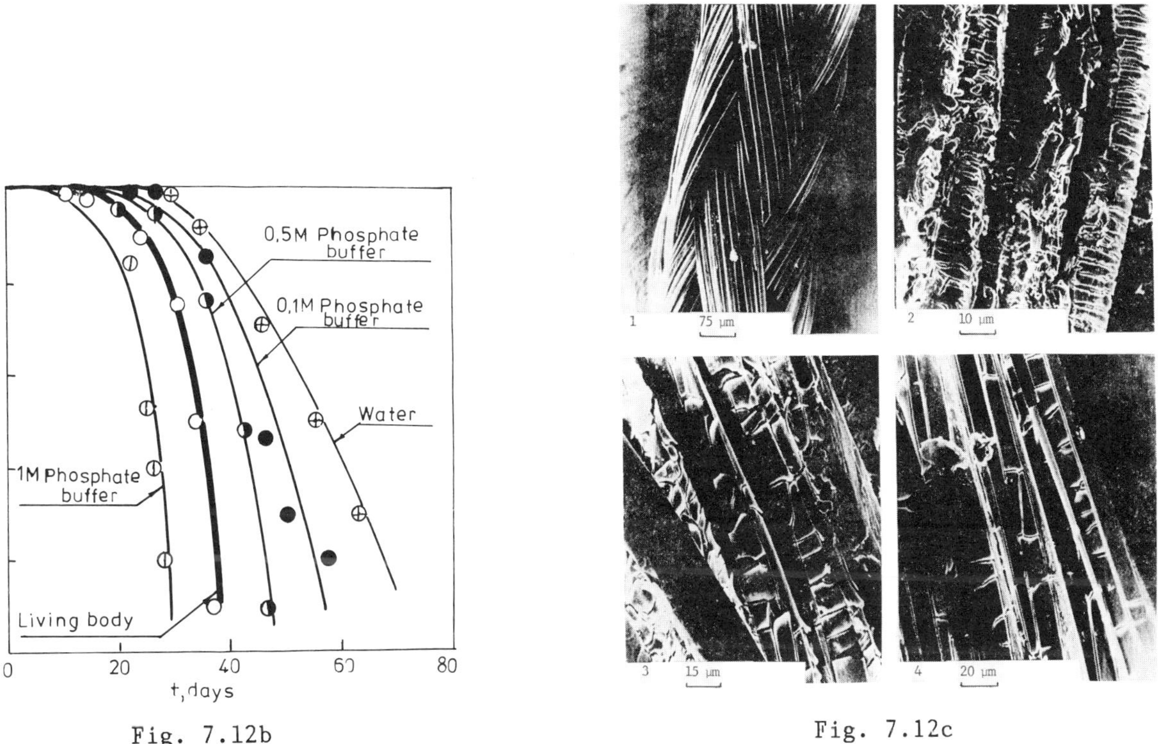

Fig. 7.12b

Fig. 7.12c

Fig. 7.12b. Kinetic curves of the change in relative mass of Dexon fibers at 37°C: ⊕) in water; ●) in 0.1 M phosphate buffer; ◐) in 0.5 M phosphate buffer; ○) in the living organism; ⊕) in 1 M phosphate buffer [27c, d].

Fig. 7.12c. Electron micrographs of Dexon fibers. Original fibers (1); treated fibers (2-4); 2) with water for 5 h at 100°C; 3) with aqueous solution of chymo-trypsin, 20 days at 37°C; 4) with 1.0 N solution of H_2SO_4, 5.5 days at 80°C [27c-g].

there is an increase in the dimensions of the microvoids between the curved surfaces of the crystallites:

$$\bar{c}_{H_2O}^{V-S} = \bar{c}_{H_2O}^{(V-S)_0}\left\{1 - \frac{m}{m_0} - 0{,}2\left[1 - \exp\left(-k_{\mathbf{rand}}t\right)\bar{c}_{H_2O}^{V}\right]^2\right\} \qquad (7.18)$$

where $\bar{c}_{H_2O}^{(V-S)_0}$ is the limiting value of the surface concentration of water for large degrees of dissociation.

The values for k_{eff}^{S-V} and k_{rand} at 37°C are, respectively, $(5 \pm 0.5)\cdot 10^{-2}$ days$^{-1}\cdot$cm$^{-2}\cdot$g and $(1.3 \pm 0.2)\cdot 10^{-2}$ day^{-1}.

The degradation of PG fibers in acidic media was investigated with a view to determining the possibility of autocatalysis of the hydrolysis of the ester groups by the carboxyl groups which form on their decomposition.

As Fig. 7.10 shows, the degradation of PG fibers in buffer solutions with pH values of 4.05 and 2.01 takes place at a rate similar to that of the degradation of the polymer in water; i.e., within the particular pH range the acidic medium has no significant influence on the hydrolysis of the ester groups. Similar behavior occurs in the hydrolysis of various low-molecular esters, e.g., β-lactones [28]. The kinetic curves of the degradation of PG filaments in acid media are described satisfactorily by Eq. (7.16).

The degradation of PG fibers in alkaline media occurs in a complex manner. At pH > 11 the kinetic curves have the same shape as the degradation of PETP filaments in bases (Fig. 7.12); i.e., there is zero order with respect to the polymer, while the mass change is described by Eq. (7.13).

In these solutions, the degradation from the surface of the filaments takes place far more rapidly than under the action of water and hydroxide ions sorbed in the amorphous regions of the polymer. With pH < 11, the degradation from the surface of the filaments and within the amorphous regions takes place at commensurate rates, and this leads to a change in the shape of the kinetic curves (Fig. 7.12). Figure 7.12a shows the kinetic curves for mass loss in acidic solutions, in physiological solutions, and in the organism [27d-h].

The general expression for the change in mass of the polymer takes the form

$$\frac{m}{m_0} = 1 - 0{,}2\left[1 - \exp\left(-k_{\mathbf{rand}}t\right)\bar{c}_{H_2O}^{V}\right]^2 - 0{,}8\left[\frac{2k_{\mathbf{eff}}^{S-V}t\bar{c}_{H_2O}^{S-V}}{N\bar{b}_0\rho}\right.$$

$$\left. - \left(\frac{k_{\mathbf{eff}}^{S-V}t\bar{c}_{H_2O}^{S-V}}{N\bar{b}_0\rho}\right)^2\right] - \frac{2k_{\mathbf{act}}^{S}tc_{OH^-}}{K_{\mathbf{eq}}\bar{r}_0\rho} + \left(\frac{k_{\mathbf{act}}^{S}tc_{OH^-}}{K_{\mathbf{eq}}\bar{r}_0\rho}\right)^2 \qquad (7.19)$$

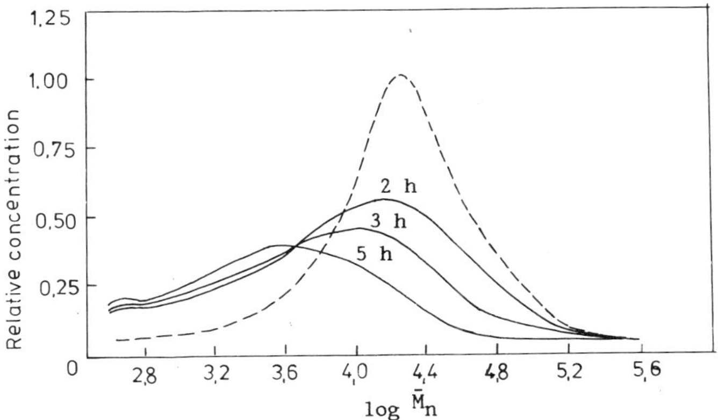

Fig. 7.13. Dependence of the MMD on the duration of degradation of F-2 polyarylate at 130°C in 60%
H_2SO_4. The broken curve corresponds to the
initial polymer [30].

The values of k^S_{act}/K_{eq} of the degradation of PG filaments in
aqueous basic solutions at various temperatures are as follows:

Temperature, °C	50	60	70
k^S_{act}/K_{eq}, g/min·sec²	$(3.3 \pm 0.3)\cdot10^{-3}$	$(7.9 \pm 0.3)\cdot10^{-3}$	$(1.6 \pm 0.2)\cdot10^{-2}$

The effective activation energy is 71 ± 1 kJ/mole.

The degradation of PG in aqueous salt solutions is investigated
in detail with a view to clarifying the mechanism of decomposition
of this polymer in the living organism [14]. It is shown that at
a pH of 7.4 salts of strong monobasic acids have practically no influence on the kinetic parameters: in the presence of phosphates
there is a considerable rise in the rate of degradation.

It may be suggested that in the case of PG there is bifunctional catalysis of the ester group by the anion $H_2PO_4^-$, as shown by
the scheme on p. 226.

Since the kinetic curves are described by Eq. (7.16), the degradation of PG filaments in the presence of phosphates proceeds
under internal kinetic conditions.

$$\underset{\underset{\|}{O}}{\sim C}-O\sim \;\underset{H_2O}{\rightleftharpoons}\; \underset{\underset{OH}{|}}{\overset{\overset{OH}{|}}{\sim C}}-O\sim \;\underset{H_2PO_4^-}{\rightleftharpoons}\; \overset{\overset{OH}{|}}{\sim C}-O\sim$$

Figure 7.12b shows the kinetic curves of the change in relative mass of Dexon fibers at 37°C in phosphate buffer and in the living organism [27c]. It may be seen that the decomposition of the PG in the organism is effected primarily by hydrolysis catalyzed by phosphate ions.

The decomposition of esters under the action of enzymes has been the subject of numerous investigations, and there are various known mechanisms of hydrolysis of esters [28a]. Many consider that the enzyme does, however, play a significant role in the decomposition of PG [28b-d]. At the same time, the experimental data on the rates of degradation of polymers in model media containing enzymes are quite contradictory. According to [27c], the addition to a solution of 1 M phosphate buffer of 0.2% of trypsin does not accelerate but in fact slows down the degradation of Dexon as compared with the corresponding process in a solution with no enzyme (and, as compared with decomposition in the organism). This is brought about, it appears, by adsorption of this enzyme on the surface of the filaments and by screening of the ester groups relative to the phosphate ions. Nevertheless, it is not ruled out that the degradation on the surface of the Dexon is catalyzed by enzymes, even if not as effectively as by the phosphate ions. Electron microscope investigation of Dexon which has been acted on by an enzyme shows that, as with tests in aqueous solutions and in the organism, the destruction of the elementary filaments is preceded by the formation of transverse defects which lead then to their breakdown into fragments (Fig. 7.12c). The diameter of the filaments remains practically unchanged in every case. The decomposition of Dexon in the organism is catalyzed mainly by the phosphate ions.

The degradation of a polyarylate (F-2) (there is a PVDF F-2) produced by polycondensation of phenolphthalein and terephthalic acid has been investigated in detail in acidic and basic media.

In aqueous basic solutions the degradation of F-2 polyarylate takes place by the same mechanism as that of PETP. The change in mass of the polymer films in various basic solutions and at various temperatures is described by the equation

$$m_n = m_n^0 - 6\cdot10^3 \exp\left(-\frac{13\,500}{RT}\right) a_{H_2O}^{b_0} st \tag{7.20}$$

In aqueous basic solutions the degradation of F-2 polyarylate films of thickness 45 ± 5 μm takes place under internal diffusion-kinetic conditions but at high temperatures (>130°C) even after 1 h the degradation takes place under internal kinetic conditions.

To clarify the type of decomposition of the macromolecules of F-2 polyarylate in the solid phase MMD curves were obtained using gel permeation chromatography (Fig. 7.13). During the process there is a shift of the maximum on the differential curve toward lower masses, and peaks become evident in the region of MMs around 1000 and 1500, i.e., the ester groups in the dimer and trimer have lower reactivities than those in the polymer.

The degradation of other polyesters has been effected, as a rule, under homogeneous conditions in order to study the influence of the structure of the acid and glycol components on the reactivity of the ester groups [29-32], to clarify the role of the hydrolytic processes in thermal degradation [33, p. 57], and to study the influence of microtacticity on the reactivity of the ester groups [34].

Some of the data are given in Appendix 7.2.

7.3. CELLULOSE AND ITS DERIVATIVES

The acid-catalyzed degradation of cellulose has been studied by numerous investigators [35-37]. It has been studied mainly in concentrated mineral acids (50% H_2SO_4, 40% HCl, and 85% H_3PO_4) at low temperatures. The degradation is accompanied by the formation of oxonium compounds and esterification. The main degradation product is glucose.

We shall now consider the main rules governing the degradation of cellulose and its derivatives under homogeneous conditions in acidic solutions.

Freudenberg [44] and others [38-43] consider that the degradation of cellulose and its derivatives occurs randomly. To support this mechanism there are adduced the constant value of $P_m/P_n = 2$ during the process of degradation [36, 40] and the change in the degree of polymerization of the polymer with time according to Eq. (4.2) [38-44].

In a number of investigations it is, however, noted that the terminal glycoside group of the macromolecule is more reactive than the other groups. This is confirmed by an increase in the rate constant of degradation of amylose [45], methylcellulose [46], and cellulose [47] with a reduction in the degree of polymerization, and also by the difference in reactivity of the glycoside groups in cellotriose [48-50].

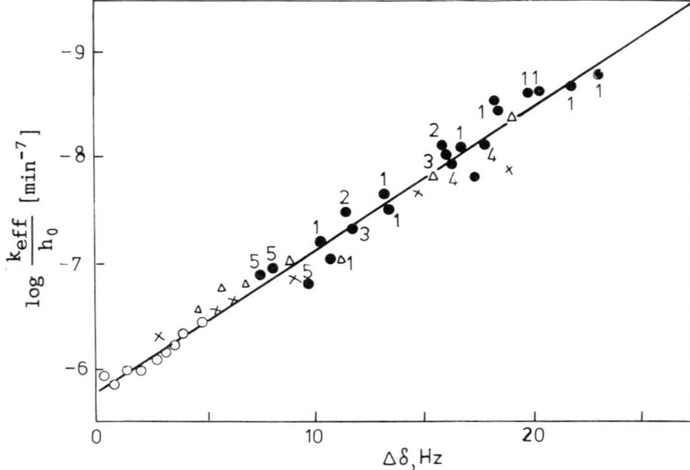

Fig. 7.14.   Relationship between log (k_{eff}/h_0) and $\Delta\delta$ for
             the degradation of methylcellulose in aque-
             ous solutions of HCl (O), $HClO_4$ (×), H_2SO_4
             (Δ), and of cellulose in solutions of H_2SO_4
             (●); data from various sources: 1) [51];
             2) [52]; 3) [53]; 4) [54]; 5) [44].

However, the difference in reactivity of the terminal and nonterminal glycoside bonds is not great (approximately 1.5-fold), and
it can be assumed that the degradation of cellulose and its derivatives under homogeneous conditions in solution proceeds practically according to a random process.

The degradation of cellulose and its derivatives occurs according to the same mechanism as was established for oligomers (see
Chap. 2). Figure 7.14 shows log k_{eff}/h_0 as a function of $\Delta\delta$ for
the degradation of methylcellulose in aqueous solutions of HCl,
$HClO_4$, and H_2SO_4 at 25°C [37] and for the degradation of cellulose
in solutions of H_2SO_4 [44, 51-54].

The degradation of derivatives of cellulose in nonaqueous acid
solutions has been investigated in a number of studies in connection with the acylation of cellulose [55-61]. With acylation there
is, as a result of the simultaneous degradation of the cellulose
and of its esters, a reduction in the degree of polymerization.

In [60] there is a detailed study of the kinetics of degradation of cellulose triacetate in acylating mixtures of various compositions.

Figure 7.15 shows the relationship between log k_{eff} and H_0,
with the slope close to unity. This means that, in this case, Eq.

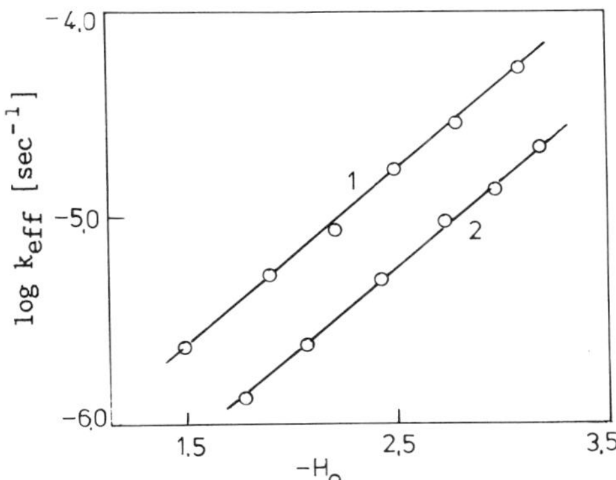

Fig. 7.15. Relationship between log k_{eff} and
H_0 for the degradation of cellulose
triacetate at 25°C in various acylating mixtures CH_2Cl_2, CH_3COOH, and
$(CH_3CO)_2O$ [60]: 1) H_2SO_4; 2) $HClO_4$.

(2.13) is satisfied. For the sulfuric and perchloric acids we get
parallel straight lines, which shows that the anion of the acid plays
a significant role in the formation of the reactive form.

In dilute or moderately concentrated aqueous acid solutions
the degradation of cellulose takes place under heterogeneous conditions. The cellulose is subjected to acid-catalyzed degradation
in the form of nap or fibers. It may be assumed that in either case
there are no diffusion limitations since the diameter of the elementary filament is about 10 μm, and the duration of the degradation
several hours.

Cellulose is a crystallizing polymer with an amorphous phase
content from 5 to 30%, which has a significant influence on the kinetics of degradation.

Figure 7.16 shows a typical kinetic curve of the change in
mass of an insoluble cellulose in degradation in 6 N HCl at 100°C
[41]. The kinetic curve, presented on semi-log coordinates, has
two characteristic stages: an initial stage, in which the rate of
degradation is high and the process takes place in the amorphous
phase, and a final stage in which the rate of degradation is considerably lower and there is breakdown of the crystallites.

In a number of studies [62, p. 60 of Russian translation], it
is shown that in the first stage there is a considerable reduction
in the degree of polymerization, whereas in the second stage the degree
of polymerization remains practically unchanged (see Fig. 7.16).

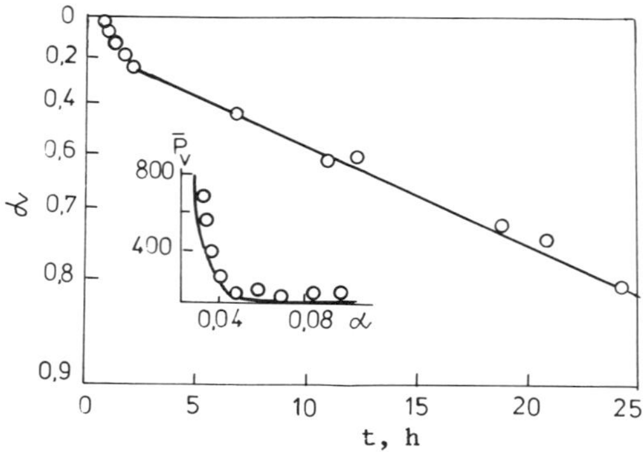

Fig. 7.16. Kinetics of the change in the degree
            of dissociation by mass of insoluble
            cellulose (α) in the degradation of
            cotton nap in 6 N HCl at 100°C, and
            the dependence of the degree of poly-
            merization P̄v on α for the same pro-
            cess [41].

The degradation of cellulose in the amorphous phase takes place
under internal kinetic conditions according to a random process,
as is shown, for instance, by the data in [41]. The rate constants
of the reaction can be calculated from the equation

$$\frac{2}{\overline{P}_w} - \frac{2}{\overline{P}_{w_0}} = kNt \tag{7.21}$$

where N is the proportion of glycoside groups in the amorphous phase.

The change in mass of the polymer in this stage of the kinetic
curve can be described by Eq. (4.17), assuming that the oligomers
of cellulose with degree of polymerization lower than 8 are soluble
[41, 63], and that diffusion restrictions in the desorption of these
oligomers from the polymer are not present. The activation energy,
according to [41], is 118 ± 7 kJ/mole, which differs from the data
of other authors [64].

The only slight change in the MM of the polymer in the second
stage of the kinetic curve [41, 65-68] shows that the degradation
takes place from the surface of the crystallites. However, it is
found experimentally that the change in mass of the polymer up to
degree of dissociation 0.8-0.9 is described by a first-order reaction, the effective rate constant depending on the type of cellulose
and the external conditions [66-68]. The activation energy corresponding to this stage of the kinetic curve is considerably higher,
at 161 ± 8 [41] and 180 kJ/mole [69].

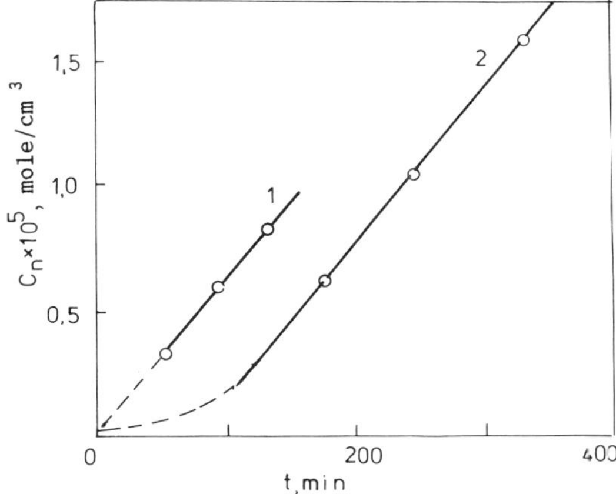

Fig. 7.17.   Dependence of the number of scissions
             of glycoside groups on the time of
             degradation of films of ethylcellu-
             lose of various thicknesses in vapor
             of 28% HCl at 35°C (45 torr) [72]:
             1) 30 ± 2 μm; 2) 180 ± 10 μm.

An explanation of the kinetic behavior in the degradation of
the crystallites was first put forth by Sharples [41]. The degradation proceeds from the faces of the crystallites at a rate proportional, to a first approximation, to their average length. On the
basis of experimental data [70], Sharples assumes an exponential
distribution for the length of the crystallites (x_3):

$$n_0 = a_0 \exp\left(-bx_0\right) \tag{7.22}$$

where b is a distribution constant, depending on the type of cellulose.

The final expression for the change in mass of the crystallites
with time takes the form

$$-\ln \frac{m}{m_0} = 2bk_{\text{eff}}t\,\frac{1}{\rho} \tag{7.23}$$

where k_{eff} is the zero-order rate constant.

Manley [71] considers that with the degradation of the crystallites there passes into solution fragments with a degree of polymerization of 8. This is due to the folded structure of the fibrils.

The degradation of fibers and of thinner films (less than 200
μm) of cellulose esters takes place under internal kinetic conditions and is described by Eq. (6.20).

Figure 7.17 gives experimental data for the degradation of films of various thicknesses of ethylcellulose in hydrochloric acid vapor. The identical slopes of the plots of c_n vs. t after some time from the beginning of the process indicates that the degradation takes place under internal kinetic conditions over the whole mass of the amorphous polymer; practically no degradation takes place from the surface of the film [72]. This kind of degradation is characterized by a low activation energy and a negative value of the entropy of activation. The following are the kinetic parameters of the acid-catalyzed degradation of ethylcellulose under heterogeneous conditions [72] and of methylcellulose under homogeneous conditions [37]:

Polymer	k_{eff} 35°C, cm³/(mole·min)	E, kJ/mole	$\Delta S^{\neq}$, J/(mole·°K)
Methylcellulose	$5 \cdot 10^{-6}$	126	17 ± 5
Ethylcellulose	$1.2 \cdot 10^{-1}$	23	-210 ± 8

The rate constants are found by dividing the effective rate constant by the catalyst concentration in the polymer (ethylcellulose) or in a dilute solution (methylcellulose). The reduction in the activation energy as we pass from homogeneous to heterogeneous conditions may be brought about by two factors. On the one hand, the dissolving of the aggressive medium in the ethylcellulose (about 2 mass % of water and HCl dissolve) unfreezes the molecular movements to a sufficient extent, and the kinetics of the chemical reaction in the solid polymer will not greatly differ from the kinetics in the liquid phase (a similar effect has been found for radical reactions [73]). On the other hand, if we consider the acid-catalyzed degradation of ethylcellulose as a process taking place in solution, it may be assumed that the major factor responsible for the lowering of the activation energy is the permittivity of the medium ε.

Figure 7.18 shows the change in the activation energy of the degradation of cellulose esters as a function of the reciprocal of ε. The satisfactory correlation leads one to hope that as a model of chemical degradation in solid polymers there might be used the corresponding reaction in a solvent with identical physical parameters. The future gathering of experimental data will show the correctness of this approach.

7.4. POLYACETALS

Investigation of the acid-catalyzed degradation of polyacetals was begun in the 1930s by Staudinger [74, 75] in connection with

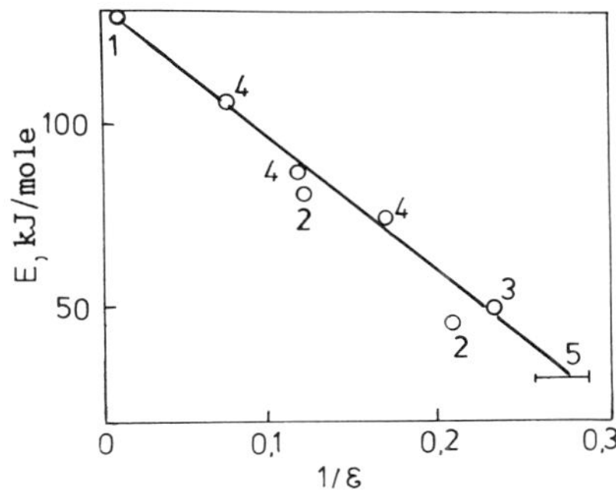

Fig. 7.18. Dependence of the activation energy
            of the degradation of cellulose esters
            on the reciprocal of the permittiv-
            ity of the medium: 1) [37]; 2) [58];
            3) [61]; 4) [60]; 5) [72].

the development of the technology of the production of stable high-
molecular polyoxymethylene (POM). Degradation was studied both
under homogeneous and heterogeneous conditions.

Degradation under homogeneous conditions was investigated
either in the melt or in organic solvents. As in the case of low-
molecular acetals (Chap. 2), it was assumed that the limiting stage
of the process was decomposition of the protonated form [76]. For
polyacetals it has been established that the first-order rate constant is a function of the acid concentration [77-80], and the rate
of degradation in acetolysis depends on the dissociation constant
of the acids in question (Table 7.2). The relationship found is
described by the Brønsted equation

$$\log w = 2.1 - 0.2 \mathrm{p}K_a \tag{7.24}$$

The activation energy of the acid-catalyzed degradation of POM
varies from 71 to 105 kJ/mole [82-83], which is close to the activation energy of decomposition of low-molecular acetals (see Chap. 2).

The principal decomposition products are formaldehyde and cyclic
oligomers [83]:

$$\sim O-CH_2-O-(CH_2-O)_n-\overset{+}{C}H_2 \longrightarrow \sim O-CH_2-\overset{+}{O}-CH_2-O \longrightarrow$$
$$CH_2-(O-CH_2)_n$$
$$\longrightarrow \sim O-\overset{+}{C}H_2 + O\underset{(CH_2-O)_n}{\overset{CH_2-O}{\diagup\diagdown}}CH_2$$

TABLE 7.2. Dependence of the Rate of Degradation of Molten Dimethyl
            Ester of POM in the Melt at 190°C on the Dissociation
            Constant of the Acid [81]

Acid	pK_a	w, h^{-1}	Acid	pK_a	w, h^{-1}
Picric	0,87	0,4	β-Naphthol	0,11	8,0
Salicylic	0,36	3,0	Hydroquinone	0,013	10,0
p-Hydroxybenzoic	0,17	4,5			

In acid media the cyclic oligomers undergo further dissociation
and decompose to the monomer, particularly when there are traces
of water in the system [84].

The mechanism of the acid-catalyzed degradation of POMs of differing MMs with differing terminal groups at low temperatures is
studied in [77] (see also Chap. 4). In sulfuric acid solutions of
hexafluoroacetone the depolymerization reaction proceeds stepwise,
from the terminal bonds, and, as was found, the rate of degradation,
as determined from the accumulation of formaldehyde, does not depend
on the nature of the terminal group and the initial degree of polymerization.

At high temperatures the depolymerization reaction at the terminal bonds takes place at considerable rates (the kinetic length
of the chain approximates the degree of polymerization). This means
that POM-OHs (hydroxyl-terminated POMs) degrade at higher rates than
polymers with comparable degrees of polymerization but with ester
terminal groups [86-88].

The catalyst concentration in the system may have a significant
influence on the ratio of the rates of decomposition randomly and
of decomposition at the terminal bonds. Thus, in the degradation
of a molten POM with terminal acetate groups in the presence of 20
mass % citric acid the change in mass of the polymer is described
by the equation [89]

$$-\frac{dm}{dt} = k_{\text{eff}}\, c_{\text{p}}\, c_{\text{HA}}^{1/2}\, \overline{M}_n \tag{7.25}$$

where c_{p} and c_{HA} are, respectively, the concentrations of polymer
and acid.

With low concentrations of acid (around 2 mass %) the rate of
the process is described by the equation

$$-\frac{dm}{dt} = k_{eff}\overline{M}_n \qquad (7.26)$$

In the opinion of the authors of [89], in the first case the rate of the degradation process is limited by the random decomposition reaction; Eq. (7.26) describes degradation where depolymerization determines the rate of the whole process.

Nevertheless, the data obtained in degradation in the melt should be treated with caution since, at the very least, two factors can distort the kinetic behavior in the acid-catalyzed process:

the nonisothermal conditions of the degradation process, which are brought about by the low thermal conductivity of POM and the high rate of the process itself, which takes place with absorption of heat;

the viscosity of the melt, which influences the rate of liberation of formaldehyde from the specimen into the gaseous phase.

Degradation under heterogeneous conditions has been investigated in aqueous inorganic acid solutions, taking as examples POM powder and slabs. At the present time there is a lack of data on the diffusion coefficients of aqueous inorganic acid solutions in POM. There is known only the diffusion coefficient of water in films and slabs of POM-OOCCH$_3$, DH$_2$O = $2.5 \cdot 10^{-8}$ cm^2/sec at 25°C [90]. On the basis of these data, and also taking into account that the powder particles do not exceed $1.0 \cdot 10^{-2}$ cm in diameter, it may be assumed that equilibrium concentrations of water and acid are established within a short time (less than 1 min).

It was shown in Chap. 5 that the equilibrium concentrations of water and particularly of inorganic acids in hydrophobic polymers are exceptionally low. Taking into account that the granules of POM have a heavily convoluted surface, it may be imagined that degradation proceeds from the surface of the granules and only to a very slight degree within them. A strict mathematical solution is impossible in this case since the acid concentration within the granules is not known, while the granules themselves do not have a definite geometrical shape. If we assume, as a first approximation, that the granules have a spherical shape, then in the case of degradation proceeding from the surface, the change in mass of the polymer will be described as follows:

$$m^{1/3} = m_0^{1/3} - 1.6k_{eff}\rho^{-2/3}\,t \qquad (7.27)$$

Figure 7.19 shows a typical kinetic curve drawn on the coordinates of Eq. (7.27). The dependence of keff on the degree of polymerization of the initial polymer is analyzed in [92]. Two reactions take place on the surface of the polymer granules:

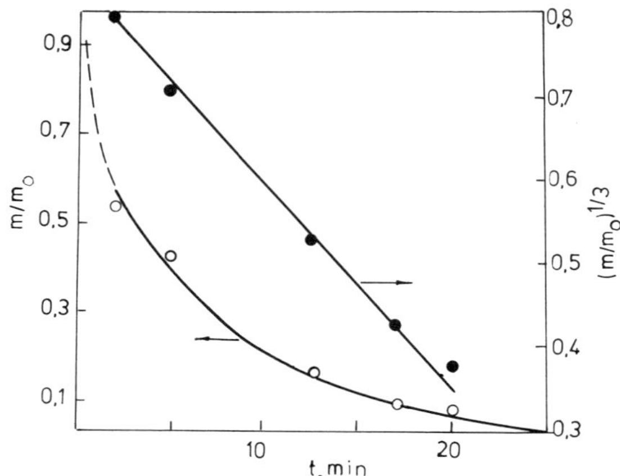

Fig. 7.19. Kinetic curve of the change in mass
            of POM-OH powder of $\bar{M}_v = 7000$ at
            35°C in 35% H_2SO_4, drawn on the
            coordinates of Eq. (7.27) [92].

depolymerization at a terminal bond, leading to a reduction in
mass of the polymer (bearing in mind that the active sites of depolymerization are the hemiacetal groups);

chain breakdown by a random process, leading to the formation
of hemiacetal groups.

It was shown in Chap. 4 that in aqueous acid solutions there
is practically instantaneous depolymerization of fragments of the
polymer with hemiacetal groups. Thus, degradation on the surface of the polymer may be represented as follows. The accumulation of hemiacetal groups is compensated for by a reduction in the
number of these groups on account of the "burning up" of the polymer
chains in depolymerization. Using the method of quasi-steady-state
concentrations, we get

$$\frac{dc_{N(surf)}}{dt} = k_{rand}c_{p(surf)} - k_d c_{N(surf)} = 0 \qquad (7.28)$$

where k_d is the destruction constant of active sites for depolymerization as a result of the "burning up" of the polymer chain,
and $c_{N(surf)}$ and $c_{p(surf)}$, respectively, the concentrations of hemiacetal and acetal groups on the surface of the polymer granules.

The concentration, at any time, of hemiacetal groups on the
surface of the POM-OH granules is in the general case equal to the
sum of the concentrations of the initial hemiacetal groups (for
POM-OH ~ $1/\bar{P}_n$) and of the hemiacetal groups formed according to Eq.
(7.28), i.e.,

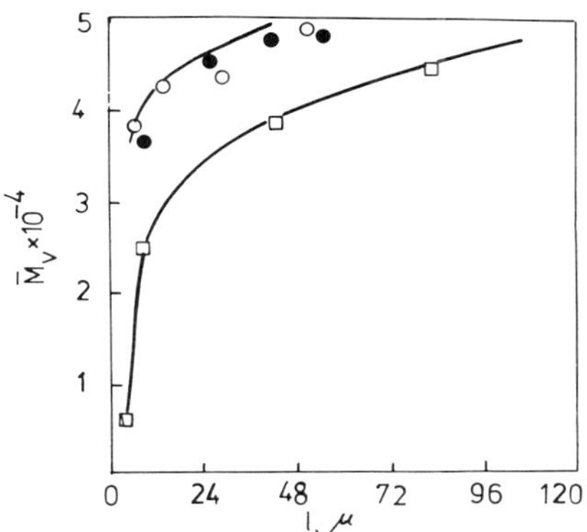

Fig. 7.20. Change in $\bar{M}_V$ in sections from the
surface of slabs of POM (Delrin 500)
in degradation in 50% H_2SO_4 for 40
min (●) and 300 min (○), and in
30.5% H_2SO_4 for 300 min (□) [92].

$$k_{eff} = k_{dep}\left(\frac{1}{\bar{P}_n} + \frac{k_{rand}}{k_d}\right) \qquad (7.29)$$

For POM-OCH$_3$, as might be expected, k_{eff} is practically independent of the initial degree of polymerization. The changes in
k_{eff} of the degradation of powdered POMs of various MMs and with
differing terminal groups in acid solutions of various concentrations was considered in Chap. 4. The acid-catalyzed degradation
of slabs of POM is an example of the process taking place not only
from the surface but also throughout the volume of a polymer article [92].
There is evidence of this in the change in MM in sections taken from
the surface of the slabs, with the size of the reaction zone depending on the concentration and nature of the acid (Fig. 7.20), and
also in the linear dependence of the logarithm of the rate of degradation on the acidity function H_0.

It is not possible to obtain an analytical solution for this
type of degradation and, therefore, the following approach has to
be considered.

The rate of accumulation of formaldehyde during degradation
taking place from the surface and throughout the volume of a polymer
article without taking into account diffusion limitations is

$$w = \frac{\Delta m}{\Delta t} = k_{surf}c_{p(surf)}c_{H_S^+(surf)}S + k_{(vol)}c_{p(vol)}c_{H_S^+(vol)}V \qquad (7.30)$$

Taking into account that the ratio $c_p(vol)/c_p(surf) = z$ is the number of monolayers per unit length, we get

$$\frac{\Delta m}{\Delta t} = \frac{1}{c_p V}\left[k_{surf}\, c_{H_S^+}(surf)\frac{1}{zl} + k_{(vol)}\, c_{H_S^+}(vol)\right] \qquad (7.31)$$

where l is the average length of the reaction zone.

The scission of the POM chain is as given by mechanism A-1 (see Chap. 2) and, consequently, $c_{H_S^+}^+(surf) \sim h_0$. Acids dissolve in POM in very low concentrations (see Chap. 5) and $c_{H_S^+}^+(surf) > c_{H_S^+}^+(vol)$.

The nonlinear character of the relationship between log w and H_0 is due to the considerable contribution of the degradation taking place in the volume of the polymer in moderately concentrated solutions of sulfuric acid ($\sim 50\%$ H_2SO_4) and, accordingly, the two terms in Eq. (7.24) are commensurate.

In more concentrated solutions of H_2SO_4, H_0 increases considerably more than $c_{HS(vol)}^+$ and

$$k_{surf}\, h_0 \frac{1}{zl} \gg k_{(vol)}\, c_{H_S^+}(vol) \qquad (7.32)$$

The degradation of slabs of POM-OOCH$_3$ takes place practically from the surface, as is shown by the only slight change in the MM in sections taken from the surface of the slabs.

When the degradation of powders of POM-OOCH$_3$ is effected in solutions of sulfuric acid the relationship between log w and H_0 is linear, with a slope of unity since the value of l is not great and condition (7.25) is satisfied for powdered POM-OOCH$_3$ in more dilute solutions of sulfuric acid than is the case for slabs of POM.

It must be noted that if the degradation is as described by mechanism A-2, the contribution of the reaction taking place in the volume will be predominant since $c_{HS(surf)}^+ \sim c_{HS(solv)}^+$. The latter value increases with increase in the acid concentration far less sharply than h_0, and over practically the whole range of acids the relationship

$$k_{surf}\, c_{H_S^+}(solv)\frac{1}{zl} \ll k_{(vol)} \qquad (7.33)$$

is valid.

It is specifically because of this that the degradation of polyesters in acidic media practically does not take place from the surface of specimens.

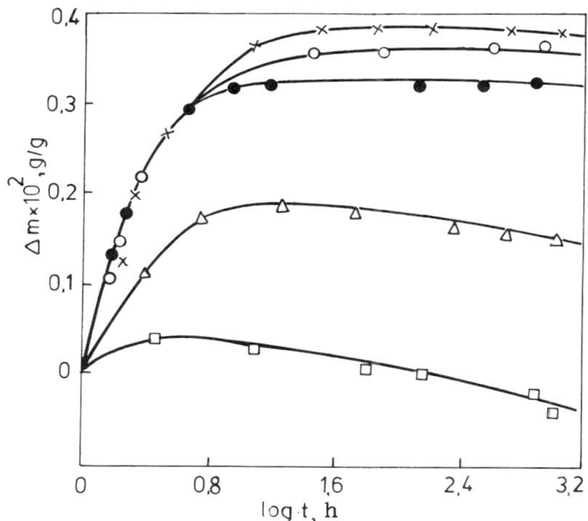

Fig. 7.21. Change in the mass of specimens of
            Diflon polycarbonate as a function
            of the immersion time at 70°C in water
            (×) and solutions of 3% (○), 10%
            (●), 30% (Δ), and 50% (□) H_2SO_4 [93].

Some data are given in Appendix 7.3.

7.5. POLYCARBONATES

The chemical degradation of polycarbonates has been investigated in acidic and basic salts, particularly that of the polycarbonate based on diphenylolpropane and phosgene ("Diflon").

The degradation of Diflon in acidic media is investigated in
detail in [93].

It is shown using permittivity and indicator methods that sulfuric acid is not sorbed by Diflon in appreciable amounts. Figure
7.21 shows the change in mass of Diflon slabs in contact with solutions of sulfuric acid of different concentrations. Two features
should be noted.

The sorption equilibrium increases with reduction in the concentration of sulfuric acid. Since sulfuric acid is practically
not sorbed by Diflon, it may be assumed that in this particular case,
as with the majority of hydrophobic polymers, there is selective
sorption of water from the electrolyte solution.

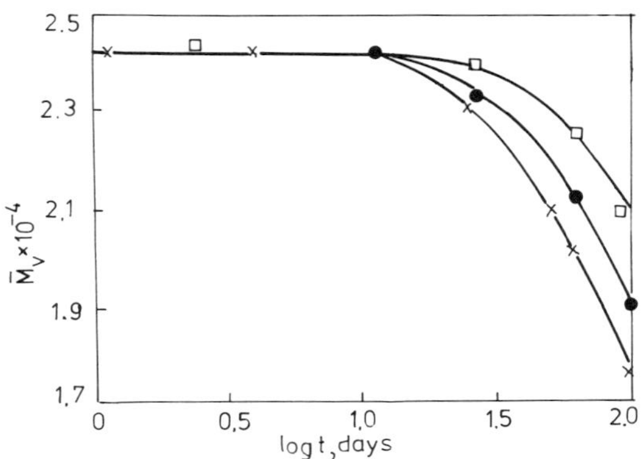

Fig. 7.22. Change in $\bar{M}_V$ of Diflon polycarbonate
as a function of the immersion time
at 70°C in solutions of sulfuric
acid (symbols as in Fig. 7.21).

The sorption isotherms are S-shaped, and the equilibrium values
for the sorption of water increase considerably with increase in temperature. The diffusion of water in Diflon is described by the equation

$$D_{H_2O} = 3.3 \cdot 10^{-3} \exp\left(-\frac{6500}{RT}\right) \tag{7.34}$$

In concentrated solutions of sulfuric acid the mass of the
polymer first increases due to the sorption of water, and then decreases as a result of the degradation.

To ascertain under which kinetic conditions the degradation
takes place, we found the dependence of the change in the viscosity-
average MM on the time of degradation (Fig. 7.22). On the basis
of the resulting data there was proposed a mechanism according to
which the process of degradation takes place both from the surface
(specific acid catalysis) and in the volume of the polymer article
(catalysis by water molecules).

In the former case, degradation takes place under external diffusion-kinetic conditions and the total change in mass of the polymer is described by the equation

$$m_P = m_P^0 + \Delta m_{H_2O}^\infty \left[1 - \frac{8}{\pi^2}\exp\left(-\frac{\pi^2 \bar{D} t}{l^2}\right)\right] - k^s h_0 s t \tag{7.35}$$

where $m_{H_2O}^\infty$ is the amount of water in the molecule which corresponds
to equilibrium sorption; k^s the rate constant (independent of the
concentration of the catalyst) of the degradation of the polymer
article from the surface.

TABLE 7.3.   Influence of Structure Differentials of Diflon Polycarbonate on k_{eff} of Degradation in 30% KOH at Various Temperatures

t, °C	$k_{eff} \cdot 10^6$, g/(min·cm^2)			
	Amorphous Film	Molded Film	Crystalline Film	Orientated Film
48.5	1.9	1.7	1.8	1.9
59.5	4.1	4.3	4.1	4.1
70.5	10.2	11.0	8.8	9.2

In the latter case degradation takes place under internal kinetic conditions, and the degree of dissociation of the carbonate groups (decomposition by a random process) is satisfactorily described by Eq. (6.20).

The degradation of Diflon in basic solutions takes place intensively even with low concentrations of catalyst. As in the case of PETP, the degradation of slabs and films of Diflon takes place in a thin surface layer [94]; nevertheless, the effective rate constant of degradation over the whole range of concentrations (11.5-53.0% KOH) is described by the equation

$$\log k_{eff} + B_0 = \text{const} \tag{7.36}$$

This equation may be regarded as a special case of Eqs. (7.8) and (7.10). Thus, on the basis of the kinetic data it can be asserted that water molecules do not take part in the limiting stage of the decomposition of the carbonate group. It remains, however, unclear whether in the limiting stage there is a direct attack of the hydroxide ion on the carboxyl group of the carbonate group or if this stage is a fast one, while the slow stage is subsequent decomposition of the ionized form of the carbonate group into a phenolate ion and ion of CO_3^{2-}:

The change in mass of films and slabs of Diflon in the course of degradation is expressed by the equation

$$m_\text{p} = m_\text{p}^0 - 10^4 \exp\left(-\frac{16\,500}{RT}\right) st \qquad (7.37)$$

where m is in g, t in min, and s in cm^2.

For this type of degradation, as already observed, a change in structure has practically no influence on the kinetic parameters of the process.

Table 7.3 shows the values of k_eff for the degradation of amorphous specimens of Diflon prepared by injection molding or by casting a solution in chloroform with subsequent rapid evaporation of the solvent, and of film with degree of crystallinity about 30%, and oriented film with degree of orientation about 3.

As Table 7.3 shows, within the limits of experimental error, the structure differentials do not influence the value of k_eff; the effective activation energy of the degradation processes of these specimens is likewise identical, 69 ± 3 kJ/mole.

Glöckner [95] investigated the degradation of polycarbonates based on diphenylolpropane in chloroform under the action of various alcohols at 20°C. The rate of degradation, as determined from the change in $\bar{M}_v$, rises with rise in the alcohol concentration in the mixture, with the catalytic activity decreasing as follows:

$$\text{MeOH} > \text{EtOH} > n\text{-PrOH} > n\text{-BuOH}$$

There are various investigations of the influence of the chemical structure of polycarbonates on their reactivity in alkaline media.

The following are the reactivities of polycarbonates of various structures in 10% NaOH at 100°C [96]:

Elementary structural unit	$k_\text{eff} \cdot 10^4$, g/(cm²·h)
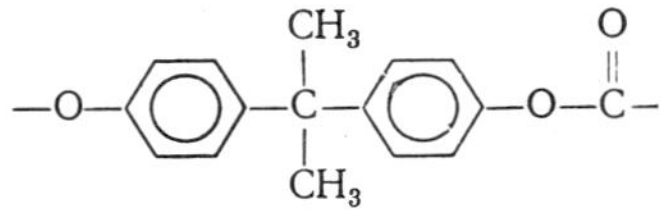 $-\mathrm{O}-\!\!\big\langle\!\!\bigcirc\!\!\big\rangle\!\!-\overset{\overset{\text{CH}_3}{\vert}}{\underset{\underset{\text{CH}_3}{\vert}}{\text{C}}}-\!\!\big\langle\!\!\bigcirc\!\!\big\rangle\!\!-\mathrm{O}-\overset{\text{O}}{\overset{\Vert}{\text{C}}}-$	1.5
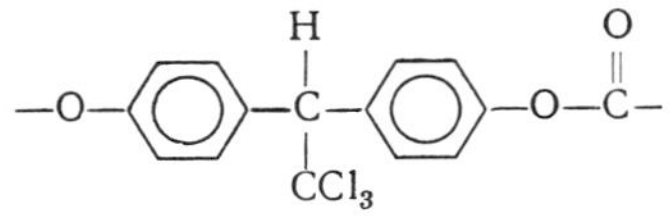 $-\mathrm{O}-\!\!\big\langle\!\!\bigcirc\!\!\big\rangle\!\!-\overset{\overset{\text{H}}{\vert}}{\underset{\underset{\text{CCl}_3}{\vert}}{\text{C}}}-\!\!\big\langle\!\!\bigcirc\!\!\big\rangle\!\!-\mathrm{O}-\overset{\text{O}}{\overset{\Vert}{\text{C}}}-$	4.3

(continue)

Elementary structural unit	$k_{eff} \cdot 10^4$ g/(cm$^2 \cdot$h)
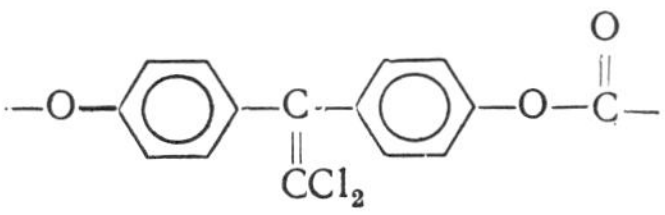 (structure: $-O-\langle\bigcirc\rangle-\overset{\overset{O}{\parallel}}{C}(CCl_2)-\langle\bigcirc\rangle-O-\overset{\overset{O}{\parallel}}{C}-$)	4.6
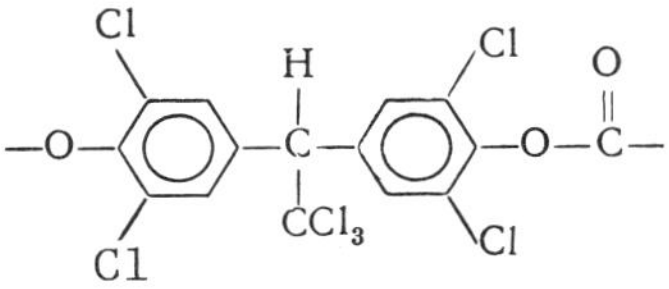 (structure: tetrachloro bisphenol carbonate with $\overset{H}{\underset{CCl_3}{C}}$ bridge)	0.023
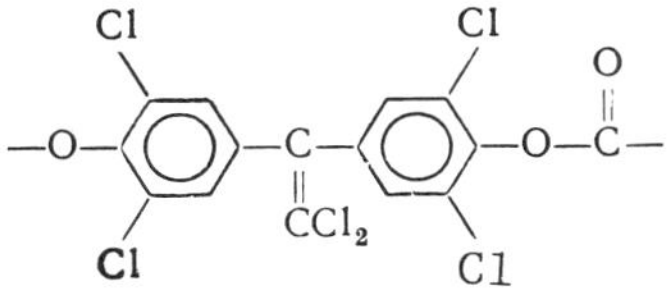 (structure: tetrachloro bisphenol carbonate with CCl_2 bridge)	0.028

The following are the reactivities of polycarbonates of various
structures in 30% KOH at 20°C [97]:

Elementary structural unit	$k_{eff} \cdot 10^7$, g/(cm$^2 \cdot$min)
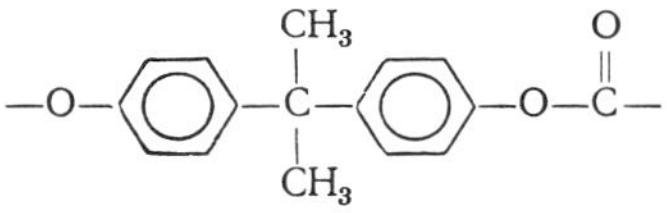 (structure: $-O-\langle\bigcirc\rangle-\overset{\overset{CH_3}{\mid}}{\underset{CH_3}{C}}-\langle\bigcirc\rangle-O-\overset{\overset{O}{\parallel}}{C}-$)	1.2
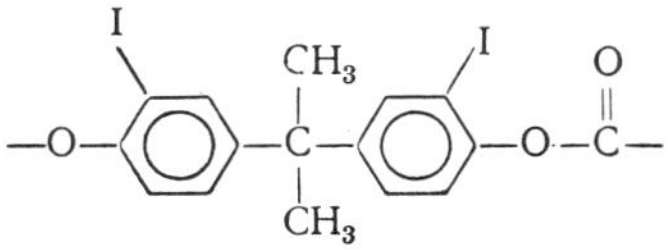 (structure: diiodo bisphenol A carbonate, I substituents, $C(CH_3)_2$ bridge)	0.6
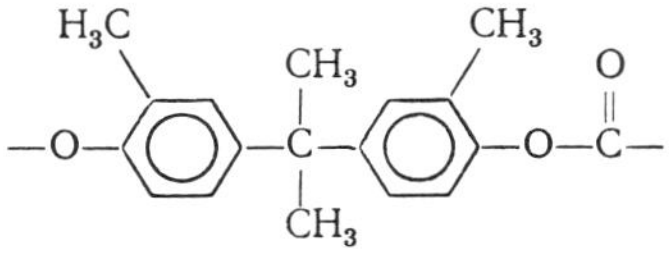 (structure: dimethyl bisphenol A carbonate, H_3C and CH_3 substituents)	0.5
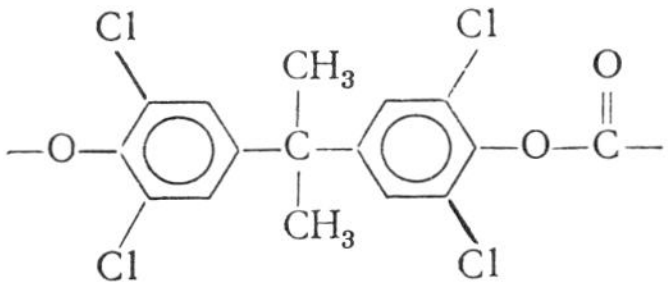 (structure: tetrachloro bisphenol A carbonate, Cl substituents, $C(CH_3)_2$ bridge)	0.016

The data obtained in [97] were treated according to the Taft
equation:

$$\log k_{eff} = \lg k_{eff}^0 - 0.35 \sum \sigma^* + 0.21 E_S \tag{7.38}$$

where σ^* and E_S are, respectively, the induction and steric constants of the substituents.

As the reaction site we selected the carbonate group, and the influence of the adjacent benzene ring via the group $CH_3-\overset{\textstyle|}{\underset{\textstyle|}{C}}-CH_3$ was not taken into account.

7.6. POLYSILOXANES

At the present time it is accepted that polysiloxanes are unstable at high temperatures (>100°C) in water, alcohols, acids, and bases. However, in contact with these media there occur two processes, which sometimes greatly hinder the interpretation of the data on the kinetics of degradation.

In the first place, there is the interconversion of cyclic and linear siloxanes. The equilibrium between these compounds and the polymer is influenced by the temperature, pressure, nature of the solvent, and also the structure of the elementary unit of the polymer. In [98] there is a detailed consideration of the influence of thermodynamic, kinetic, and statistical factors on the polymerization of cyclic siloxanes and on the depolymerization of linear polymers with the formation of cyclic degradation products. The occurrence of these processes makes it impossible, for instance, to give a clear interpretation of the data on the kinetics of accumulation of silanol groups.

In the second place, commercial polysiloxanes contain fillers (e.g., SiO_2), low-molecular cyclic compounds, and stabilizers. During prolonged contact with aggressive media these compounds may be desorbed from the polymer and, accordingly, any conclusion as to the course of degradation based solely on a reduction in the mass of a polymer article is risky.

Lewis [99] presents copious data on changes in mass, volume, and the breaking load (TS) of polydimethylsiloxane rubber (PDMS) over 1 week at 25°C. In acids and in salt solutions and weak bases there is sorption accompanied by a reduction in the TS, while in concentrated solutions of strong bases there is a reduction in the mass and volume of the polymer.

The kinetics of degradation of PDMSs with various terminal groups has been investigated in detail in the presence of 20 mass % of H_2O at 200-300°C [100]. Comparison of the data on mass loss and the change in the number of scissions during degradation shows that the most probable is a degradation mechanism including two stages: random decomposition of the polymer chain and subsequent depolymerization of fragments with silanol groups:

$$\sim \overset{|}{\underset{|}{Si}} - O - \overset{|}{\underset{|}{Si}} - O \sim \quad \xrightarrow{H_2O} \quad \overset{\diagdown\diagup}{\underset{\vdots\quad\vdots}{Si}} \cdots O - \overset{|}{\underset{|}{Si}} - O \sim$$

There is no mention here of interconversions of linear and cyclic siloxanes, but IR data are presented which show an increase in the concentration of silanol groups with time. This shows that the given scheme of degradation is evidently correct under these conditions. According to [100], the effective activation energy of degradation of PDMS in the presence of water is 71 ± 4 kJ/mole. Martellock [101] found an effective activation energy for thermal decomposition in vacuum of 97 kJ/mole, and in the presence of acidic or basic catalysts 25 kJ/mole.

The most correct data on the kinetics of degradation of PDMS in aqueous solutions of acids and bases were obtained in [102]. Using a Langmuir—Adam balance, the authors determined, from the straight sections of the isotherms with constant surface pressure (F), the area appertaining to the elementary unit (a). A PDMS with $\bar{M}_v = 5.7 \cdot 10^4$ was in a monolayer on the surface of the aqueous solution of aggressive medium. It was shown by special experiments that in the aqueous solutions it is practically only monomeric dimethyl-silanediol which dissolves.

Taking this into account, and assuming that the degradation is a random process, from the equation linking the concentration of oligomers with the degree of polymerization $\bar{P}_X$, the time of degradation, and the initial concentration c_0 of the macromolecules with the degree of polymerization $\bar{P}_0$ [103, p. 372]

$$c_X = c_0 \left(1 - \exp\left(-kt\right)\right) \exp\left[-(\bar{P}_X - 1)\, kt\right] \left[2 + (\bar{P}_0 - \bar{P}_X - 1)\left(1 - \exp\left(-kt\right)\right)\right] \quad (7.39)$$

we get, with X = 1,

$$c_X = c_0 \bar{P}_0 \left(1 - \exp\left(-kt\right)\right)^2 \quad (7.40)$$

The proportion of monomer units transported from the monolayer into the solution is

$$\alpha = \frac{c_X}{c_0 \bar{P}_0} = \frac{a_0 - a_t}{a_0} \quad (7.41)$$

where a_0 and a_t are the surface areas found by extrapolating to F = 0 the straight sections of the isotherms F − a which correspond to condensation of the actual monolayer of PDMS and which relate to the beginning of the experiment and the time of hydrolysis t.

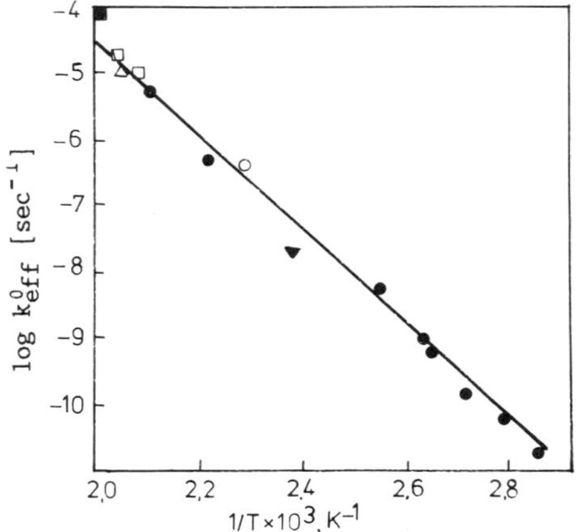

Fig. 7.23. Dependence of log k$_{eff}^{0}$ on 1/T for
noncatalytic dehydrochlorination of
PVC in vacuo according to various
authors: $\triangle$) [106]; $\blacksquare$) [107]; $\square$)
[108]; $\circ$) [109]; $\blacktriangledown$) [110]; $\bullet$)
[111].

From (7.40) and (7.41) we get

$$\ln\left[1 - \left(\frac{a_0 - a_t}{a_0}\right)^{1/2}\right] = -kt \qquad (7.42)$$

which gives a good description of the experimental data.

Data from various studies are given in Appendix 7.4.

7.7. POLY(VINYL CHLORIDE)

In the thermal degradation of PVC hydrogen chloride is formed,
which strongly catalyzes the dehydrochlorination reaction. Dehydrochlorination is the principal reaction at temperatures below 200°C
in the thermal degradation of PVC. It can either occur noncat-
alytically or else be catalyzed by a number of compounds, of which
the most active catalyst is the said hydrogen chloride.

7.7.1. Noncatalytic Dehydrochlorination

Noncatalytic dehydrochlorination occurs when the rate of diffusion of hydrogen chloride from the polymer specimen is much higher
than the rate of the reaction itself. For instance, this condition
applies at temperatures below 25°C for finely powdered PVC.

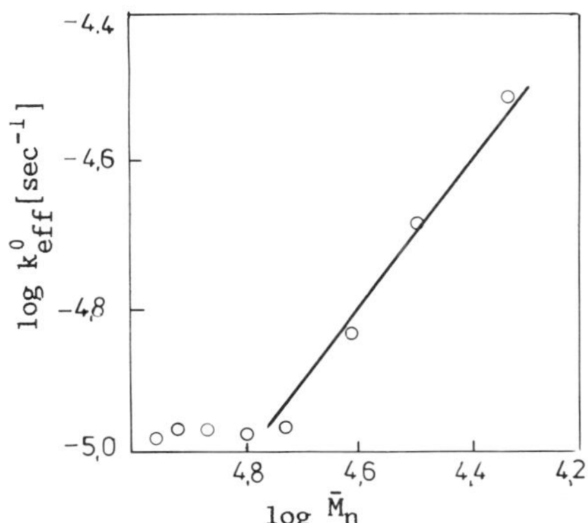

Fig. 7.24.   Dependence of $\log k^0_{eff}$ on $\log \bar{M}_n$ for
the dehydrochlorination of PVC at 220
220°C in vacuo [107].

The reaction is described by a first-order equation with respect to the polymer; correct data are obtained in solutions [104].
In the course of thermal degradation of solid polymers there is a
change in the chemical structure of the polymer, which leads sometimes to distortion of the order of the reaction [105]. Figure
7.23 shows the rate constants of the process at various temperatures,
as obtained by various authors [106-111]. The effective rate constants of noncatalytic dehydrochlorination k^0_{eff} are calculated
from the equation

$$\alpha = \frac{n_{HCl}}{n^\infty_{HCl}} = k^0_{eff} t \tag{7.43}$$

where n_{HCl} and n^∞_{HCl} are, respectively, the amounts of HCl formed
in time t and with complete dissociation.

For the noncatalytic dehydrochlorination of commercial polymer
specimens the effective activation energy is 134 ± 8 kJ/mole.

If we assume that the transition of the polymer from the high-
elastic to the glassy state has no significant influence on the
activation energy, we can estimate k^0_{eff} for temperatures below T_g.
For instance, at 25°C, $k^0_{eff} \approx 10^{-14}$ sec^{-1}. A degree of dissociation of 0.01 is reached in 10^7 days.

One of the reasons for the lower thermal stability of PVC as
compared with model compounds (see Chap. 4) is the presence of unstable groups or active sites in the polymer. At present the existence of the following active sites has been shown.

β-Chloroallyl Groups ~CH$_2$—CH=CH—CHCl~.   Depending on the nature of the adjacent groups these have various reactivities (Tables 2.3 and 2.4).

It is known that PVC molecules may contain terminal double bonds and that their concentration may reach about 10^{-3} mole/mole [109, 112]. Figure 7.24 shows the dependence of log k_{eff}^0 on log $\bar{M}_n$ for specimens of PVC, the degradation of which was investigated at 220°C under vacuum [107]. It may be noted that for the polymers with $\bar{M}_n$ from 21,000 to 61,000 there is satisfactory correlation between k_{eff}^0 and the number of terminal double bonds (the slope is close to unity). For polymers with higher MMs the experimental values for the rate constants are higher than is suggested by the straight-line relationship in Fig. 7.24.

From this it evidently follows that for polymers with $\bar{M}_n$ < 60,000 the terminal double bonds are among the principal active sites; for polymers with higher MMs other active sites start to play a considerable part.

The number of internal double bonds was determined in [109, 113]. In commercial polymers the concentration of these is $(0.6-2.0) \cdot 10^{-4}$ mole/mole, i.e., lower than that of the terminal double bonds.

According to Braun [114], for some PVC specimens there is a correlation between the rate of dehydrochlorination and their content of internal double bonds.

Oxidized Structures. It is known that compounds with double bonds oxidize readily and, accordingly, various oxidized structures may form in the PVC [115], among which there predominate ketoallyl groups, ~CO—CH=CH—CHCl~.

The presence of ketoallyl groups in the macromolecules has been shown by a) alkaline hydrolysis (a specific reaction for unsaturated ketones); b) a reduction in the MM of the PVC analogous to that which takes place in oxidative ozonization of the polymer [113]; c) the presence, in the IR spectra, of thick films of PVC, of absorp-tion bands at 1675 cm^{-1}, which are characteristic of $\sim\overset{\displaystyle O}{\overset{\displaystyle \|}{C}}$—CH=CH—CHCl~ groups; d) the quantitative closing, under mild conditions, of the internal —CH=CH— bonds in the reaction of PVC with organic phosphites [116, 117], and, also, e) a number of other, independent, experiments.

The concentration of ketoallyl groups is $(0.1-2.5) \cdot 10^4$ mole × (mole)$^{-1}$. The value of k_{eff}^0 for PVC specimens with $\bar{M}_v$ = 50,000-300,000 depends linearly on the concentration of ketoallyl groups (Fig. 7.25), the straight line being extrapolated to the origin of the coordinates

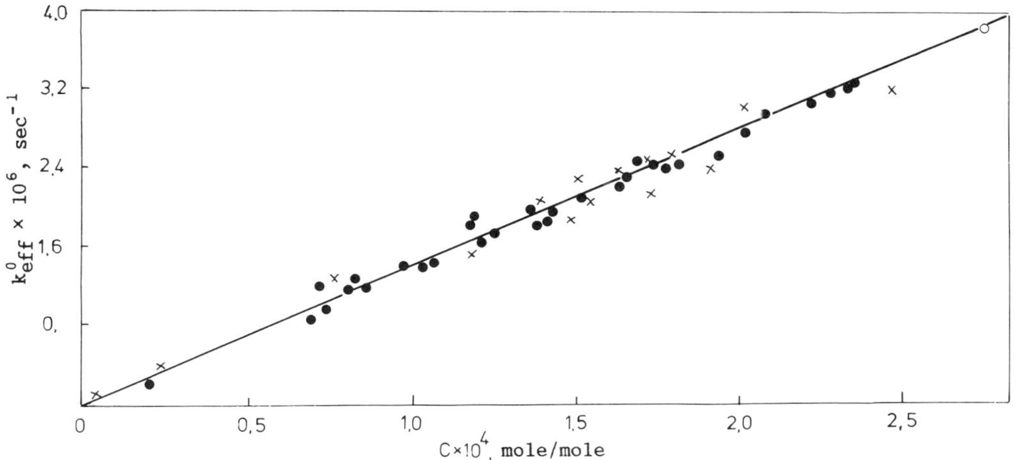

Fig. 7.25. Dependence of k_{eff}^0 of the dehydrochlorination of PVC
on the content of $-CO-CH=CH-CHCl-$ groups for various
specimens of PVC ($\bar{M}_v$ = 50,000-300,000) at 175°C and
10^{-4} torr [117].

[108, p. 20]. In the opinion of Minsker, for PVC specimens with
high MMs it is indeed these groups which are the principal active
sites for the dehydrochlorination process.

There have also been observed in PVC up to $5 \cdot 10^4$ mole/mole of
peroxide groups [118].

<u>Chlorovinyl Groups with Tertiary Carbon Atoms</u>. It has been
shown experimentally that branched structures eliminate HCl considerably more readily than linear structures do. According to some
data [119], one tertiary carbon atom, on the average, occurs every
70 monomer units, while according to other data [120] it is practically impossible to detect branching in commercial PVC.

In PVC specimens the number of unstable groups depends on the
method of production, finishing, and storage of the polymer. The
presence of unstable groups brings about complex dehydrochlorination
of the PVC, including a number of reactions which can occur either
in series or in parallel.

1. Statistical elimination of HCl from the normal chloroalkane groups of the PVC, leading to the formation of β-chloroallyl
groups with rate constant k_c; the value of this constant is assessed
in [113, 121]. For instance, at 175°C, $k_c = (0.8 \pm 0.1) \cdot 10^{-7}$ sec^{-1}
[113], which practically coincides with the rate constant of dehydrochlorination of the model compound, 8-chlorohexadecane (Table 3.5).
As was to be expected, k_c is a fundamental characteristic of PVC,

depending neither on the MM nor on the method of production of the polymer [113].

2. Formation of polyene systems which are initiated by β-chloroallyl and ketoallyl groups. The presence of a double bond considerably assists the process of elimination of HCl. This is confirmed by data showing that with mild chlorination of PVC there is a considerable reduction in the rate of its dehydrochlorination [109, 113]. Supporting this, we should also adduce the experimental data obtained for model compounds in the gaseous and liquid phases (see Chap. 3), from which it is possible to estimate the rate constants of formation of diene (k_{1p}) and triene (k_{2p}) systems from fragments containing a β-chloroallyl group. At 175°C, k_{1p} is within the limits $(1-0.5) \cdot 10^{-4}$ sec^{-1} , and k_{2p} within the limits $(3-4) \cdot 10^{-2}$ sec^{-1}. At the present time it is, unfortunately, not clear what the rate constant for the formation of polyenes with a large number of conjugated double bonds is.

In fragments of PVC with ketoallyl groups an electron-withdrawing carbonyl group considerably facilitates the process of elimination as compared with fragments containing β-chloroallyl groups [122, p. 245 of Russian translation]. The rate constant k_{1p}^{CO} can be determined from the data in Fig. 7.25; at 175°C, k_{1p}^{CO} is 7.5 × 10^{-3} sec^{-1}.

3. Reactions of terminating the growth of the polyene chain as a consequence of parallel reactions of intermolecular or intramolecular cyclization of the polyenes being formed.

The equation describing the dehydrochlorination of PVC, while taking into account the proposed reactions of HCl elimination, is complex. Minsker and Berlin [108, p. 41] obtained a kinetic equation describing the dehydrochlorination of a PVC containing ketoallyl groups:

$$\alpha = \left(2k_c + \frac{k_{1p}^{CO}k_c}{k_d} \right) t - \frac{k_c}{k_{1p}} \cdot \frac{k_d - k_{1p}^{CO} - k_{1p}}{k_d - k_{1p}} (1 - \exp(-k_{1p}t))$$
$$+ \frac{k_{1p}^{CO}}{k_d} \left(c_0 - \frac{k_c}{k_d} + \frac{k_c}{k_d - k_{1p}} \right) (1 - \exp(-k_d t)) \qquad (7.44)$$

where k_d is the rate constant for the destruction of the polyene chain.

For the initial stages of the dehydrochlorination of PVC with $k_{1p} \cdot t$ and $k_d t \ll 1$ the following equation holds:

$$\alpha = \left(k_c + k_{1p}^{CO}c_0 \right) t = k_{eff}^0 t \qquad (7.45)$$

which defines the linear dependence of the yield of HCl with time, which in fact occurs in the majority of experiments [106-111, 123].

The mechanism of noncatalytic dehydrochlorination is not clear at the present time although there have been attempts to describe it [108]. Generally, the formation of olefins from halogen alkyls may occur as indicated by any of three mechanisms [124, p. 96].

1. The monomolecular mechanism (E_1):

$$\sim CH_2-CHCl\sim \; \rightleftharpoons \; \sim CH_2-\overset{+}{C}H\sim + Cl^- \qquad (slow)$$

$$\sim CH_2-\overset{+}{C}H\sim + S \longrightarrow \sim CH=CH\sim + SH^+ \qquad (fast)$$

where S is the solvent.

Ordinarily this mechanism applies in ionized solvents (there is no foundation for proposing the possibility of this mechanism in a polymer matrix with a permittivity around 4).

2. The bimolecular mechanism (E_2):

$$\sim CH_2-CHCl\sim + B \; \rightleftharpoons \; \underset{\underset{H\cdots B}{|}}{\sim CH-CHCl\sim} \longrightarrow \sim CH=CH\sim + Cl^- + BH^+$$

where B is a base.

The elimination reaction suggested by this mechanism takes place in the presence of strong bases, but the occurrence of this mechanism in commercial polymers is of low probability.

3. The cyclic mechanism. Cyclic elimination is a monomolecular single-stage process (unlike mechanism E_1), which involves a chlorine atom which is eliminated, taking with it a β-proton:

$$\underset{\sim CH-CH\sim}{\overset{\overset{\displaystyle H \quad Cl}{|\quad\;\, |}}{}} \longrightarrow \sim CH=CH\sim + HCl$$

This process is not ionic since it does not proceed via the formation of ions in the free state. But, in individual cases, it is possible on examination to adduce an analogy with ionic processes. For instance, in the pyrolysis of hydrogen alkyls there is the same dependence of the reactivity on the structure of the compound as there is with mechanism E_1. This means that the transition state is somewhat reminiscent, in structure, of the ion pair R^+Cl^-. For instance, it may be imagined that the elimination of HCl from the chloroallyl group proceeds via the formation of an allyl zwitterion:

$$\sim CH{=}CH{-}CHCl{-}CH_2\sim \longrightarrow \sim \overset{\delta+}{C}H{=}CH{-}\overset{\delta-}{C}H{-}CH_2\sim$$
$$\underset{\displaystyle Cl}{\mid}$$

It would appear that the cyclic mechanism is the most applicable
for the noncatalytic dehydrochlorination of PVC.

7.7.2. Catalytic Dehydrochlorination

Catalytic dehydrochlorination of PVC occurs when the rate of
diffusion of the hydrogen chloride is commensurate with the rate of
noncatalytic dehydrochlorination or when the reaction occurs in an
atmosphere of hydrogen chloride.

Solubility of Hydrogen Chloride in PVC. Hydrogen chloride dissolves in PVC; evidence of this is the kinetic data which show that
the rate of the catalytic reaction is proportional to the volume
of the polymer article [125]. However, attempts to obtain quantitative
data by a gravimetric method (the McBain balance) proved fruitless [125].

In [126], the solubility of hydrogen chloride in the polymer
was determined by a manometric method, from the reduction in the
pressure of HCl in a vessel of volume v_0, containing m grams of the
polymer. The solubility coefficient σ was determined from the equation

$$\sigma = \frac{p_0 v_0 - p_1 (v_0 - \Delta v)}{p_1} \; \frac{\rho}{m} \tag{7.46}$$

where p_0 and p_1 are, respectively, the initial and final (after introduction of the polymer) pressures of HCl, and Δv the change in
volume of the system.

The found solubility ($\sigma_{20^\circ} = 5.5$) is, as far as can be seen,
on the high side because of strong adsorption of the HCl. Nevertheless, these data are of considerable interest since they show
that the solubility of the HCl rises in the course of the degradation:

$$\sigma = \sigma_0 (1 + \beta\alpha) \tag{7.47}$$

where β is a coefficient of proportionality, 63 at 20°C, and α the
degree of dehydrochlorination.

Active Sites of Catalytic Dehydrochlorination of PVC. Generally, these include chloroalkane, β-chloroallyl, and ketoallyl groups,
and also chlorovinyl groups containing tertiary carbon atoms.

TABLE 7.4. Dehydrochlorination Rate Constants of 4-Chloro-2-hexene
in Various Solvents in the Presence of 0.8 M HCl at 60°C

Solvent	Permittivity	$K \cdot 10^{-4}$, min^{-1}	Solvent	Permittivity	$K \cdot 10^{-4}$ min^{-1}
Methanol	32.6	8	Dioxane	2.2	1.0
1,2-Dichloroethane	10.3	3.5	Heptane	2.0	0.8
Tetrahydrofuran	7.2	0	2-Ethylhexyl phthalate (DOP)	11.0	1.5

There has been practically no demonstration of the role of
chloroalkane groups as active sites of catalytic dehydrochlorination.
Pocker [127] found that HCl is a powerful electrophilic catalyst
in media with low permittivity, and considerably raises the rate
of dehydrochlorination of tert-butylchloride in these media.

β-Chloroallyl groups are π-bases and consequently are capable
of combining with HCl. The existence of complexes of β-chloroallyl
groups with hydrogen chloride was first shown by Schlimper [128]
and Onozuka et al. [129].

The role of these complexes in catalytic dehydrochlorination
is examined fully in [130]. It is found that the rate of dehydrochlorination of 4-chloro-2-hexene increases with increase in the
polarity of the solvent (a favorable condition for complex formation) (Table 7.4).

According to the authors in question, the complex has the characteristics of a charge transfer complex. The most rigorous proof
that β-chloroallyl groups are active sites for catalytic dehydrochlorination are obtained in [131], in which there are determined
the number of scissions in the PVC macromolecules with the elimination of $P(RO)_3$ of ozonized specimens of the polymer after degradation.

If the active sites of catalytic dehydrochlorination are β-
chloroallyl groups, then $\bar{M}_n$ of specimens which have been subjected
to degradation in the presence or absence of HCl is bound to be practically identical since the removal of polyene fragments has no influence on $\bar{M}_n$ (the length of the polyene fragments is much less than
that of the "splinters" of PVC which include not fully reacted chloroalkane groups).

If the active sites are chloroalkane groups, then the $\bar{M}_n$ of
the PVC specimens is bound to change during degradation in the presence of HCl since the formation of double bonds in the PVC molecules
is by a random process. Experiments at temperatures close to room
temperature show that the change in the number of scissions with
time in specimens undergoing degradation in the presence or absence

of HCl remains practically identical, i.e., under the conditions
in question the reactivity of the β-chloroallyl groups is far higher
than that of the chloroalkane groups.

It is to be expected that the ketoallyl groups will also serve
as active sites in catalytic dehydrochlorination.

The chlorovinyl groups with tertiary carbon atoms are π-bases
and, consequently, can be active sites of catalytic dehydrochlorination. It is suggested in [132] that active sites of the general
formula $R_1R_2R_3CCl$ are formed in the interaction of HCl with polyenes with subsequent formation of the hydrochlorination product.
In point of fact, with the addition of Ph_3Cl the rate of dehydrochlorination of the PVC rises considerably [132].

<u>Macrokinetics of Catalytic Dehydrochlorination of PVC</u>. The
general equation for catalytic dehydrochlorination takes the form

$$\frac{dc_{HCl}}{dt} = \frac{dc_=}{dt} = \sum_i k^k_{eff_i}\, c_{HCl}\, c(a.s.)_i \tag{7.48}$$

where c_{HCl} is the concentration of HCl in the polymer; $c_=$ the concentration of double bonds forming, and $c_{a.s.}$ the concentration of
active sites in the polymer.

At temperatures which are close to room temperature, the active
sites are the β-chloroallyl and ketoallyl groups present in the polymer since noncatalytic dehydrochlorination occurs at a very low rate.
With low degrees of dissociation it may be assumed that the concentration of active sites will be practically constant. If the pressure of HCl in the system remains unchanged during the reaction,
then the degree of dissociation, as determined from the increase
in the number of double bonds in the polymer, is found from the equation

$$\alpha = k_{eff}t \tag{7.49}$$

where $k_{eff} = k^k_{eff} c^0_{HCl} c^0_{a.s.}$

An equation like this was used to express the experimental data
on the catalytic dehydrochlorination of thin PVC films (diffusion
limitations were not present) in HCl gas and vapor at 60°C [125].

In [106, 133] the reaction of catalytic dehydrochlorination
was investigated in a closed volume at high temperatures. During
the course of the reaction the rise in pressure of the hydrogen
chloride in the system was followed. At high temperatures (>185°C)
noncatalytic dehydrochlorination took place at a perceptible rate
and the concentration of active sites, e.g., β-chloroallyl groups,

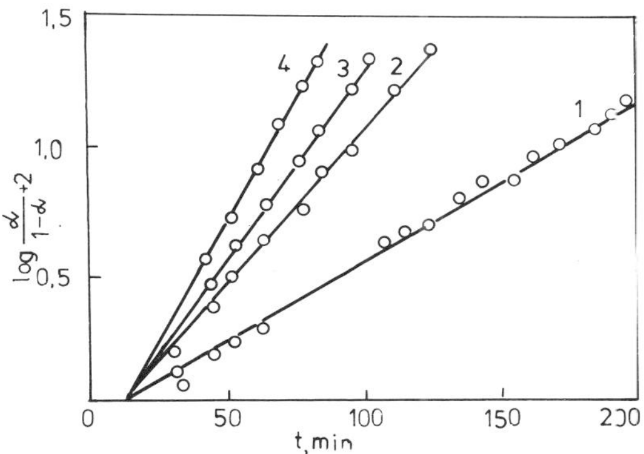

Fig. 7.26. Experimental data for the catalytic
dehydrochlorination of PVC at 210°C
at various pressures plotted according to Eq. (7.50) [106]: 1) vacuum;
2) 72 torr; 3) 137 torr; 4) 233 torr.

rose during the course of the reaction. Thus, the autocatalytic
character of dehydrochlorination of PVC is brought about, in the
case in question, by two factors: a rise in the concentration of
hydrogen chloride in the polymer, and an increase in the number of
active sites for catalytic dehydrochlorination.

The general kinetic equation of the reaction is complex in form
[125]; nevertheless, the experimental data obtained in [106, 133]
are, at fairly low degrees of dissociation, expressed by the equation (Fig. 7.26)

$$\log \frac{\alpha}{1-\alpha} = \log \alpha_0 + 0.434 k^k_{eff} c^0_{a.s.} t \tag{7.50}$$

where α is a parameter connected with the pressure of HCl in the
system.

The values of the products $k^k_{eff} c^0_{a.s.}$ calculated from Eq.
(7.50) at various initial pressures of HCl for the dehydrochlorination of powdered PVC at 210°C are as follows [106]:

p_{HCl}, torr	0	72	137	233
$k^k_{eff} c^0_{a.s.} \cdot 10^4$, sec^{-1}	1.0	2.0	2.3	2.8

The change in $k^k_{eff} c^0_{a.s.}$ with rise in p^0_{HCl} may be brought
about by the following factors: nonobservance of Henry's law, together with a change in the concentration of the complexes of the

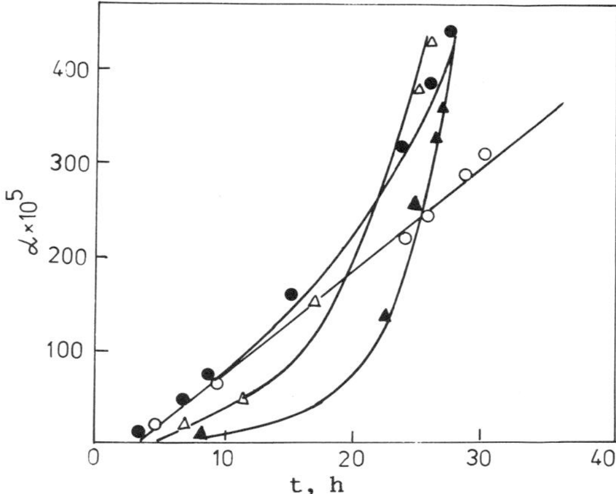

Fig. 7.27. Dependence of the degree of dissociation of HCl on the time of dehydrochlorination of PVC disks of various
thicknesses: o) 0.05 cm; ●) 0.1
cm; ▲) 0.2 cm; △) 0.4 cm at 140°C
[136].

active sites with HCl, with increase in the concentration of the latter in the polymer.

The formation of polyene systems with catalytic dehydrochlorination of PVC according to the kinetic scheme has received practically no consideration. According to Minsker [131], an increase
in dehydrochlorination in the presence of HCl is brought about
principally by an increase in the rate of growth of polyene systems,
initiated by ketoallyl groups. The length of the polyene systems
is greatly influenced by the temperature and concentration of HCl
in the polymer [134].

However, according to [132] there is no significant difference
in the polyene structure of PVC specimens degraded in the presence
of HCl or in vacuum.

The kinetic pattern of dehydrochlorination of PVC films with
the simultaneous occurrence of noncatalytic and catalytic reactions
is studied in [135, 136].

When dehydrochlorination occurs under diffusion-kinetic conditions, the change in the concentration of HCl in the polymer may
be expressed by means of the equation

$$\frac{dc_{HCl}}{dt} = D\,\frac{d^2c_{HCl}}{dX^2} + k_{eff}^0 + k_{eff}^k\,c_{a.s.}c_{HCl} \tag{7.51}$$

TABLE 7.5. Values of $k_{eff}^{k} c_{a.s.}^{0}$ and D_{HCl} at Various Temperatures

t, °C	$k_{eff}^{k} c_{a.s.}^{0} \cdot 10^{4}$, sec^{-1}	$D \cdot 10^{8}$, cm^{2}/sec	Reference	t, °C	$k_{eff}^{k} c_{a.s.}^{0} \cdot 10^{4}$, sec^{-1}	$D \cdot 10^{8}$, cm^{2}/sec	Reference
120	0.075 ± 0.005	3.5 ± 0.5	[136]	185	1.1	—	[133]
130	0.19 ± 0.02	5.5 ± 0.5	[136]	200	1.7	33	[135]
140	0.44 ± 0.02	7.5 ± 0.5	[136]	210	1.7	—	[110]
180	1.0	—	[137]				

where D is the diffusion coefficient of the HCl in the polymer.

The equation has an analytical solution if it is assumed that the concentration of active sites for catalytic dehydrochlorination remains unchanged ($c_{a.s.} \approx c_{a.s.}^{0}$).

The solution of Eq. (7.51) takes the form

$$c_{HCl} = -\frac{2k_{eff}^{0}}{\pi} \sum_{n=1}^{\infty} \frac{1-\cos n\pi}{n\left(k_{eff}^{k} c_{a.s}^{0} - Dn^2\pi^2 l^{-2}\right)} \left[1 - \exp t\left(k_{eff}^{k} c_{a.s.} - Dn^2\pi^2 l^{-2}\right)\right] \sin \frac{n\pi x}{l}$$

$$(7.52)$$

where l is the thickness of the film.

The amount of HCl which has diffused through unit surface in time dt is

$$dm = -D\left[\frac{dc_{HCl}}{dx}\right]_{x=e} dt$$

$$= \frac{2Dk_{eff}^{0}}{l} dt \sum_{n=1}^{\infty} \frac{(1-\cos n\pi)\cos n\pi}{\left(k_{eff}^{k} c_{a.s.}^{0} - Dn^2\pi^2 l^{-2}\right)} \left[1 - \exp t\left(k_{eff}^{k} c_{a.s.}^{0} - Dn^2\pi^2 l^{-2}\right)\right] \quad (7.54)$$

The final solution for the degree of dissociation α may be presented as follows:

$$\alpha \approx \frac{4Dk_{eff}^{0}}{l^2 m_{\infty}} \sum_{n=1}^{\infty} \frac{1-\cos n\pi}{\left(k_{eff}^{k} c_{a.s.}^{0} - Dn^2\pi^2 l^{-2}\right)^2} \left[\exp t\left(k_{eff}^{k} c_{a.s.}^{0} - Dn^2\pi^2 l^{-2}\right)\right.$$

$$\left. - t\left(k_{eff}^{k} c_{a.s.}^{0} - Dn^2\pi^2 l^{-2}\right) - 1\right]$$

$$(7.53)$$

where m_{∞} is the theoretical amount of HCl which can be evolved from unit volume of the polymer.

Examination of Eq. (7.54) shows that there can be two limiting cases:

$k^k_{eff}c^0_{a.s.} \gg D\pi^2 l^{-2}$, the dehydrochlorination reaction takes place under autocatalytic conditions, and there is a sharp increase in the rate of the process;

$k^k_{eff}c^0_{a.s.} < D\pi^2 l^{-2}$, acceleration of the dehydrochlorination reaction is not observed, and the maximum rate is reached at a time when the exponential term is infinitely small. Figure 7.27 shows a typical change in the degree of dissociation α with time for the dehydrochlorination of PVC specimens of various thicknesses.

Figure 7.27 shows that the rate of the process increases with increase in the thickness of the specimens. Pudov [135] was the first to introduce the concept of a critical thickness of the specimen, i.e., the thickness l at which there is a transition of the degradation process from steady-state to autoaccelerated. The critical value l_{cr} is found from the equation

$$\frac{\pi^2 D}{l^2_{cr}} = k^k_{eff}c^0_{a.s.} \qquad (7.55)$$

Table 7.5 gives the values of $k^k_{eff}c^0_{a.s.}$ and D obtained by various authors.

The activation energy of catalytic dehydrochlorination is determined in [136] as 117 ± 8 kJ/mole. At the present time there is a lack of any data, except the work already cited [130], on the mechanism of catalytic dehydrochlorination of PVC. It may be suggested that the reactive entity is the π-complex

$$[H^+Cl]$$
$$\sim HC{=}CH{-}CHCl{-}CH_2\sim$$

Such a π-complex, above all, will not form a carbonium ion, but will promote the elimination of HCl. At high temperatures the possibility of a direct attack of the HCl on the chloroalkane group is not excluded [138]

$$\sim CH_2{-}CH{-}CH{-}CHCl\sim$$
$$\qquad \quad | \quad \ |$$
$$\qquad \quad Cl \ \ H$$
$$\qquad \quad H{-}Cl$$

The dehydrochlorination reaction is accelerated not only in the presence of HCl, but also in the presence of various other acids.

For instance, degradation of PVC takes place in the presence of sulfuric acid; nevertheless, in the presence of H_2SO_4—HNO_3 PVC is practically stable [139]. Apparently the nitric acid molecules combine with the double bonds and thus suppress catalytic dehydro-chlorination.

In the presence of bases PVC undergoes double dissociation [140]. In the first place, there is an elimination reaction

$$\underset{B:\longrightarrow H}{\overset{\overset{\textstyle Cl}{|}}{\sim CH-CH\sim}} \longrightarrow \sim CH=CH\sim + BH^+ + Cl^-$$

and, second, a substitution reaction

$$\underset{\overset{\uparrow}{B:}}{\overset{\overset{\textstyle Cl}{|}}{\sim CH_2-CH}} \longrightarrow \underset{\overset{|}{B^+}}{\sim CH_2-CH} + Cl^-$$

In recent years a method of utilizing PVC waste has been developed, by treatment with aqueous strong basic solutions at high temperatures [141].

APPENDIX 7.1. Kinetic Parameters of Degradation of Polyamides in Aggressive Media

Polymer	Aggressive medium	Temp., °C	Kinetic parameters	References
PCA ($\bar{P}_V = 160$)	40% H_2SO_4 in water	50	Under homogeneous conditions, in solution. Random decomposition, $k_{rand} = 1 \cdot 10^{-3}$ h^{-1}	[6]
PCA ($\bar{P}_V = 160$)	40% H_2SO_4 in water	25	Under homogeneous conditions, in solution. Random decomposition, $k_{rand} = (0.50 \pm 0.6) \times 10^{-3}$ h^{-1}	[7]
Poly(hexamethylene adipamide) ($\bar{M}_n = 3000$)	60% H_2SO_4 in water	40–60	Under homogeneous conditions, in solution. Random decomposition, k_{rand} (h^{-1}): 40°C, $0.44 \cdot 10^{-4}$; 50°C, $1.10 \cdot 10^{-4}$; 60°C, 2.03×10^{-4}	[8]
PCA ($\bar{P}_n = 15$)	50% H_2SO_4 in water	30–50	Under homogeneous conditions, in solution. Random decomposition	[9]
	25% HCl in water	72	Under homogeneous conditions, in solution. Random decomposition, $k_{rand} = 0.36$ h^{-1}	[5]
PCA	25% HCl in water	110	Under homogeneous conditions, in solution. Random decomposition	[4]
PCA ($\bar{M}_n = 12{,}500$)	40–60% H_2SO_4 in water	110–140	Under homogeneous conditions, in solution. Random decomposition $$\kappa_{eff} = 2.5 \cdot 10^9 \exp\left(-\frac{23\,000}{RT}\right) \frac{c_{H_S^+}}{1 + \dfrac{k_A}{5.0}} \quad (\text{sec}^{-1})$$	[3]
PCA ($\bar{M}_V = 20{,}000$)	40 and 92% H_2SO_4 in water	98–118	Under homogeneous conditions, in solution, random decomposition. In 40% H_2SO_4, at 98°C, k_{rand} (min^{-1}) $(2.5 \pm 0.2) \cdot 10^{-3}$; at 108°C $(10.7 \pm 0.8) \cdot 10^{-3}$. In 92% H_2SO_4 at 108°C degradation does not occur within 7 h	[10]

PCA ($\bar{M}_V$ = 20,000)	0.05-0.2 H_2SO_4 in water	96-101	Under heterogeneous conditions, in fiber. The reaction is internal kinetically controlled. With increase in the degree of orientation the degradation rate falls	[11]
PCA ($\bar{M}_n$ = 8,800)	4 N HCl in water	96	Under heterogeneous conditions, in film. The reaction takes place under internal kinetic control. Degree of accessibility of the amide groups ~0.7	[12]
PCA	50% H_2SO_4 in water	30-50	In solution, under homogeneous conditions. Random decomposition	[142]
PCA	H_3PO_4 and its salts	255	Under homogeneous conditions, in the melt. The reaction product is a caprolactam	[143]
PCA	H_3PO_4	310	Under homogeneous conditions, in the melt. The reaction product is a caprolactam	[141]
Poly(amido acid) based on anilinefluorene and 3,3',4,4'-tetra-carboxydiphenyl-2,2'-propane	pH 0-7	25	Under homogeneous conditions, in solution $$k_{\text{eff}} = \frac{k^0_{\text{act}}}{1 + K_{\text{diss}} h_A^{-1}}$$	[13]
PCA ($\bar{M}_V$ = 20,000) Polydodecanamide ($\bar{M}_V$ = 24,000)	Subcutaneous tissue of rabbits	37	In accessible regions. Random, internal kinetically controlled, $k_{\text{rand}} \cdot c_{\text{cat}}$ for the PCA and polydodecanamide = $(2.3 \pm 0.2) \times 10^{-3}$ day^{-1} and $(2.5 \pm 0.3) \cdot 10^{-3}$ day^{-1}, respectively; from the surface of the specimens, k_{eff} for the PCA and polydodecanamide = $(2.7 \pm 0.3) \cdot 10^{-5}$ and $(5 \pm 1) \cdot 10^{-5}$ g $\times$ (day$\cdot$cm^2)$^{-1}$ respectively	[14]

APPENDIX 7.2. Kinetic and Diffusion Parameters of Degradation of Polyesters in Aggressive Media

Polymer	Aggressive medium	Temp., °C	Kinetic and diffusion parameters	References
PETP ($\bar{M}_V$ = 18,000)	H_2O, vapor	60-175	Under heterogeneous conditions, in film (ℓ = 125 μm), internal kinetic control, k_{H_2O} = 10^9 exp $(-26,000/RT)$ min^{-1}	[16]
PETP ($\bar{M}_V$ = 18,000)	H_2O	25-100	Under heterogeneous conditions, internal control, E_{eff} = 109-113 kJ/mole	[18, 19]
PETP ($\bar{M}_V$ = 21,000)	H_2O	100	Under heterogeneous conditions, internal kinetic control. At 100°C degradation takes place under autocatalytic conditions, k_{eff} = $(7.5 \pm 1.5) \cdot 10^{-3}$ $cm^3 \cdot mole^{-1}$, min^{-1}	[20]
PETP ($\bar{M}_V$ = 18,600-21,800)	1-7 M HCl in water	50-70	Under heterogeneous conditions, in fiber form with differing r, internal kinetic control. at 70°C, c_n = $k_{eff}(c_{HCl}K_{diss} \times K_{distr})^{1/2}t$, where k_{eff} = $(1.1 \pm 0.1) \times 10^{-6}$ g/mole·h. k_{eff} does not depend on the degree of crystallinity in the range amorphous—48% crystallinity, and decreases with increase in the fiber orientation	[21]
PETP ($\bar{M}_V$ = 19,000)	53-70% H_2SO_4 in water	90-126	Under heterogeneous conditions, in film of various thicknesses (5-80 μm). Internal diffusion-kinetic control. At 90°C, $D_{H_2SO_4}$ = $(1.5 \pm 0.5) \cdot 10^{-12}$ cm^2/sec	[15]
PETP ($\bar{M}_V$ = 18,000)	20% KOH in water	25	Under heterogeneous conditions, in film form, external diffusion-kinetic control	[145]
PETP ($\bar{M}_V$ = 19,000)	8.2 ± 39.1% KOH	27-93	Under heterogeneous conditions, in film of various thicknesses. External diffusion-	[25]

			kinetic control $m_n = m_n^0$ $$-\frac{6,4\cdot10^6\exp\left(-\dfrac{16\,500}{RT}\right)a_{H_2O}st}{1+4\cdot10^2 b_0}$$ (m_n in g; s in cm², and t in min), where b_0 is the average size of the crystal	
PETP ($\bar{M}_V$ = 19,000)	10-35% KOH, 10-25% NaOH, 3-9.8% LiOH in water	32-92	Under heterogeneous conditions, in fiber, external diffusion-kinetic control $$m_n = m_n^0\left[1-\frac{6,4\cdot10^6\exp\left(-\dfrac{16\,500}{RT}\right)a_{H_2O}t}{(1+4\cdot10^2 b_0)\overline{r_0}\rho}\right]^2$$	[24]
Polyesters prepared by polycondensation of dicarboxylic acids and diols ($\bar{M}_V$ = 1000)	0.5 N KOH in a water–acetone mixture	20	Under homogeneous conditions, in solution. Determination of k_{eff} for the various polyesters	[31]
Polyesters prepared by polycondensation of dicarboxylic acids and diols	$1.4\cdot10^{-2}$-1.7 × 10^{-1} N HCl in water	105	Under homogeneous conditions, in the melt. At 105°C, $k_{eff} = 1.5\cdot10^{-1}c_{HCl}$ (min^{-1})	[32]
Poly(1,4-cyclohexylene dimethyleneterephthalate)	0.001-0.2% H_2O in the melt	295-330	Under homogeneous conditions, in the melt. Random decomposition, E ~ 172 kJ/mole	[146]
PMMA	13.6% H_2SO_4 in a mixture of CH_3COOH–H_2O (90:10)	115	Under homogeneous conditions, in solution. The reactivity of the ester groups in isotactic triads is 16 or 43 times as high as in hetero- and syndiotactic triads	[147]
Copolymers of methyl methacrylate and diphenylmethyl methacrylate with methacrylic acid	0.5-0.2 M KOH in water. Pyridine–H_2O (95:5)	145	Under homogeneous conditions, in solution. Determination of the rate constants of hydrolysis of the ester groups in the triads	[34]

APPENDIX 7.2 (continued)

Polymer	Aggressive medium	Temp., °C	Kinetic and diffusion parameters	References
Copolymers of vinyl acetate with styrene, methyl methacrylate, and vinyl phthalimide	0.2 M H_2SO_4 in water	90	Under homogeneous conditions, in solution. In the course of the process autoacceleration takes place. Determination of the rate constants of hydrolysis of copolymers of differing composition	[148]
Polyesters prepared by polycondensation of diethyleneglycol with dicarboxylic acids	Ethanol—benzene (1:1)	20	Under homogeneous conditions, in solution. Determination of the rate constants of solvolysis of polyesters of differing composition	[149]
Polyglycolid	pH = 2.01-14.0	37-100	Under heterogeneous conditions, in solution. Internal kinetic control and, at pH < 10, from the surface of crystallites and at pH > 10 from the surface of the fiber $$\frac{m}{m_0} = 1 - 0.2[1 - \exp(-k_{\text{rand}}t)\,\overline{c}^v_{H_2O}]^2 - 0.8\left[-\frac{2k^{s-v}_{\text{eff}}t\overline{c}^{s-v}_{H_2O}}{N\overline{b}_0\rho} - \left(\frac{k^{s-v}_{\text{eff}}t c^{s-v}_{H_2O}}{N\overline{b}_0\rho}\right)^2\right] - \frac{2k^s_{\text{act}}t c_{OH^-}}{K_p r_0\rho} + \left(\frac{k^s_{\text{act}}t c_{OH^-}}{K_p \overline{r}_0\rho}\right)^2$$	[14]
Polyacrylate prepared by polycondensation of phenolphthalein and terephthalic acid	5-25% KOH 60% H_2SO_4 in water	25-60 100-130	In basic solutions, external diffusion-kinetic control $$m_n = m_n^0 - 6\cdot10^3\left(\exp-\frac{13\,500}{RT}\right)a_{H_2O}b_0 st$$ In acid solutions, internal kinetic control	

APPENDIX 7.3. Kinetic Parameters of Degradation of Polyacetals in Aggresive Media

Polymer	Aggressive medium	Temp., °C	Kinetic parameters	References
Polyoxymethylene (POM)-OH	n-Butylacetate	20	Under homogeneous conditions, in solution, intrinsic viscosity decreases 2.5-fold in 300 h	[79]
Poly-1,3-dioxalane ($\bar{M}_W = 17,000$)	Picric acid, 1 mass %	140-250	Under homogeneous conditions, in solution, $\log (\bar{M}_W/\bar{M}_{W_0}) \sim \alpha$ for all temperatures, $E_{eff} = 71 \pm 8$ kJ/mole	[78]
Poly-1,3-dioxalane ($\bar{M}_W = 1,200-40,000$)	$10^{-5}-10^{-3}$ mole × (liter)$^{-1}$ HCl in water	22-75	Under homogeneous conditions in solution, $$k_{eff} = 5 \cdot 10^6 \exp\left(-\frac{17\,500}{RT}\right) c_{HCl}\,(c^{-1})$$	[80]
POM-OCH$_3$	Picric, salicylic, and hydrobenzoic acids, β-naphthol	190	In the melt, $\log w = 2.1-0.2\,pK_a$, where w is the rate of reduction in the mass of the polymer (in h^{-1})	[81]
Poly-3,4-acrolein	$3.15 \cdot 10^{-3}$ mole × (liter)$^{-1}$ H$_2$SO$_4$ in dioxane	15.5-35	Under homogeneous conditions, in solution. Random decomposition with depolymerization; at low degrees of dissociation $$m = m_0\left(1 - \frac{k_{rand}k_{dep}}{k_{form}}t\right)$$	[86]
POM-OH ($\bar{M}_V = 390,000$) POM-OOCCH$_3$ ($\bar{M}_V = 50,000$)	Hexafluoroacetone + 2% triethylamine	25	Under homogeneous conditions, in solution, increase in $\bar{M}_V$ in the course of the experiment, $\log \alpha \sim k_{eff}t$	[150]
POM-OH POM-OOCCH$_3$	Phenol, p-chlorophenol, p-nitrophenol	90-130	Under homogeneous conditions, random decomposition with depolymerization $$\bar{P}_n = \frac{\bar{P}_{n_0}\exp(-at)}{1 + \frac{k_{rand}}{a}\bar{P}_{n_0}[1-\exp(-at)]}; \quad a = \frac{k_{rand}\,k_{dep}}{k_{form}}$$	[88]

Polymer	Aggressive medium	Temp., °C	Kinetic parameters	References
POM-OOCH$_3$	0.9-3.41 mole × (liter)$^{-1}$ di-chloroacetic acid in methylene chloride	30-33	Under homogeneous conditions, in solution, reduction in mass $k_{eff} = 2.5 \cdot 10^{10} \exp\left(-\dfrac{19\,100}{RT}\right) c_{HA}(c^{-1})$	[87]
POM-OOCH$_3$ ($\bar{M}_V$ = 50,000)	2-20% citric acid	Low-temp. cond., heat rate 3°/min	In the melt, random decomposition with de-polymerization, $\log \alpha = k_{eff} t c_{HA}^{0.5}$ for 20% acid. Rate-determining step is random decomposition, $\alpha = k_{eff} t$ for 2% acid. Rate-determining step is depolymerization	[89]
POM-OCH$_3$ ($\bar{M}_V$ = 1,500, 6,500, 10,500, and 19,500) POM-OOCH$_3$ ($\bar{M}_V$ = 50,000 and 100,000) POM-OH ($\bar{M}_V$ = 7,000 and 220,000)	4.4·10^{-3} mole × (liter)$^{-1}$ HCl, 8.9 mole/liter H$_2$O in hexa-fluoroacetone	30	Under homogeneous conditions, random decom-position with depolymerization $\alpha = 1 - \exp\left\{\left(\dfrac{2k_{dep}}{k_{rand}} + 1\right)\ln[2 - \exp(-k_{rand} \cdot t)] - k_{rand} \cdot t\right\}$ $k_{dep} = 30 \pm 10,\ k_{rand} = 2.3 \pm 0.2$ liter/(mole·min)	[77]
POM-OH ($\bar{M}_V$ = 7,000 and 220,000) POM-OOCH$_3$ ($\bar{M}_V$ = 100,000) POM-OCH$_3$ ($\bar{M}_V$ = 6,600 and 19,500)	5-50% H$_2$SO$_4$ in water	25-70	Powder, diffusion-kinetic control $(1-\alpha)^{1/3} = k_{dep} c_{HS}^{+} t \left(\dfrac{1}{\bar{P}_{n_0}} + \dfrac{k_{rand} h_0}{k_r c_{HS}^{+}}\right)$ At 25°C, $k_{dep} = (1.2 \pm 0.3) \cdot 10^{-1}$ min^{-1}, $k_{rand} \sim 10^{-8}$ min^{-1}	[92]
POM-OOCH$_3$ ($\bar{M}_V$ = 50,000)	5-50% H$_2$SO$_4$ 26-32% HCl 55-73% H$_3$PO$_4$	25-80	Under heterogeneous conditions, slab form, diffusion-kinetic control, $\alpha \sim 1 - k_{eff} t h_0$ for $h_0 > 100$; σ is a nonlinear function of α	[151]

APPENDIX 7.4. Kinetic Degradation of Polysiloxanes in Aggressive Media

Polymer	Aggressive medium	Temp., °C	Kinetic and diffusion parameters	References
Polydimethylsiloxane	H_2O, alkali, acid	150	Under homogeneous conditions, random decomposition, internal-kinetic control $$-\frac{dc}{dt} = k_1 c^2 - k_2 (c_0 - c)\, c_{H_2O},$$ where c and $(c_0 - c)$ are, respectively, the contents of silanol and siloxane groups, $E_{noncat} = 97$ kJ/mole, $E_{cat} = 25$ kJ/mole	[101]
Polydimethylsiloxane with SiOH and SiOK terminal groups	20 mass % of H_2O	200–300	Under homogeneous conditions, random decomposition with depolymerization of fragments containing silanol groups, internal-kinetic control $$\frac{\overline{P_n}}{\overline{P_{n_0}}} \sim (1-\alpha)^{1/2} \quad E_{eff} = 71 \pm 4 \text{ kJ}$$ for the initial rates of change of mass	[100]
Polydimethylsiloxane $(\overline{M}_V = 5.7\cdot10^4)$	H_2O, pH = 1	20	Random decomposition in a monolayer on the surface of an aqueous solution $$\ln\left| 1 - \left(\frac{a_0 - a_t}{a_0}\right)^{1/2} \right| = -kt,$$ where a_0 and a_t are the areas of an elementary unit of the polymer in the condensation of an actual monolayer of PDMS, $k = 7.2\cdot10^{-3}$ min^{-1}	[102]

REFERENCES

1. Z. A. Rogovin et al., Zh. Obshch. Khim., 17, 1316 (1947).
2. V. A. Bershtein and L. M. Egorova, Vysokomol. Soedin., A, 19, No. 6, 1260 (1977).
3. P. P. Nechaev, Yu. V. Moiseev, and G. E. Zaikov, Vysokomol. Soedin., A, 14, No. 5, 11048 (1972).
4. D. Heikins, P. H. Hermans, and H. A. Veldhoveh, Makromol. Chem., 30, No. 2, 154 (1959).
5. D. Heikens, J. Polym. Sci., 35, 277 (1959).
6. A. Matthes, J. Prakt. Chem., 162, 245 (1943).
7. A. Matthes, Makromol. Chem., 5, 165 (1950).
8. A. M. Lignori and A. Mele, Gazz. Chim. Ital., 82, 828 (1952).
9. K. Hoshino and M. Watanabe, J. Am. Chem. Soc., 73, 4816 (1951).
10. V. A. Myagkov and A. B. Pakshver, Kolloidn. Zh., 14, 172 (1952).
11. E. K. Mankash and A. B. Pakshver, Zh. Fiz. Khim., 25, 468 (1951).
12. L. I. Razumovskii et al., Vysokomol. Soedin., A, 19, No. 6, 1358 (1977).
13. P. P. Nechaev, Candidate's Dissertation, Inst. Khim. Fiz. Akad. Nauk SSSR, Moscow (1975).
14. Yu. V. Moiseev et al., 17th Symp. on Macromolecules, Prague (1977), p. 27.
15. T. E. Rudakova et al., Vysokomol. Soedin., A, 16, No. 6, 1356 (1974).
16. H. C. Golike and S. W. Zasoski, J. Phys. Chem., 64, 895 (1960).
17. T. Davies et al., J. Phys. Chem., 66, 175 (1962).
18. D. A. S. Ravens and J. M. Ward, Trans. Faraday Soc., 57, 150 (1961).
19. W. McMahon et al., J. Chem. Eng. Data, 4, 57 (1959).
20. H. Zimmermann, 2nd International Symposium on Degradation and Stabilization of Polymers, Dubrovnik (1978).
21. D. A. S. Ravens, Polymer, 1, 375 (1960).
22. E. M. Arnett, in: Contemporary Problems in Physical Organic Chemistry, M. E. Vol'pin (ed.), Mir, Moscow (1967), p. 195.
23. T. E. Rudakova et al., Vysokomol. Soedin., A, 14, No. 2, 449 (1972).
24. T. E. Rudakova et al., in: Nauchno-Issled. Tr., Litov. Nauchno-Issled. Inst. Teks. Promst., 3, 265 (1974).
25. T. E. Rudakova et al., Vysokomol. Soedin., A, 17, No. 8, 1797 (1975).
26. Z. Pekkarinen, Ann. Acad. Sci. Fenn., A, 11, 62 (1954).
27. R. W. Taft, J. Am. Chem. Soc., 74, No. 11, 3120 (1952).
27a. K. Hirono, G. Wasai, T. Saegusa, and J. Furukawa, J. Chem. Soc. Jpn., Ind. Chem. Sect., 67, 604-612 (1964).
27b. Y. Chatani, K. Suehiro, G. Okita, H. Todokoro, and K. Chojo, Makromol. Chem., 113, 215-225 (1968).
27c. L. G. Privalova, H. Kus, and G. E. Zaikov, Polim. Med., 11, No. 1, 3-25 (1981).

27d.   Yu. V. Moiseev, T. T. Daurova, O. S. Voronkova, K. Z. Gumar-
       galieva, and L. G. Privalova, J. Polym. Sci., Polym. Symp.,
       66, 269-276 (1979).
27e.   L. G. Privalova, T. T. Daurova, and G. E. Zaikov, in: 10.
       Donauländergespräch., Vienna (1977), p. 22.
27f.   L. L. Razumova, T. T. Daurova, G. E. Zaikov, A. A. Vereten-
       nikova, L. G. Privalova, et al., Polim. Med., 9, No. 2, 119-
       125 (1979).
27g.   L. G. Privalova, K. Z. Gumargalieva, T. T. Daurova, O. S.
       Voronkova, Yu. V. Moiseev, L. L. Razumova, and G. E. Zaikov,
       Vysokomol. Soedin., A, 22, No. 8, 1891-1899 (1980).
27h.   L. G. Privalova, K. Z. Gumargalieva, L. L. Razumova, and
       G. E. Zaikov, in: Chemie Kunstst. Aktuel. Sondernummer. 2.
       Donauländergespräch. Die natürliche und künstliche Alterung
       von Kunststoffen, Oct. 2-3, 1978, Dubrovnik. Fak.-Verlag,
       Vienna (1979), pp. 37-39.
28.   H. Johanesson, Lunds Univ. Arrskr., 12, 8 (1916).
28a.   E. Kosower, Molecular Biochemistry, McGraw-Hill, New York
       (1962).
28b.   J. Salthouse, J. Biomed. Mater. Res., 10, 197-229 (1976).
28c.   A. B. Dav'dov, Khim. Tekhn. Vysokomol. Soedin., VINITI,
       Moscow, No. 10, 5-29 (1976).
28d.   T. E. Lipatova and G. A. Pkhakadze, Use of Polymers in
       Surgery, Naukova Dumka, Kiev (1977).
29.   C. K. Ingold and W. S. Nathan, J. Chem. Soc., No. 1, 222 (1936).
30.   E. W. Timm and C. N. Hinshelwood, J. Chem. Soc., No. 3, 862
       (1938).
31.   G. Schulz, P. Fijolka, and H. Kriegsmann, Plaste Kautsch.,
       15, 816 (1968).
32.   E. Szabo-Rethy and A. Vancso-Szmercssanyi, Chem. Zvesti, 26,
       390 (1972).
33.   B. M. Kovarskaya, A. B. Blyumenfel'd, and I. I. Levantovskaya,
       Thermal Stability of Hetero-Chain Polymers, Khimiya, Moscow
       (1977).
34.   A. D. Litmanovich et al., Vysokomol. Soedin., A, 17, No. 5,
       1112 (1975).
35.   L. Jörgensen, Studies on the Partial Hydrolysis of Cellulose,
       Moestue, Oslo (1950).
36.   L. F. McBurney, in: Cellulose and Cellulose Derivatives,
       E. Ott et al. (eds.), Wiley-Interscience, New York (1954),
       Part 1, Chap. 3C.
37.   Yu. V. Moiseev, N. A. Khalturinskii, and G. E. Zaikov, Carbo-
       hydr. Res., 51, 23, 39 (1976).
38.   A. Meller, Holzforschung, 9, 149 (1955).
39.   M. L. Wolfrom, J. C. Sowden, and E. A. Metcalf, J. Am. Chem.
       Soc., 63, No. 6, 1688 (1941).
40.   A. Sharples, Chem. Ind. (London), No. 32, 870 (1953).
41.   A. Sharples, Trans. Faraday Soc., 53, No. 7, 1003 (1957).
42.   G. G. Gibbons, J. Text. Inst., 43, 25 (1952).
43.   H. Mark and R. Simha, Trans. Faraday Soc., 36, No. 7, 611 (1940).

44. K. Frendenberg and G. Blomqvist, Ber., **68**, 2070 (1935).

45. J. N. Be Miller, Adv. Carbohydr. Chem., **22**, 25 (1967).

46. M. L. Wolfrom, J. Am. Chem. Soc., **59**, No. 2, 282 (1937); **60**, No. 5, 1026 (1938); **61**, No. 5, 1072 (1939).

47. G. V. Schula and E. Husemann, Z. Phys. Chem., **52**, 1, 23 (1942).

48. R. W. Jones, R. J. Dimler, and C. E. Rist, J. Am. Chem. Soc., **77**, No. 6, 1659 (1955).

49. M. S. Feather and J. F. Harris, J. Am. Chem. Soc., **89**, No. 22, 5661 (1967).

50. J. Hollo et al., Stärke, **16**, 211 (1964).

51. L. I. Novikov and A. A. Konkin, Zh. Prikl. Khim., **32**, 1081 (1959).

52. A. Ekenstam, Chem. Ber., **69**, 549, 553 (1936).

53. J. Sakurada and S. Okamura, Z. Phys. Chem. (Leipzig), A, **187**, 289 (1940).

54. P. N. Odintsov and Ya. V. Vitol, Izv. Akad. Nauk Latv. SSR, Ser. Khim., No. 3, 375 (1966).

55. A. J. Rosenthal, J. Polym. Sci., **51**, 111 (1961).

56. W. C. Frith, Tappi, **46**, 739 (1963).

57. C. J. Malm, Tappi, **47**, 533 (1964).

58. E. P. Bytenskii, E.P. Kuznetsova, and N. I. Klenkova, Zh. Prikl. Khim., **39**, 2319 (1966); **40**, 410 (1967).

59. E. P. Kuznetsova and N. I. Klenkova, Zh. Prikl. Khim., **37**, 1073 (1964); **39**, 478 (1966).

60. N. K. Pyatakina, Yu. V. Moiseev, and G. E. Zaikov, Vysokomol. Soedin., A, **13**, No. 1, 200 (1971).

61. M. A. Bhatti and P. Howard, Preprint Intern. Symp. Macromol., Helsinki, **5**, 43 (1972).

62. T. Alfrey, in: Chemical Reactions of Polymers, E. M. Fettes (ed.), Wiley, New York (1964).

63. M. L. Wolfrom and J. C. Dacons, J. Am. Chem. Soc., **74**, No. 21, 5331 (1952).

64. K. Frendenberg, Trans. Faraday Soc., **32**, 74 (1936).

65. G. F. Davidson, J. Text. Inst., **34**, 87 (1943).

66. R. F. Nickerson and J. A. Habrle, Ind. Eng. Chem., **39**, 1507 (1947).

67. H. J. Philipp, M. L. Nelson, and H. M. Ziifle, Text. Res. J., **17**, 585 (1947).

68. M. A. Millett, W. E. Moore, and J. E. Saeman, Ind. Eng. Chem., **46**, 1493 (1954).

69. J. E. Saeman, Ind. Eng. Chem., **37**, 43 (1945).

70. B. Ranby, PhD Thesis, Institute of Physics, Uppsala (1952).

71. M. Chang, T. C. Pound, and R. St. J. Manley, J. Polym. Sci., A-2, **11**, 399 (1973).

72. M. I. Artsis et al., Vysokomol. Soedin., A, **17**, No. 1, 128 (1975).

73. V. A. Roginskii and V. B. Miller, Dokl. Akad. Nauk SSSR, **215**, No. 5, 1164 (1974).

74. H. Staudinger and M. Züthy, Helv. Chim. Acta, **8**, 41 (1925).

75. H. Staudinger, Ann., **474**, 145 (1929).

76. V. Jaacks, W. Kern, and H. Baader, Makromol. Chem., _83_, 56 (1965).
77. L. V. Ivanova et al., Vysokomol. Soedin., A, _16_, No. 8, 1831 (1974).
78. E. N. Kumpanenko et al., Vysokomol. Soedin., A, _12_, No. 1, 229 (1970).
79. J. C. Bevington and R. S. W. Norrish, Proc. Roy. Soc., A, _196_, 363 (1949).
80. A. A. Berlin et al., Vysokomol. Soedin., A, _10_, No. 7, 1496 (1968).
81. N. S. Enikolopyan and S. A. Vol'fson, Chemistry and Technology of Polyformaldehyde, Khimiya, Moscow (1968).
82. G. Delzenne and G. Smets, Makromol. Chem., _18_, 82 (1956).
83. J. Furukawa and T. Saegusa, Polymerization of Aldehydes and Oxides, Wiley, New York (1963).
84. V. Jaacks, Makromol. Chem., _101_, 33 (1967).
85. L. Leese and M. W. Baumber, Polymer, _6_, 269 (1965).
86. R. C. Schulz and G. Wegner, Makromol. Chem., _104_, 185 (1967).
87. J. Mejzlik and I. Pak, Makromol. Chem., _82_, 226 (1965).
88. W. Fukuda and H. Kakiudu, Kogyo Kagaku Zasshi, J. Chem. Soc., Jpn., Ind. Chem. Sect., _66_, 387 (1963).
89. D. E. Cagliostro, S. Riccitiello, and J. A. Parker, J. Makromol. Sci., A, _3_, 1601 (1969).
90. G. F. Hardy, J. Polym. Sci., A-2, _5_, No. 5, 671 (1967).
91. M. Popova, Khim. Ind., _3_, 117 (1970).
92. L. V. Ivanova, Yu. V. Moiseev, and G. E. Zaikov, Vysokomol. Soedin., A, _14_, No. 5, 1057 (1972).
93. A. A. Khokhlov, Candidate's Dissertation, Moscow Chemical Technology Institute, Moscow (1979).
94. N. N. Pavlov et al., Vysokomol. Soedin., A, _18_, No. 7, 1591 (1976).
95. G. Glöckner, Plaste Kautsch., _17_, No. 116 (1970).
96. S. Boranowska and Z. Wielgosz, Polim. Tworz. Wielkoczast., _15_, No. 1, 12 (1970).
97. A. A. Khokhlov et al., Vysokomol. Soedin., B, _20_, No. 3, 231 (1978).
98. H. R. Allcock, J. Macromol. Sci., C., _4_, No. 2, 149 (1970).
99. F. M. Lewis, Rubber Chem. Technol., _35_, 1222 (1962).
100. I. A. Metkin, K. B. Piotrovskii, and Yu. A. Yuzhelevskii, Zh. Prikl. Khimii, _48_, No. 5, 1108 (1975).
101. G. Martellock, Polym. Prepr., Am. Chem. Soc., Div. Polym. Chem., 10 (1950).
102. V. M. Rudov, Candidate's Dissertation, Moscow State University, Moscow (1976).
103. N. M. Emanuel' and D. G. Knorre, Course in Chemical Kinetics, Vyssh. Shkola, Moscow (1974).
104. W. J. Bengouth and H. M. Sharpe, Makromol. Chem., _66_, 31 (1963).
105. R. R. Stromberg, S. Straus, and B. G. Achhammer, J. Polym. Sci., _35_, No. 129, 355 (1959).

106. G. Talamini, G. Cingue, and G. Palma, Mater. Plast. Elasto-
     meri, $\underline{30}$, 317 (1964).
107. A. Grosato-Arnaldi et al., J. Appl. Polym. Sci., $\underline{8}$, No. 2,
     747 (1964).
108. K. S. Minsker and G. G. Fedoseeva, Degradation and Stabiliza-
     tion of Poly(vinyl chloride), Khimiya, Moscow (1979).
109. B. Baum and L. H. Wartman, J. Polym. Sci., $\underline{28}$, 537 (1958).
110. G. Talamini, Makromol. Chem., $\underline{39}$, No. 1/2, 26 (1960).
111. G. Palma and M. Carenza, J. Appl. Polym. Sci., $\underline{14}$, No. 7, 1737
     (1970).
112. F. Danusso, G. Pajaro, and D. Sianesi, Chim. Ind., $\underline{41}$, 1170
     (1959).
113. K. S. Minsker et al., Vysokomol. Soedin., A, $\underline{15}$, No. 4, 866
     (1973).
114. D. Braun, Pure Appl. Chem., $\underline{26}$, 173 (1971).
115. W. C. Geddes, Rubber Chem. Technol., $\underline{40}$, 177 (1967).
116. K. S. Minsker, Plaste Kautsch., $\underline{24}$, No. 6, 375 (1977).
117. K. S. Minsker et al., Fiz.-Khim. Osn. Sint. Pererab. Polim.,
     No. 2, 52 (1977).
118. J. Landler and P. Lebel, J. Polym. Sci., $\underline{48}$, 477 (1960).
119. J. Cotman, Ann. N. Y. Acad. Sci., $\underline{57}$, 417 (1953).
120. A. A. Caraculacu and O. Wichterle, J. Polym. Sci., C, No. 16,
     459 (1967).
121. B. B. Troitskii, PhD Thesis, Moscow State University, Moscow (1977)
122. P. M. Sykes, Guidebook to Mechanism in Organic Chemistry,
     Wiley, New York (1965).
123. Z. Vymazal et al., J. Appl. Polym. Sci., $\underline{18}$, 2361 (1957).
124. W. H. Saunders, Jr., Ionic Aliphatic Reactions, Prentice-Hall,
     Englewood Cliffs, New Jersey (1965).
125. M. I. Artsis, Candidate's Dissertation, Institute of Chemical
     Physics, Academy of Sciences of the USSR, Moscow (1976).
126. R. A. Panko and V. S. Pudov, Vysokomol. Soedin., A, $\underline{18}$, No.
     17, 864 (1976).
127. A. Pocker, Rep. Chem. Soc. (London), $\underline{57}$, 165 (1960).
128. R. Schlimper, Plaste Kautsch., $\underline{13}$, 196 (1966).
129. M. Onozuka and M. Ashina, J. Macromol. Sci., C, $\underline{3}$, 236 (1969).
130. A. Michil, Tran-van-Hoang, and A. Guyot, Intern. Symp. on
     Degradation and Stabilization of Polymers, Brussels (1974),
     p. 17.
131. K. S. Minsker et al., Dokl. Akad. Nauk SSSR, $\underline{223}$, No. 1, 138
     (1975).
132. V. N. Myakov, Candidate's Dissertation, Dzerzhinsk (1978).
133. L. S. Troitskaya, V. N. Myakov, and B. B. Troitskii, Vysoko-
     mol. Soedin., A, $\underline{9}$, No. 10, 2119 (1967).
134. M. Thallmaier and D. Braun, Makromol. Chem., $\underline{108}$, No. 3, 241
     (1967).
135. V. S. Pudov and R. A. Papko, Vysokomol. Soedin., B, $\underline{12}$, No.
     3, 218 (1970).
136. M. Carenza, Yu. V. Moiseev, and G. Palma, J. Appl. Polym. Sci.,
     $\underline{17}$, 2685 (1973).

137. K. S. Minsker, V. P. Malinskaya, and A. A. Panasenko, Vyso-
     komol. Soedin., A, $\underline{12}$, No. 5, 1151 (1970).
138. D. Braun and R. F. Bender, Eur. Polym. J., Supplement, 269
     (1969).
139. Z. Wolkober, J. Polym. Sci., $\underline{58}$, 1311 (1962).
140. J. P. Roth and P. Rempp, J. Polym. Sci., C, No. 4, 1347
     (1963).
141. M. Ikariwa and S. Takeshita, Nippon Kagaku Kaishi, J. Chem.
     Soc. Jpn. Chem. Ind. Chem., No. 3, 555 (1975).
142. A. B. Thomson and D. W. Woods, Trans. Faraday Soc., $\underline{52}$, No.
     10, 1383 (1956).
143. S. Otani, Kogyo Kagaku Zasshi, J. Chem. Soc. Jpn., Ind. Chem.
     Sect., $\underline{65}$, 1641 (1962).
144. T. Ohtsubo et al., Nippon Kagaku Zasshi, J. Chem. Soc. Jpn.,
     . Chem. Ind. Chem., No. 3, 1834 (1975).
145. E. Waters, J. Soc. Dyers Colour., $\underline{66}$, 609 (1950).
146. F. C. Wampler and D. R. Gregory, J. Appl. Polym. Sci., $\underline{16}$,
     3253 (1972).
146a. M. R. Padhyc and N. Nadaf, Indian J. Text. Res., $\underline{4}$, No. 3,
     99-110 (1979).
147. A. B. Robertson and H. J. Harwood, Isot. Radiat. Technol.,
     $\underline{9}$, 56 (1971).
148. K. V. Belogorskaya, N. I. Seliverstova, and A. F. Nikolaev,
     Vysokomol. Soedin., B, $\underline{16}$, No. 8, 619 (1974).
149. R. S. Barshtein et al., Plast. Massy, No. 5, 54 (1974); trans-
     lated in: Int. Polym. Sci. Technol., No. 10, 8 (1974).
150. S. Iwabuchi et al., Makromol. Chem., $\underline{100}$, 276 (1967).
151. L. V. Ivanova, Candidate's Dissertation, Inst. Khim. Fiz. Akad.
     Nauk SSSR, Moscow (1973).

Chapter 8

INFLUENCE OF AGGRESSIVE MEDIA ON THE MECHANICAL PROPERTIES OF POLYMERS

The mechanical properties of polymers, as a whole, determine the mechanical behavior of polymers under the action of external forces. Polymer articles deform under the action of a force field, and fail at particular mechanical stresses and times. The changes in the mechanical properties of polymer articles in force fields and with temperature are considered in detail in a number of monographs [1-9]. In the following chapter we shall consider those mechanical properties the changes in which are most frequently determined by the action of aggressive media on polymer articles. The most important deformational properties are the creep (by which solids slowly increase their deformation under the action of constant stresses) and the viscosity (the way in which the solids resist irreversible changes in shape). The most important strength properties are the strength, characterized by the stress at which there is failure of the polymer under conditions of loading carried out with a particular pattern of increase in deformation, and the durability, which is determined as the time from the moment of loading up to the failure of the polymer article under constant stress.

There are now known to be various quantitative relationships which relate the aforementioned mechanical properties to other mechanical properties of the polymers [10].

8.1. STRENGTH OF POLYMERS IN AGGRESSIVE MEDIA

Polymers may be called "viscoelastic materials," stressing their intermediate position between viscous liquids and elastic solids [9, p. 23 of Russian translation]. The strength of polymers depends greatly on the conditions of testing, the temperature,

275

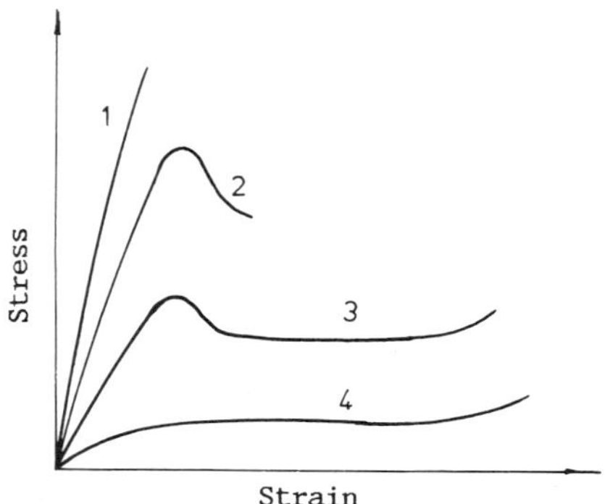

Fig. 8.1.   Stress—strain curves at various tem-
            peratures: 1) brittle failure; 2)
            plastic failure; 3) cold drawing; 4)
            rubbery state.

the rate of application of the stress, and the nature of the aggressive medium. Figure 8.1 shows the influence of the temperature
on the form of the stress—strain relationship. At temperatures
considerably below T_g (curve 1) the stress increases linearly with
increase in strain and rupture ensues at low strains of the specimen. At high temperatures (curve 4) the polymer is rubbery and
the stress changes with strain according to an S-shaped relationship, while rupture takes place at very high strains (from
30 to 1000%). At temperatures below T_g (curve 2) the stress—strain
relationship is characterized by the existence of a stress flow
limit, at which there is necking even before the onset of failure.
At higher temperatures (curve 3) there develops, after the formation of the neck, leathery ("forced") elastic deformation, which
may reach tens or hundreds of percent.

The strength of polymers depends on the duration of action
of the stress. As a rule, the strength increases with increase
in the rate of tensile deformation.

8.1.1. Theoretical Strength of Solid Polymers

Several approaches to the calculation of the theoretical
strength of polymers have been proposed. Since for solid polymers
strict theoretical calculations of strength are impossible, semiempirical methods have been used, the structure of the polymers
being regarded as an ideal monocrystal, and the polymer body being
subjected to uniaxial stretching.

The approach of Orovan [11] takes the form of the calculation
of the change in the quasi-elastic force F in the simultaneous

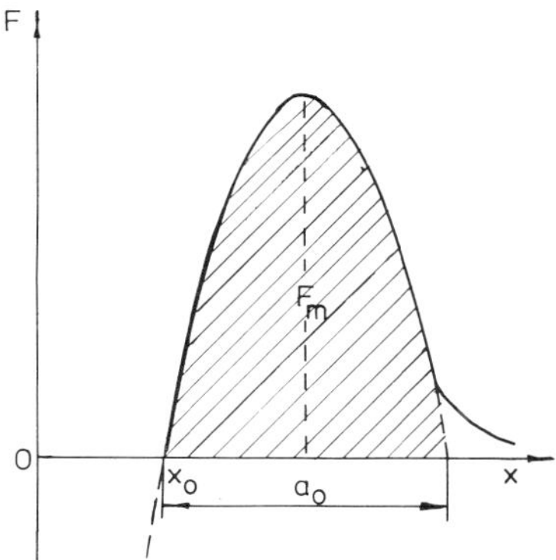

Fig. 8.2. Dependence of the quasi-elastic
force on the distance between the
particles [11].

parting of individual surfaces of a solid polymer. He proposed
an approximate equation expressing the dependence of the absolute
magnitude of the quasi-elastic force on the distance x between
the particles:

$$F = F_m \sin\left[\frac{\pi}{a_0}(x - x_0)\right]$$

(8.1)

where a_0 is twice the distance from the equilibrium position to
the position corresponding to the maximum of the quasi-elastic
force F_m (Fig. 8.2) and x_0 the equilibrium distance between the
particles.

With slow parting of the surfaces

$$\sigma_0 = NF$$

(8.2)

where σ_0 is the stress applied to the surface, and N the number
of particles belonging to this surface.

On substituting (8.2) into (8.1), we get

$$\sigma_0 = \sigma_{th}\sin\left[\frac{\pi}{a_0}(x - x_0)\right]$$

(8.3)

where σ_{th} = NF_m is the theoretical strength.

The work of the external forces which is expended on moving
a surface by distance a_0 is found by integrating σ between the
limits x_0 and $x_0 + a_0$

$$\int_{x_0}^{x_0+a_0} \sigma_{\text{th}} \sin\left[\frac{\pi}{a_0}(x-x_0)\right] dx = \frac{2a_0}{\pi}\sigma_{\text{th}} \qquad (8.4)$$

It is assumed, moreover [11], that the calculated work is fully converted into potential energy in the surfaces which are formed

$$\frac{2a_0\sigma_{\text{th}}}{\pi} = 2\gamma_{\text{surf-}\Gamma} \qquad (8.5)$$

where $\gamma_{\text{surf-}\Gamma}$ is the free surface energy of the solid polymer.

With only slight deviations from the equilibrium position it is assumed that Hooke's law holds:

$$\sigma_0 = E\,\frac{x-x_0}{x_0} \qquad (8.6)$$

where E is Young's modulus.

Taking into account these conditions from Eq. (8.3), it becomes possible to determine the constant

$$a_0 = \frac{\sigma_{\text{th}}\pi x_0}{E} \qquad (8.7)$$

Substitution of (8.7) into (8.5) leads to the expression

$$\sigma_{\text{th}} = \sqrt{\frac{\gamma_{\text{surf-}\Gamma}E}{a_0}} \qquad (8.8)$$

From this relationship there follows the very important conclusion that the strength of solid polymers depends on the surface energy at the polymer-medium interface, i.e., on a parameter which is bound to change when testing polymer articles in aggressive media.

As a rule there is used for the estimation of the theoretical strength an approximate formula which emerges from (8.7). The maximum of the quasi-elastic force is reached within 10-20% elongation of the bonds; consequently, $a_0/2$ is equal to 0.1-$0.2x_0$ and on the average $\sigma_{\text{th}} \approx 0.1E$.

For calculating the strength <u>Kobeko</u> [1] used the Morse formula

$$U = -D\exp\left[-B_0(x-x_0)\right]\{2 - \exp\left[-B_0(x-x_0)\right]\} \qquad (8.9)$$

where U is the potential energy of interaction, calculated per particle; D the dissociation energy associated with one particle; and

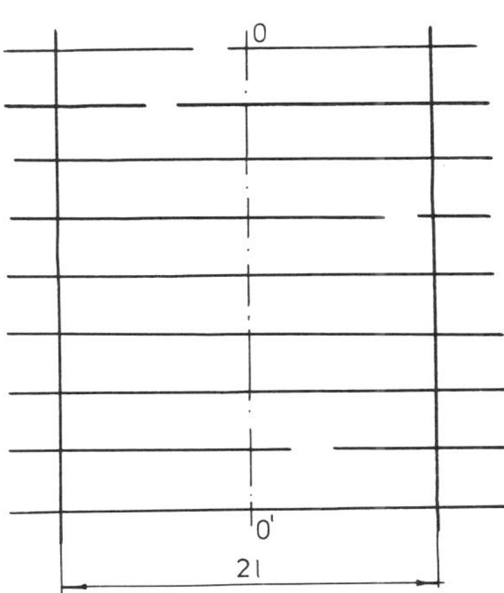

Fig. 8.3.   Diagram of the formation of defects
on a model polymer [12].

$B_0 = \left(\dfrac{w}{D}\right)^{1/2} \dfrac{1}{A_0}$ (where w is the frequency of oscillation of the particles and A_0 the amplitude).

The maximum value of the quasi-elastic force F_{th}, as found from (8.9), is equal to $B_0 D/2$, whence it follows that

$$\sigma_{th} = NB_0 D/2 \tag{8.10}$$

where N is the number of chains belonging to unit area of the cross section of the unstressed specimen of orientated polymer.

The value of (8.10) lies in the fact that it makes it possible to estimate σ_{th} by way of physical parameters which can be determined experimentally. For instance, the dissociation energy D can be calculated from the heats of combustion, and the constant B_0 calculated from spectroscopic data. Kobeko found by calculation that the relationship $\sigma_{th} \approx 0.1E$ holds well for different types of bonds. However, the use of this relationship has shown that the theoretical strength is always higher than the technical strength of polymer materials. An exception to this rule is presented by superfine filaments (whiskers) made from inorganic polymers and metals for which the theoretical and technical strengths practically coincide.

<u>The approach of Gubanov and Chevychelov</u> [12] is based on three premises: the sole form of defects in the orientated polymer lies in the bonds between the ends of the molecules; the intermolecular forces are more than ten times weaker than the intramolecular forces;

with scission of the polymer in the plane OO' (Fig. 8.3) there takes
place main chain scission of those molecules whose ends lie beyond
plane OO' by more than the distance of the monomers. The rest of
the molecules will pull their ends out from the mass of the polymer
when the specimen ruptures.

The probability of incidence of an end of a molecule which is
x monomers long in a layer 2ℓ is $2\ell/x$; the probability of scission
of the polymer chain γ in this same layer can be found from the
equation

$$\gamma = \sum_{x=1}^{2l-1} g\,(x) + 2l \sum_{x=2l}^{\infty} \frac{g\,(x)}{x} \tag{8.11}$$

where $g(x)$ is the probability of the incidence of a molecule having x monomers in the layer 2ℓ:

$$g\,(x) = \frac{n\,(x)}{n_0} = \frac{1}{\overline{P}_n}\left(1 - \frac{1}{\overline{P}_n}\right)^{x-1} \tag{8.12}$$

where $n(x)$ and n_0 are, respectively, the number of molecules with
x monomers and the total number of molecules in the layer 2ℓ. The
distribution of defects in the layer under consideration is described by the function

$$k\,(S) = n_0\gamma^{l-1}\,(1 - \gamma)^{2\,(\pi S)^{1/2}} \tag{8.13}$$

where S is the surface area of the defect, equal to iS_0, where S_0
is the cross-sectional area of the polymer chain.

An analysis carried out by the present authors showed that
for polymers with $\overline{M}_n < 10^5$

$$\sigma_0 = \frac{3}{2}\left(\frac{F\gamma_{\text{surf-}\Gamma}}{d_0}\right)^{1/2}(1 - \gamma) \tag{8.14}$$

i.e., the strength is governed by the reduction in the cross section of the specimen on account of the ruptured bonds.

For polymers with $\overline{M}_n > 10^5$

$$\sigma_0 = \frac{3}{2}\left(\frac{E\gamma_{\text{surf-}\Gamma}}{d}\right)^{1/2} \tag{8.15}$$

where $d = 2(S/\pi)^{1/2}$, and the strength is governed by the stress
concentration in the region of the biggest defect with surface area
S. Table 8.1 shows results of calculations for three polymers of
differing MMs.

TABLE 8.1.   Theoretical Strength of Polymers with Differing MMs [17]

Polymer	$E \cdot 10^{-3}$, MPa	$\gamma_{\text{surf-}\Gamma}$, J/m^2	$\sigma_0^* \cdot 10^{-2}$, MPa			σ_0 exp $\cdot 10^{-2}$, MPa
			$\overline{M}_n=10^7$	$\overline{M}_n=10^6$	$\overline{M}_n=10^5$	
PMMA	6	17.5	3.1	2.8	2.5	2.1—3.6
PVA	18	52.5	12.2	11.2	9.8	12.0—15.5
PCA	9	64.0	10.0	9.2	8.0	6.8—8.0

*Calculated for layer $2\ell = 10$, where ℓ is the length of the monomer.

It may be seen that with high $\overline{M}_n$ the strength differs only slightly from the experimental values. The Gubanov and Chevychelova approach makes it possible to assess the influence of aggressive media on the strength [12]. For instance, physically active media may weaken the intermolecular interaction within the matrix of the polymer which, in the present context, is equivalent to an increase in ℓ (see Fig. 8.3). Increasing the measurements of the layer twofold leads to a 10% reduction in σ.

When chemically active media, causing degradation of the main chains of the molecules, act upon a polymer, the reduction in the MM leads to a reduction in the strength of the specimen (Table 8.1).

Thus, if we regard the structure of the polymer as an ideal monocrystal, the strength of the polymer body in uniaxial stretching will depend on the surface energy at the polymer—medium interface, the energy of the chemical bonds in the molecules, and the MM of the polymers.

8.1.2. Influence of Aggressive Media on the Strength of Polymers

Aggressive media, as already indicated, may be divided into two groups: physically and chemically active media.

Physically Active Media. Physically active media can be either adsorbed on the surface or else sorbed by the polymer article.

The adsorption of the components of the aggressive medium leads to a change in the surface energy $\gamma_{\text{surf-solv}}$ at the polymer—medium interface. If the physically active medium is multicomponent, the components which are soluble in the solvent may be surface-active ("surfactants") or surface-inactive according to whether they reduce or increase the surface energy $\gamma_{\text{surf-solv}}$.

Surfactants include the majority of aqueous solutions of strong electrolytes. The basic concepts of the mechanism of action of surfactants on the strength of solids have been stated by Rebinder [13, 14]. By reducing the free surface energy at the polymer—medium interface, surfactants reduce the work of formation of a new surface and the nucleation and development of surface defects. Their molecules penetrate into the apices of microcracks as a result of surface diffusion, which is described by the equation [15]

$$c = 4\pi m D_n t \exp\left(\frac{-r^2}{4D_n t}\right) \qquad (8.16)$$

where c is the concentration of the diffusing substance at a point a distance r from the source, and D the coefficient of surface diffusion.

The motive force of surface diffusion is the reduction in the surface energy:

$$\Delta\gamma = \gamma_0 - \gamma_\Gamma \qquad (8.17)$$

where γ_0 is the surface energy of the polymer in a vacuum; γ_Γ the surface energy of the polymer which is covered by an adsorption layer Γ.

The adsorption effect may be exhibited in its pure form in the case of polymers which practically do not swell in physically active media. One such system is polystyrene with aqueous alcohol solutions. It is known that on contact with methanol, polystyrene retains its rigid structure, while there is either adsorption on the outside surface of the polymer, or else within its volume according to a mechanism of filling up the volume of the micropores [16].

Bakeev et al. [17] investigated in detail the change in the leathery elastic limit of specimens of polystyrene of various thicknesses in aqueous solutions of a homologous series of alcohols. The action of these media on the leathery elastic limit increases with increase in the number of methylene groups in the alcohol molecule, in accordance with the Duclos—Traube rule (the surface activity increases threefold with each addition of a methylene group to the homolog). As might be expected, there was a critical thickness of the specimen (about 1.9 mm), above which aqueous alcohol solutions had practically no influence on this limit (Fig. 8.4).

Thus it may be assumed that in the case of those surfactants which do not cause swelling of polymers the adsorption effect plays a predominant role in the reduction of polymer strength.

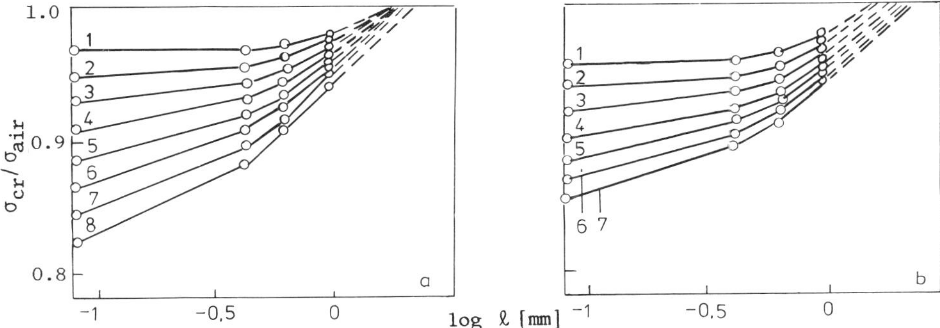

Fig. 8.4. Influence of the thickness of specimens of polystyrene
on the relative value of the limit of leathery elastic-
ity in aqueous solutions of propyl (a) and amyl (b) alcohols [17]:   a: 1) 0.2; 2) 0.4; 3) 0.6; 4) 0.8; 5)
1.0; 6) 1.2; 7) 1.4; 8) 1.6.   b: 1) 0.04; 2) 0.06; 3)
0.08; 4) 0.10; 5) 0.12; 6) 0.14; 7) 0.16.

If there is swelling of the polymer when components of the
aggressive medium are sorbed by polymer articles, generally two effects are exhibited: a) reduction in the intermolecular interaction
which brings about the formation of microcracks within the volume
of the specimen and, b) more uniform stress distribution, increase
in the flexibility of the molecules, and possible increase in the
degree of crystallinity.

The first effect leads to reduction in the strength of the
polymer articles, the second to an increase.

These effects occur simultaneously, but the intensity of each
depends on the magnitude of the force field and the duration of
contact of the polymer article with the physically active medium. Most
often there is an extremal dependence of the strength on time (Fig.
8.5) [18].

The relationship between the microcracking and increase in
flexibility of the molecules or increase in the degree of crystallinity at identical temperatures, loading, and duration of contact
with the aggressive media is determined by the following factors.

The physical state of the polymer. Usually there is a predominance of the first effect in the case of glassy polymers, and
with sorption of physically active media their strength characteristics suffer, with cracking proceeding actively.

In elastic polymers there is practically no cracking of the
polymers with the sorption of components of aggressive media. In
this case the second effect plays the main part.

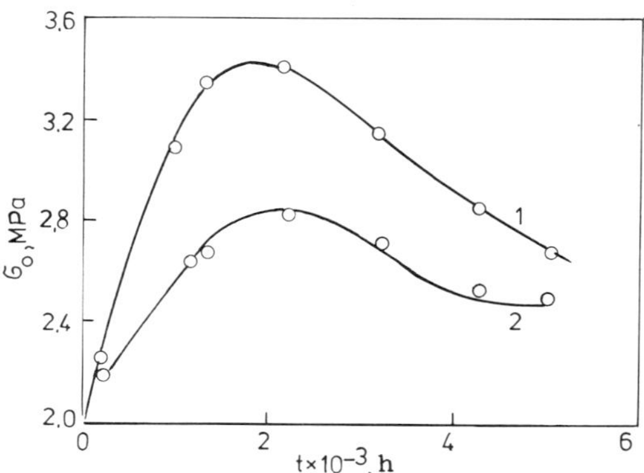

Fig. 8.5.   Dependence of the strength of films
of polyethylene at 20°C on the time
of action of water without loading
(1) and with loading (2) [18].

<u>Degree of crystallinity of the polymer</u>.   Under conditions of
simultaneous action of loading and an aggressive medium the strength
of amorphous polymers mainly increases, whereas with polymers with
a high degree of crystallinity there is some reduction in strength
[18].

<u>Interaction of a physically active medium with the polymer</u>.
There is much data showing the influence of various media on
strength.   If the solubility of the components of the aggressive
medium in the polymer is sufficiently high, microcracks are formed
in the specimen during the diffusion.   These microcracks lie in a
plane perpendicular to the axis of diffusion and the stresses developing within a nonuniformly swollen specimen (there being a
rigid core) mean tearing of the polymer.   The microcracks act as
a nucleus of formation of macrocracks due to osmotic flow in the
swollen polymer or due to other regroupings [19].

All this leads to a reduction in strength of the polymer articles.

<u>Pressure</u>.   With increase in pressure there is a sharp reduc-
tion in the strength of polymers in physically active media as a
result of an increase in the rate of diffusion of the liquid phase
through the porous structure of the microcracks and also because
of dissociation of the clusters of water into individual molecules,
which act more effectively on the growth of the microcracks [20].

<u>Chemically Active Media</u>.   In the presence of chemically active
media there is degradation leading to reduction in strength of the
polymer article.

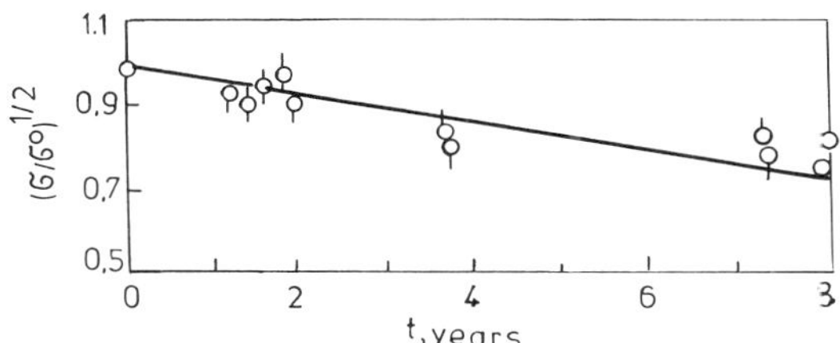

Fig. 8.6. Dependence of the TS of PETP filaments on the time of implantation
[treated according to Eq. (8.20)]
[21].

Unlike the experiments on physically active media, those on
the action of chemically active media on strength were for the most
part carried out in two stages: to begin with, the article was
held for a given time in the aggressive medium under the given conditions, and then the article was washed free of the aggressive
medium and its strength in air determined from the stress–strain
plot.

In these experiments we did not determine the actual strength
of the polymer article in the aggressive medium under the given conditions, but the data obtained did correctly reflect the relative change
in strength under the action of chemically active media.

If the degradation takes place under external diffusion-
kinetic conditions (from the surface of the polymer article), while diffusion of the components of the aggressive medium within the polymer
does not bring about irreversible changes in structure, then for the pure
form of the stress–strain relationship the reduction in the tensile
strength (TS) (σ) is proportional to the change in the cross section (S_c) of the polymer article at the point of scission:

$$\sigma = \sigma_0 S_c \qquad (8.18)$$

On using (6.27) and (6.34), we get for the change in the TS in the
case of a film of thickness ℓ_0 and a filament of radius r_0 the following relationships:

$$\sigma = \sigma^0 \left(1 - k_{\text{surf}}\frac{M}{A}\, c^0_{\text{p(surf)}}\, \frac{1}{z\rho l_0}\, c_{\text{cat(surf)}}\, c_{\text{solv(surf)}}\, t\right) = \sigma^0 \left(1 - \frac{t}{\tau}\right) \qquad (8.19)$$

$$\sigma = \sigma^0 \left(1 - k_{\text{surf}}\frac{M}{A}\, c^0_{\text{p(surf)}}\, \frac{1}{z\rho r_0}\, c_{\text{cat(surf)}}\, c_{\text{solv(surf)}}\, t\right)^2 = \sigma^0 \left(1 - \frac{t}{\tau}\right)^2 \qquad (8.20)$$

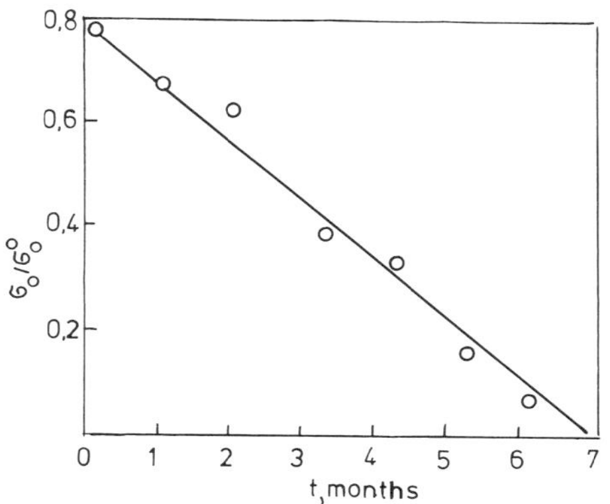

Fig. 8.7. Change in the strength of filaments
of PCA as a function of the time of
implantation in the subcutaneous
tissue of rabbits [21].

where σ_0 is the TS for $t = 0$. For instance, by means of Eq. (8.20)
we describe the reduction in the TS of PETP filaments on the time
of implantation of a PETP mesh in human subcutaneous tissue [21]
(Fig. 8.6) and also that of PETP fabric on the time of contact with
aqueous basic solutions [22]. In both series of experiments there
was no appreciable change in the surface roughness of the specimens.
When degradation takes place under internal kinetic and diffusion-
kinetic conditions the cross section of the polymer articles remains
practically unchanged (the mass of the article change only very slightly), and the main reason for reduction in the TS is the reduction
in the strength σ_0 as a consequence of the destruction of load-
bearing bonds.

As far back as 1945, Flory proposed an equation linking the
TS with the number-average MM of a polymer [23]:

$$\sigma_0 = A - \frac{B}{M_n} \tag{8.21}$$

where A and B are constants.

This relationship was later confirmed theoretically [12] and
experimentally [24] for a number of polymers undergoing brittle
rupture. If degradation takes place under internal kinetic conditions and the molecules decompose randomly, then for low degrees
of dissociation of the load-bearing bonds the reduction in TS is
described by an equation derived from (6.20), (8.18), and (8.21):

$$\sigma_0 = \sigma_0^0 \left(1 - B k_{\text{eff.}} c_{\text{cat}}^0 \, t \right) \tag{8.22}$$

Figure 8.7 shows the change in TS of PCA filaments as a function of the time of implantation in the subcutaneous tissue of rabbits [21]. If degradation takes place under internal diffusion-kinetic conditions, then the relationship giving the change in σ_0 with time is more complex. Unfortunately, there is at present a lack of experimental data to verify these relationships.

8.2. FAILURE OF POLYMERS IN AGGRESSIVE MEDIA

The failure of polymers in aggressive media is a complex physicochemical process; knowledge of its mechanism is necessary for the determination under service conditions of the two most important strength parameters, the strength and durability (fatigue life). We shall now consider the best-known theories of failure of polymers in air, and the use of these to describe the processes of failure in aggressive media.

8.2.1. Theories of Failure of Polymers

The failure of polymers can be brittle or viscous, depending on the conditions of the experiment. It is known that reducing the temperature (Fig. 8.1) and increasing the rate of application of the load brings about brittle failure, and vice versa.

Ludwig [26] suggested that viscous failure is caused by shearing stress, while tensile stress brings about brittle failure. Thus the type of failure is determined by the ratio of the critical stresses under the conditions of the experiment, and also by the value of the stress field in the specimen. In this chapter we consider only brittle failure.

At the prsent time investigation of the failure of polymer articles is developing mainly in two directions: study of the influence of defects on the strength (the defect theories of strength), and investigation of the molecular processes in failure (molecular-kinetic theories of strength).

<u>Defect Theories of Failure</u>. Griffith [27] proposed the first theory of the failure of actual solids, based on two postulates. First, there are microcracks of varying size in the solid. Under the action of the applied force the local stress at the ends of the crack considerably exceed the average stress in the specimen, i.e., at points some distance from the crack. Second, if the local stress at the ends of the crack becomes equal to the theoretical strength σ_{th} there is crack growth at a rate close to the speed of sound, and the specimen separates into two parts. Here the applied stress corresponds to the technically measured strength σ_{cr}.

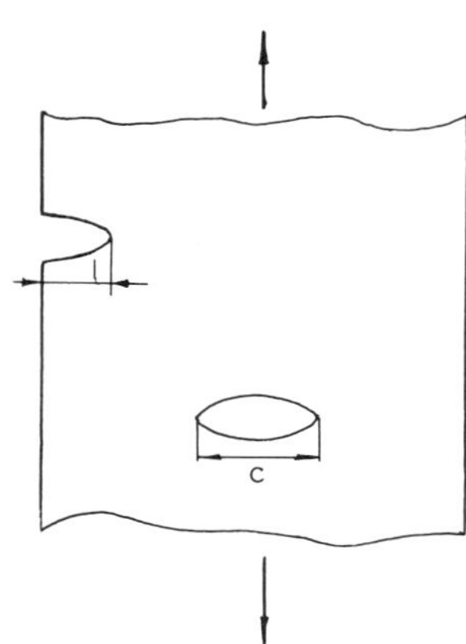

Fig. 8.8.   Elliptical cracks according to Grif-
           fith.

According to Griffith, the cracks have the shape of elongated
ellipses (Fig. 8.8). Since, other conditions being equal, the
greatest stress occurs at the least radius of curvature of the end
of the crack, the narrower the crack the greater its weakening effect on the strength of the material. The presence of such cracks
is also often responsible for an increase in the local stress by
several powers.

Griffith calculated the value of σ_{cr} based on the assumption
that the crack grows only when the reduction in the elastic energy
in the specimen during its growth is equal to or greater than the
increase in the potential energy with the formation of the surfaces
in the scission.

The energy of a specimen in the form of a thin plate which has
an internal crack of length c is, in the plane-stressed state,

$$W = W_0 - \frac{\pi c^2 \sigma^2}{4E}\, l + 2lc\gamma_{\text{surf-}\Gamma} \tag{8.23}$$

where W_0 is the elastic energy of a specimen with no crack, and l
the length of the specimen.

The second term in the equation characterizes the loss of
elastic energy in a specimen, the third term the increase in energy
of the specimen as a result of the formation of new surfaces.

According to Griffith, the conditions of failure of the specimen are found by differentiation of (8.21) with respect to c and
making $\partial W/\partial c$ equal to 0:

$$\sigma_0 = 2\left(\frac{\gamma\text{surf}_{-\Gamma}E}{\pi c}\right)^{1/2} \tag{8.24}$$

With the presence of a boundary microcrack of length δ the technical (actual) strength becomes

$$\sigma_{\text{cr}} = \left(\frac{2\gamma\text{surf}_{-\Gamma}E}{\pi\delta}\right)^{1/2} \tag{8.25}$$

The Griffith theory has been modified repeatedly. It has been shown that microcracks in a specimen may not only grow but also close up [28], with the loss of elastic energy in the process of failure going not into an increase in the surface energy but into mechanical loss (with separation of the crack walls into the work of plastic deformation as well) [29-31].

The Griffith theory has played a very important role in understanding the difference between the technical and theoretical strength of solids, and has been a prerequisite for the development of ideas on the influence of liquid media on polymer strength.

Following Griffith, several theories of the failure of brittle solids, based on the consideration of the kinetics of crack growth, have been advanced. We shall now quote theories which have been specially developed to describe the failure of polymers in liquid media although it is taken as understood that liquid media do not alter the mechanism of failure but merely accelerate crack growth.

According to Bartenev [3, p. 44 of Russian translation] the durability under constant stress is determined by the time within which a crack moving at a rate v_t reaches the critical magnitude

$$l_{\text{cr}} = z\left(1 - \frac{\sigma}{\sigma_{\text{cr}}}\right) \tag{8.26}$$

where σ_{cr} is the critical stress in the cross section of scission, and z the width of the thin strip specimen.

The durability is determined from the relationship

$$\tau_{\text{p}} = \int_0^{l_{\text{cr}}} \frac{dl}{v_t} \tag{8.27}$$

Tynnyi et al. [32, p. 182] assume that the rate of crack growth is a function of the energy of failure and of the temperature:

$$v = \frac{dl}{dt} = f_1 (\gamma - \gamma_{min}) f_2 (T) \qquad (8.28)$$

where γ_{min} is the minimum value of the energy of failure, at which slow crack growth can take place.

Assuming that the functions f_1 and f_2 are linear, the authors obtained

$$\begin{aligned} v &= v_{max} \quad \text{when} \quad \gamma = \gamma_{max} \\ v &= v_{min} \longrightarrow 0 \quad \text{when} \quad \gamma = \gamma_{min} \end{aligned} \qquad (8.29)$$

Furthermore, it is assumed that under the action of the medium a major crack develops from an edge crack and the polymer undergoes brittle scission, and then the effective energy of failure amounts to

$$\gamma = \pi l w \qquad (8.30)$$

where w is the accumulated elastic density.

Under isothermal conditions, expression (8.28) takes the form

$$\frac{dl}{dt} = \frac{v_{max}}{\gamma_{max} - \gamma_{min}} (\gamma - \gamma_{min}) \qquad (8.31)$$

The durability of the polymer is determined by integration of (8.31) within the limits ℓ_0 to ℓ_{cr}, where ℓ_0 is the characteristic size of the cracks in the polymer and $\ell_{cr} = \gamma_{max}/\pi w$ is the critical size of crack. Then

$$\tau_p = \frac{\gamma_{max} - \gamma_{min}}{\pi v_{max} w} \ln \frac{\gamma_{max} - \gamma_{min}}{\pi l_0 w - \gamma_{min}} \qquad (8.32)$$

<u>The Molecular-Kinetic Theory of Failure</u>. This theory regards the failure of polymers as the chemical reaction of the scission of the load-bearing bonds, the rate of this reaction being determined by the temperature, tensile stress (strain), and degree of dissociation.

In 1943, Tobolsky and Eyring [33] explained the dependence of durability on stress (strain) and temperature, taking as their starting point the activation mechanism of scission of the chemical bonds. Assuming that the time of retention of intactness of a solid is determined by the rate of scission of these bonds, they obtained the relationship

$$\tau_p = A \,\frac{n_0}{\sigma}\, \exp \frac{U_0 - \sigma\varepsilon}{kT} \qquad (8.33)$$

where A is a parameter linked with the properties of the polymer; n_0 the initial number of chemical bonds; U_0 the free activation energy of scission of a chemical bond in the absence of a mechanical field; and ε the deformation of the chemical bond in scission.

Further development of the activation mechanism of failure of polymers is found in the work of Bueche [34], who introduced the important assumption of a continuous increase in stress by a decreasing number of unruptured chemical bonds.

Expressing in the usual way the probability of scission of the chemical bonds, Bueche obtained the following expression for the rate of breakdown of the bonds:

$$\frac{-dn}{dt} = n_0 \,(1 - p)^{wt}\, pw \qquad (8.34)$$

where n is the number of chemical bonds broken; p the probability that the energy accumulated on a particular bond is sufficient for its scission; and w the frequency of vibration of the atoms forming the said chemical bond.

By integration of this we get

$$\frac{n}{n_0} = 1 + p\,[1 - (1 - p)^{wt}]\,[\ln(1 - p)]^{-1} \qquad (8.35)$$

Bueche assumes that p is sufficiently small for $\ln(1 - p) \approx -p$ and (8.35) takes the simpler form

$$\frac{n}{n_0} = \exp(-pwt) \qquad (8.36)$$

Using f_0 and f to denote the average tension in the chemical bonds at the initial time and at time t, we get, given a constant total loading,

$$n_0 f_0 = nf \qquad (8.37)$$

It follows that

$$f = f_0 \exp(pwt) \qquad (8.38)$$

Since f increases rapidly when pwt → 1, this condition is taken as a criterion of stability. Thus the durability is

$$\tau_p = \frac{1}{pw}\,\frac{\tau_0}{p} \tag{8.39}$$

where τ_0 is the period of vibration of the atoms.

Bueche assumes that the energy of the bond U falls, as a result of force f, by a particular amount fε.

The probability p in the case in question is determined by Boltzmann's principle:

$$p_p = \exp\left[-(U - f\varepsilon)\,\frac{1}{kT}\right] \tag{8.40}$$

With failure of the specimen there follows from (8.38) that f = f_0e and

$$p_p = \exp\left[-(U - f_0 e\varepsilon)\,\frac{1}{kT}\right] \tag{8.41}$$

The final expression takes the form

$$\tau_p = \tau_0 \exp\left[\left(U - \frac{e\varepsilon\sigma_0}{n_0}\right)\frac{1}{kT}\right] \tag{8.42}$$

The physical meaning of the parameters in the Bueche equation has been reconsidered on a number of occasions. In the most recent form of this theory, U and τ_0 characterize the movement of the segments in the polymer [35].

The best-known equation for the durability of polymers is that of Zhurkov [36]:

$$\tau_p = \tau_0 \exp\left(\frac{U_0 - \gamma\sigma_0}{kT}\right) \tag{8.43}$$

where τ_0 has the same meaning as in Bueche's equation; U_0 is the activation energy of thermal degradation; and γ a parameter of the material.

In studying the data of phenomenological investigation, Zhurkov arrived [8] at the important conclusion that the molecular mechanism of failure of polymers takes the form of thermofluctuational scission of the chemical bonds in the molecules. This means that the external applied force plays a preparatory role, reducing the potential barrier of the bond scission, while the final separation of the atoms is accomplished as a result of thermal movement. By the use of NMR, EPR, IR spectroscopy, mass spectroscopy, x-ray diffraction, and other methods, it has been found that the distribution of the stresses among the bonds is nonuniform. Under

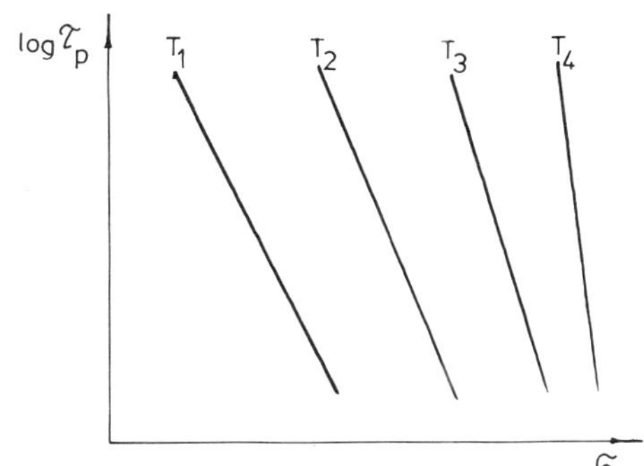

Fig. 8.9.   Dependence of log τ_p on σ at various temperatures ($T_1 > T_2 > T_3 > T_4$).

the action of thermal fluctuation a quite small proportion of strongly overstressed bonds rupture.

According to [3, p. 51 of Russian translation], $\gamma = w\beta$, where β is the coefficient of stress concentration in the region before failure, and w the fluctuation volume of an act of failure.

Equation (8.43) shows the arbitrary nature of the concept of ultimate strength. Figure 8.9 shows the dependence of log τ_p on σ at various temperatures. With lowering of the temperature the plots of log τ_p vs. σ approximate to the vertical and the durability of the polymer changes by several orders of magnitude within a quite small range of stresses. This range may be replaced by the failure stress σ_B, and it may be considered that if the applied stress is lower than σ_B the polymer can remain intact over a long period, but beyond this will fail almost instantaneously. At higher temperatures a wide range of stresses is required for any significant change in durability. It is usually assumed that U_0 and γ do not depend on the temperature; nevertheless, it has been shown in a number of studies that both of these increase to a quite small extent with lowering of the temperature [35].

Bartenev [37] suggested that failure takes place according to two different mechanisms, depending on the temperature: by a high-temperature thermofluctuation mechanism with individual scission of separate molecules, this mechanism being effective above the brittle point temperature, or by a low-temperature mechanism with groupwise scission of the molecules below the brittle point.

The difference between these mechanisms concerns the change in the number of ruptured chemically unstable bonds as a function of the test temperature.

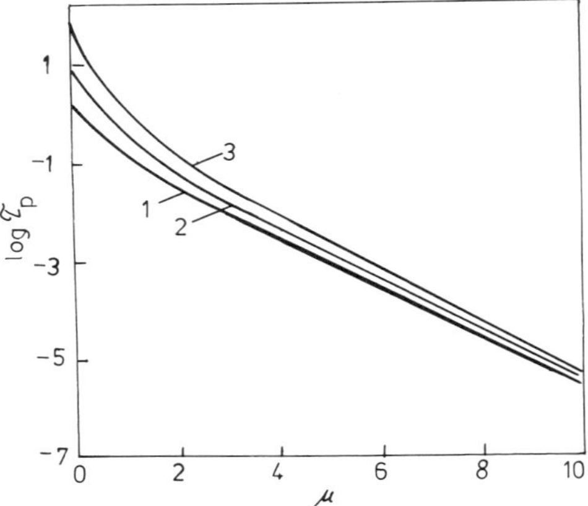

Fig. 8.10. Dependence of $\log \tau_p$ on μ: 1) $\varphi(a) =$
1; 2) $1 - \alpha$; 3) $(1 - \alpha)^2$.

Thus, extrapolation of the dependence of $\log \tau_p$ on σ to the
ordinate axis does not give U_0 (the activation energy of failure
in the absence of stress with $T \to 0K$) but gives U_0' (the activation energy of failure in the temperature range under investigation).

Manelis et al. [38] analyzed the kinetic behavior of the bulk
failure of polymers with constant σ for simple kinetic models of
degradation. They worked on the assumption that the mechanical
field, while not influencing the chemistry of the breakdown reactions of load-bearing bonds, leads to a reduction in the energetic
barrier U of the elementary process of breakdown of the chemical
bond. Generally, the dependence of U on the tension of the bond
is of a nonlinear character [39], but one which permits a linear
approximation over a comparatively wide range of changes in f.

Having assumed that the loading is distributed uniformly over
all the load-bearing bonds, Manelis et al. obtained the following
relationship:

$$U(\sigma) = U_0 - \frac{\beta \sigma n_0}{n} \tag{8.44}$$

where β is a coefficient determined by the nature and structure of
the polymer article.

Relationship (8.44) is correct as long as $\beta \sigma n_0/n < U_0$. When
$\beta \sigma n_0/n > U_0$, failure is determined by the rate of propagation of
elastic waves in the material [40].

For a simple unidirectional reaction the rate of mechanical degradation is expressed by the following dimensionless equation:

$$\frac{d\alpha}{d\tau} = \varphi(\alpha)\exp\left(\frac{\mu}{1-\alpha}\right) \tag{8.45}$$

where $\alpha = 1 - (n/n_0)$ is the degree of dissociation of the chemical bonds; $\tau = ktn^m$ is a dimensionless function of time (m is the order of the reaction); $\varphi(\alpha)$ is the kinetic expression for the chemical reaction; and $\mu = \beta\sigma_0/RT$ is a parameter characterizing the intensity of the mechanical field.

Figure 8.10 shows the dependence of $\log \tau_p$ on μ, as found by integration of (8.45), for various kinetic expressions.

When $\mu = 0$ the time is τ_0, but only for the case when a zero-order degradation reaction takes place and, in the other cases, τ_p becomes infinitely great, which agrees with the conceptions of chemical kinetics.

At low stresses ($\mu < 2$) the dependence of $\log \tau_p$ on μ is curved, and is satisfactorily expressed by the following equation [3, p. 172 of Russian translation]:

$$\tau_p = c\sigma^{-d}\exp\left(\frac{U_0}{RT}\right) \tag{8.46}$$

where c and d are parameters depending on the chemical and physical structure of the polymer.

At high stresses ($\mu > 2$) the dependence of $\log \tau_p$ on μ is linear and is expressed by the following equation [25]:

$$\tau_p = \frac{a}{k_0}\exp\left(\frac{U_0 - 1,17\beta\sigma_0}{RT}\right) \tag{8.47}$$

where the form of the parameter a is determined by the type of kinetic expression which is appropriate.

Equation (8.47) differs from those of Bueche and Zhurkov in that the pre-exponential factor in it is not the period of vibration of the atom, but the ratio of the pre-exponential factors of the elementary processes of the degradation reaction.

8.2.2. Special Features of the Failure of Polymers in Aggressive Media

The investigation of the mechanical properties of polymers, as already indicated, is being developed in two practically inde-

pendent directions: along the lines of describing the strength of
actual polymeric bodies within the framework of the Griffith elastic model, or of treating the failure of polymers as the result of
chemical reactions in a mechanical field.

Both approaches are based on well-known experimental facts:
the theoretical strength greatly exceeds the actual, which is determined by the defects in the polymer, and the strength of
polymers depends on the time for which they remain in the stressed
state.

Since there is at the present time no satisfactory theory explaining equally successfully both the macro- and microscopic as-
pects of failure, we shall now consider the special features of
the failure of polymers in aggressive media.

<u>The Role of Defects in the Failure of Polymers</u>. In rigid poly-
mers there are various defects, the most common being micropores
(or microcracks) and microregions, where the chemical bonds have
relatively high reactivity on account of the internal tensile
stresses.

The formation of these defects may be caused by differences
in the rate of cooling of the different parts of the polymer article, by
orientation and crystallization, by the conditions of shaping of
the polymer article, by degradation, or by various inclusions.

For the majority of rigid polymers the total volume of micropores does not exceed 0.1 cm^3/g. It is possible to determine experimentally the relative distribution of the micropores according
to the radius. Usually two methods are used for this purpose: the
sorption of inert gases at low temperatures, and the determination
of the volume of mercury forced into the polymer as a function of
the hydraulic pressure (the mercury porometry method) [25].

The method of small-angle scattering of x rays (diffusion diffraction) may be used to determine both the shape and the size of
micropores [41].

Although the above-mentioned methods are not without their
shortcomings, examination of the experimental data on the porosity
of polymer articles makes it possible to distinguish the following
groups of micropores [41, p. 119]:

intrafibrillar packing irregularities, measuring less than
15 Å;

interfibrillar pores, detected by the sorption, mercury porometry, or x-ray methods, whose size varies from 15 to 100 Å with the
majority about 30-40 Å;

pores formed in the swelling of polymers which have not undergone internal stress relaxation processes, with size of the order of a few hundred Å;

pores formed in the degradation of amorphous regions in partially crystalline polymers, of size from 100 to 10,000 Å.

The shape of the micropores depends on the physical structure of the polymers and in particular on the degree of orientation. For instance, with increase in the degree of orientation the micropores become more elongated [41].

Thus, rigid polymers are inhomogeneous at the micro- and macroscopic levels. As it acts on an inhomogeneous polymer, the stress field likewise becomes inhomogeneous and creates a stress concentration in the proximity of any defect.

If the stresses are sufficiently high and the accumulation of elastic energy takes place more rapidly than the increase of energy brought about by the formation of the new surfaces, then there is failure of the polymer article, with the strength being determined by the number and size of the defects.

The main postulate of the Griffith theory is that the strength of the material is determined by the size of a single defect (or ellipsoidal crack). Since an actual material contains a large number of defects of differing sizes, the strength will obviously be determined by the largest of these defects with, in the case of micropores, an important part being played not only by the dimensions but also by the shape.

Experiments show that when the size of an artificial defect (a notch) is large, the theoretical relationship in (8.25) is fully realized, but if the notch is sufficiently small there is then a deviation from the theoretical dependence. The surface of the failure does not in this case necessarily coincide with the plane of the notch, but always takes place in the region of the hairline cracks which occur in stressed glassy polymers in planes normal to the applied stress [35]. The cracks propagate preferentially along the zones of lowest density (reduced crosslinking), i.e., between crystallites, along the boundaries of domains and of spherulites, and between fibrils, i.e., in those regions where natural defects, for instance micropores, are concentrated.

The structure of hairline cracks ("crazes") was studied in detail by Kambour [43, 44] for various glassy polymers. Around 40-60% of the volume of a hairline crack is occupied by interconnected micropores, while the remainder contains polymers orientated in the direction of the acting force, in the form of fibrils of thicknesses 100-200 Å.

TABLE 8.2. Modulus of Crosslinking and Effective Surface Energy
at the Interface of PMMA with Various Media at 20°C

Medium	$k \cdot 10^4$, J/m²	γ, J/m²	Medium	$k \cdot 10^4$, J/m²	γ, J/m²
Air	128	$1.7 \cdot 10^5$	Octanal	24.2	$6.2 \cdot 10^3$
Methanol	3.3	$8.2 \cdot 10^2$	Hexane	9.8	$2.9 \cdot 10^3$
Ethanol	4.9	$9.2 \cdot 10^2$	Heptane	9.8	$2.9 \cdot 10^3$
Butanol	7.3	$1.8 \cdot 10^3$	Benzene	2.9	$7.2 \cdot 10^2$
Hexanol	14.1	$4.5 \cdot 10^3$	Toluene	2.9	$7.2 \cdot 10^2$

<u>Influence of Various Factors on the Failure of Polymers</u>. Let
us consider the influence of various factors on the failure of polymers with the simultaneous action of an aggressive medium and a
mechanical field.

<u>Stressing</u>. Where this is not large, growing cracks separate
the macromolecules to a greater degree than they rupture them, and,
as already pointed out, the cracks propagate preferentially along
zones of reduced crosslinking, i.e., in regions in which the main
mass of the sorbed aggressive medium is concentrated.

Usually the minimum stress σ_{min} at which a mainline crack
begins to grow in liquid media is recorded [32, p. 58]. The development of the crack proceeds with retardation, and after a certain time its growth ceases as a consequence of the formation of
a pencil of cracks which reduce the stress concentration.

At a particular loading σ_{cr} there occurs complete failure of
the polymer article, with the cracks developing very rapidly and
scission of the molecules taking place.

The influence of the magnitude of the stress on the failure
of polymers in a surfactant was first considered in [45]. This
question is considered below in further detail, because it is connected with the time dependence of the strength.

<u>The surface activity of the aggressive medium</u>. When polymer
articles come into contact with aggressive media, there is a change
in the surface energy at the polymer–medium interface.

Libatskii and Kovchik [46] introduced various liquid media into an artificial crack and recorded the minimum stress at which
crack growth began. Table 8.2 gives the modulus of crosslinking
k and the effective surface energy at the interface of poly(methyl
methacrylate) with various media:

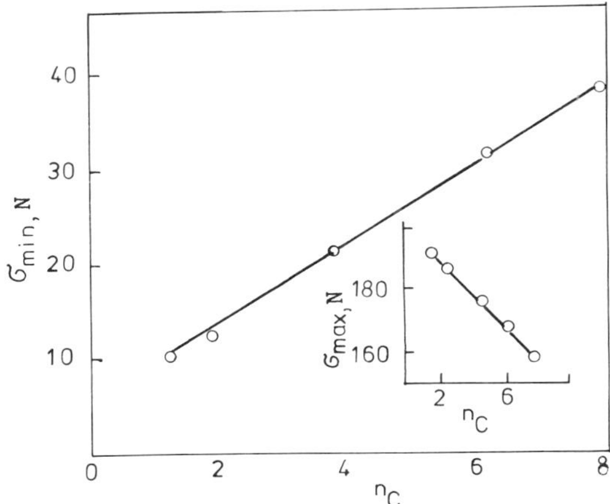

Fig. 8.11.   Dependence of σ_{max} and σ_{min} for PMMA
on the number of carbon atoms in the
alcohol [32, p. 58].

$$k = \frac{\sigma_{min}}{h\sqrt{2R}}\,\varphi\,(l,\,R)\qquad\qquad(8.48)$$

where h is the thickness of the specimen and $\varphi(l,\,R)$ a function of
the size of the crack and of the specimen, of the point of application of the loading, and of the physical constants of the material.

The effective surface energy was calculated from the formula

$$\gamma = \frac{k^2}{\pi E}\qquad\qquad(8.49)$$

The data show satisfactory correlation between the modulus of
crosslinking and the effective surface energy.

In the homologous series the surface energy cf the alcohols
in air rises with increase in the number of carbon atoms and, as
Fig. 8.11 shows, σ_{cr} decreases in the same order, but the reverse
relationship applies for σ_{min} [32, p. 58]. The magnitude of σ_{min}
is the result of interaction between the OH groups of the alcohol
and the ester groups of the polymer. With increase in the length
of the alcohol molecule there is a reduction in the number of such
contacts in the microcracks within the polymer, which in turn leads
to an increase in σ_{min}.

The scale factor. The geometrical dimensions of the specimen
play a big part in the failure of polymers in aggressive media.
However, there is at present no single viewpoint which explains the

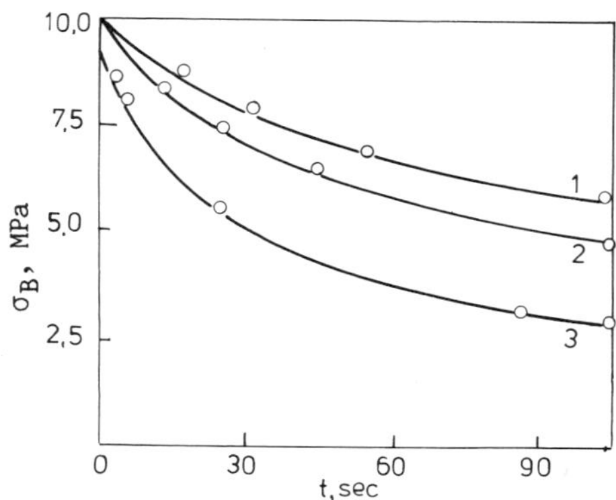

Fig. 8.12.   Dependence of σ_B on the time of immersion in ethyl alcohol for PMMA
specimens of various diameters [32,
p. 91]:   1) 5.5 mm; 2) 3.0 mm; 3)
1.5 mm.

influence of the size of the specimen on its strength in air, and
there is only a small amount of experimental data. As already
shown (Fig. 8.4), in the stretching of polymers in adsorption-
active media (swelling being practically absent) the influence of
the medium decreases as the thickness of the specimen increases and
there is a critical thickness above which the medium has no perceptible
influence on the mechanical properties. If there is sorption of
the medium by the polymer the behavior becomes more complex. For
instance, Fig. 8.12 shows the dependence of the TS, σ_B, of circular
PMMA specimens of various diameters on the time of previous immersion in alcohol. This falls monotonically with an increase in the time
of contact with the alcohol, with the specimens of large diameter
having the highest strength. If the tests on this polymer are carried out directly in ethyl alcohol the differences become insignificant since for specimens of diameter 1.3 and 16 mm, σ_B is, respectively, 66 and 87 MPa [32, p. 91].

During the time of previous immersion in ethyl alcohol there
is no time for establishment of equilibrium [32, 66], and it is
natural to expect that the strength of the specimens of larger diameter will be higher.

8.2.3.   Durability of Polymers

This is determined from plots of log τ_p vs. σ, which show the
biggest difference in shape when polymers are tested in aggressive
media (Fig. 8.13). Similar relationships were obtained in tests
of PE [49], PCA [50, 51], and hydrated cellulose [51] in aggressive media.

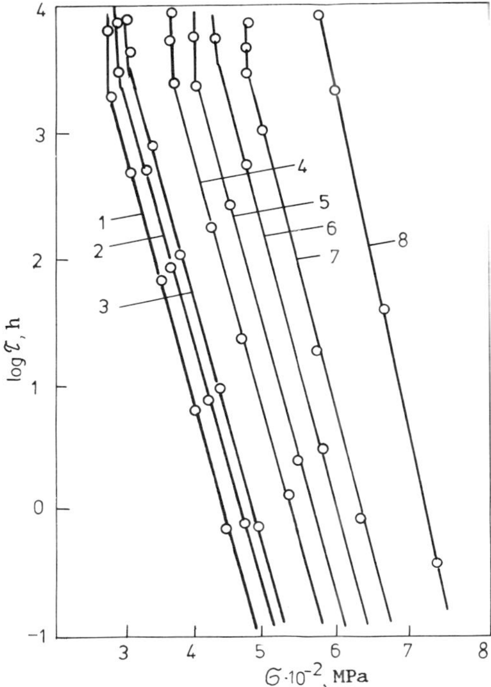

Fig. 8.13. Dependence of the durability of AG-
45 (aniline-PF, in sheet form) glass-
reinforced plastic in aggressive media on stress at 20°C: 1) 10% NaOH;
2) 30% NaOH; 3) 10% H_2SO_4; 4) 30%
H_2SO_4; 5) 3% NaOH; 6) 3% H_2SO_4; 7)
H_2O; 8) air [54].

We must now consider two extreme cases involving stress acting
on a polymer article.

In the region of low loading, the rate of surface diffusion
V_D is higher than that of crack development v_t. The aggressive
medium diffuses into the matrix of the polymer, changing, as a
rule, the physical and chemical structure of the polymer. The magnitude of σ_0 will depend not only on the applied stress, but also
on the free surface energy, the flowing-out pressure, the capillary pressure, the energy of the intermolecular bonds, and the
number of chemically unstable bonds which have been broken.

In the region of high loading, where $v_D < v_t$, the aggressive
medium is not capable of being sorbed by the polymer, and accordingly the magnitude of σ_0 is often determined solely by the action of
the applied stress [37].

TABLE 8.3.   Values of U_0 and γ for PMMA and PE in Various Physically Active Media at 20°C [32]

Medium	U_0, kJ/mole		γ, kJ/(mole · MPa)	
	PE	PMMA	PE	PMMA
Air	151	130	2.6	0.55
Hexane	—	84	—	0.13
Heptane	117	84	2.8	0.13
Toluene	122	—	3.1	—
Methanol	134	80	2.0	0.14
Ethanol	134	84	2.0	0.12
Butanol	126	84	2.0	0.12
Hexanol	126	—	2.3	—
Octanol	—	92	—	0.17

Examination of the data on the long-term strength testing of
polymers in aggressive media shows that there are cases where Eq.
(8.43) is formally applicable, and also cases where it cannot be
used. The latter cases, as a rule, are connected with the development of large deformations, and are considered in a special section.

Let us consider the influence of aggressive media on the parameters U_0, γ, and τ_0 with the formal employment of Eq. (8.43).

Tables 8.3 and 8.4 show the values of U_0, γ, and τ_0 for the
action of physically and chemically active media on various polymers.

<u>Zero activation energy of degradation, U_0</u>. As Table 8.3 shows,
the value of U_0 in tests on PMMA and PE in various media is far
lower than the energy of scission of the C—C bond, and depends on
the nature of the medium [32].

In this case, the reduction in U_0 may be brought about by a
change in the energy of the intermolecular bonds and by a reduction

TABLE 8.4.   Values of U_0, γ, and τ_0 for PCA, Hydrated
the Kinetic Parameters of Degradation [51]

Polymer	Air		
	U_0, kJ/mole	τ_0, sec	γ, kJ/(mole · MPa)
PCA	230—250	10^{-12}	0.67
Hydrated cellulose	210—230	10^{-13}	0.67
Quartz glass	420	—	—

in the free surface energy. It is not known whether, and if so, how, physically active media influence the breakdown of the chemical bonds.

In tests on polymers in chemically active media (Table 8.4), U_0 again decreases considerably as compared with tests in air and, most importantly, U_0 is close to the effective activation energy of hydrolytic degradation of these polymers.

Thus, in the case in question, the breakdown of hydrolytically unstable bonds in the molecules is the main reason for reduction in the durability of polymer articles.

The structure-sensitive coefficient γ. In the determination of $\gamma = \beta w$ the aggressive medium may also influence both the fluctuation volume of the elementary process of failure, particularly with high figures for sorption, and also the coefficient of stress concentration if there is a change in the physical or chemical structure of the polymer. Unfortunately, no specific data have been published on the influence of aggressive media on β and w separately. We can only say that, as a rule, the value of γ is not constant in the testing of polymers over a wide range of applied stresses.

The pre-exponential factor τ_0. According to the thermofluctuation theory τ_0 is the period of vibration of the atoms in the chemical bonds, which should not change to any considerable extent. However, in tests on polymers in aggressive media its value is lower by several orders than the theoretical value (10^{-12}-10^{-13} sec) [51]. According to Manelis [38], the pre-exponential factor τ_0 is determined by the activation entropy of the elementary processes of breakdown and, it would appear, this treatment of τ_0 is the most correct.

In point of fact, $\tau_0 \sim 1/A$ (see Table 8.4). For the majority of processes of chemical breakdown in the condensed phase the activation entropy is a large negative figure, an exception being

Cellulose, and Quartz Glass in Air and in Water, and

Water			Hydrolysis	
U_0, kJ/mole	τ_0, sec	γ, kJ/(mole·MPa)	E_{eff}, kJ/mole	A, sec^{-1}
105—125	10^{-5}	0.55	85—105	10^8—10^4
145—170	10^{-12} — 10^{-13}	0.55	145	—
85	10^{-4}	0.08	81—97	10^5—10^6

TABLE 8.5. Strength Parameters of EF-32-301 Glass Fabric Laminate in Various Media at 50°C and $\sigma = 118$, MPa [54]

Medium	$\tau_0 \exp\left(\dfrac{U_0}{RT}\right)$, sec	τ_p, h	Medium	$\tau_0 \exp\left(\dfrac{U_0}{RT}\right)$, sec	τ_p, h
H_2O	$1.9 \cdot 10^4$	612	NaOH, 1%	$6.9 \cdot 10^3$	82
H_2SO_4, 3%	$8.7 \cdot 10^3$	145	NaOH, 10%	$3.1 \cdot 10^2$	32
H_2SO_4, 30%	$1.9 \cdot 10^4$	612			

cellulose derivatives, for which $\Delta S^{\neq} \geq 0$ (see Chap. 2), and τ_0 is 10^{-4}-10^{-5} sec.

When polymer articles are being used under normal conditions the stressing and temperature are as a rule not constant, and determination of the durability is a difficult problem, both experimentally and theoretically.

One approach to this problem was developed by Bailey [52]. It is based on two postulates: irreversibility of failure and independence of the failure rate with respect to the mechanical prehistory of the specimen.

Both these postulates are unfulfilled under extreme conditions, but if the rates of change in stress are not high and the structure does not greatly change during testing, then the following general equation holds [53, p. 396]:

$$\int_0^{t_p} \frac{dt}{\tau_p \sigma(t)\, T(t)\, \gamma(t)} = 1 \tag{8.50}$$

where t_p is the time from the moment of application of the loading up to failure of the polymer, and τ_p is the durability at constant values of stress, temperature, and the structure-sensitive parameter; it is equal to the instantaneous values of $\sigma(t)$, $T(t)$, and $\gamma(t)$, respectively.

It is exceptionally difficult to solve this equation in the general form, and in practice we use special cases of it.

With varied loading but constant T and γ, we have

$$\int_0^{t_p} \frac{dt}{\tau_0 \exp\left[\dfrac{U_0 - \gamma\sigma\,(t)}{RT}\right]} = 1$$

With varied temperature but constant σ and γ, we have

$$\int_0^{t_p} \frac{dt}{\tau_0 \exp\left[\dfrac{U_0 - \gamma\sigma}{RT\,(t)}\right]} = 1$$

Table 8.5 gives the strength parameters calculated using Eq. (8.50), in tests of EF-32-301 glass fabric laminate.

8.3.   CREEP OF POLYMERS IN AGGRESSIVE MEDIA

With simultaneous action of mechanical stresses and aggressive media on polymer articles, considerable deformation (far greater than in air) develops, to an extent depending on time, stress, and temperature (Fig. 8.14). It is because of this that creep is one of the most inportant parameters in establishing the service properties of polymer articles in aggressive media.

We shall now consider the theoretical ideas on the creep of polymers.

8.3.1.   Theoretical Ideas on Polymer Creep

As already indicated, polymers occupy an intermediate position between viscous liquids and elastic solids, and the total creep deformation ε_{creep} is a combination of elastic, delayed elastic (nonsteady state), and viscous (steady state) (strain increasing uniformly) deformation.

With constant uniaxial stretching and constant temperature ε_{creep} is equal to [55]

$$\varepsilon_{creep} = \varepsilon_{el} + \varepsilon_{del.\,el} + \varepsilon_v \tag{8.51}$$

where $\varepsilon_{el} = \sigma(t)/E$, the elastic component determined by Hooke's law; $\varepsilon_{del.el}$ is the delayed elastic deformation; $\varepsilon_v = (1/\eta)\sigma(\tau)d\tau$ is the viscous component, determined by the Newton equation; τ is the moment of time; and $\sigma(t)$ is the stress at the moment of time t.

Delayed elastic deformation can be described qualitatively by the Kelvin–Voight–Meyer equation, which corresponds to a model consisting of elastic and viscous elements connected in parallel [53, p. 79]:

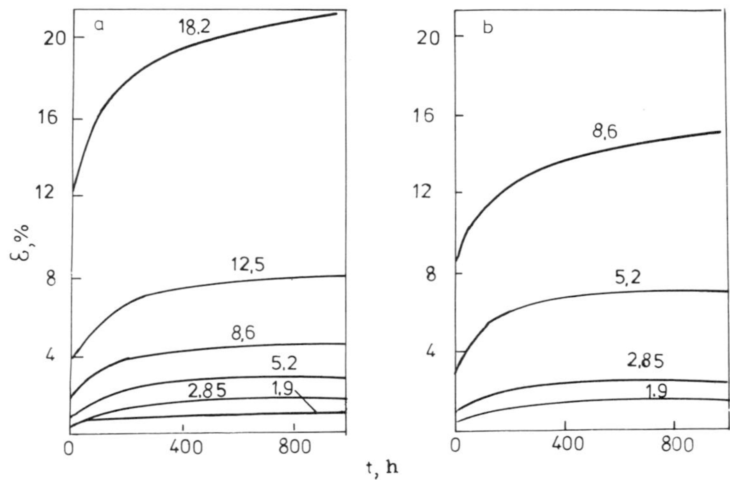

Fig. 8.14.   Creep curves of polypropylene in water under
             various stresses at 20°C (a) and 60°C (b)
             (the numbers against the curves indicate the
             stress in MPa) [54].

$$\varepsilon_{del.\ el} = \frac{1}{\eta'} \int_0^t \exp\left(-\frac{t-\tau}{\theta}\right) \sigma(\tau)\,d\tau \qquad (8.52)$$

where θ is the retardation time, i.e., the time at the end of which
the deformation amounts to $(1 - e^{-1})$ of the equilibrium value.

Since substituting Eq. (8.52) in (8.51) does not make a quantitative description of the deformation of a polymer article possible,
there have been proposed from time to time various theories of
creep, of which we shall now consider those most used in polymer
mechanics.

The Elastic History Theory. This is based on the Boltzmann
superposition principle, which assumes that the creep of a specimen depends on its stress prehistory, and that each stage of stressing makes an independent contribution to the final deformation, so
that the total deformation can be found by simple summation of all
these contributions. For reversible deformations at low stresses
the Boltzmann—Volterra equation is used [56]:

$$\varepsilon(t) = \frac{\sigma(t)}{E} + \int_0^t \varphi(t-\tau)\,\sigma(\tau)\,d\tau \qquad (8.53)$$

The function $\varphi(t - \tau)$ is called the memory function or the kernel of the Boltzmann equation. It takes into account the whole prehistory of the specimen, i.e., the influence of the preceding deformations.

For actual polymers $\varphi(t - \tau)$ may take differing forms; it is found either from experimental data or else by trial and error.

One of the expressions for the memory function, which has become popular for describing the creep of polymers in aggressive media, was proposed by Askadskii [57]:

$$\varphi\,(t-\tau) = \frac{m}{\theta}\left(\frac{t-\tau}{\theta}\right)^{m-1}\exp\left[-\left(\frac{t-\tau}{\theta}\right)^{m}\right] + \frac{1}{\eta} \qquad (8.54)$$

Substituting this function into Eq. (8.53) leads to the following expression for describing creep:

$$\varepsilon\,(t) = \frac{\sigma\,(t)}{E} + \varepsilon_{\infty}\left[1 - \exp-\left(\frac{t}{\theta}\right)^{m}\right] + \frac{\sigma\,(t)\,t}{\eta} \qquad (8.55)$$

where ε_{∞} is the equilibrium deformation when $t \to \infty$; m is an empirical constant.

The Theory of the Distribution of Retardation Times. Alfrey [2] was the first to express the notion that the mechanical properties of an actual polymer body should be described by a distribution of retardation times θ_i. In this case the creep of polymers is expressed by the equation

$$\varepsilon\,(t) = \frac{\sigma\,(t)}{E} + \sigma\int_{0}^{\infty} J\,(\theta)\left[1 - \exp\left(-\frac{t}{\theta}\right)\right]d\theta + \frac{\sigma\,(t)}{\eta}\,t \qquad (8.56)$$

where $J(\theta)$ is a function of the retardation times which, like the function $\varphi(t - \tau)$ in the Boltzmann–Volterra equation, may be found either from experimental data or by trial and error.

Equations (8.53) and (8.56) describe the process of creep under isothermal conditions. The dependence of the rate of deformation on temperature can be described by an exponential equation [56]

$$\frac{d\varepsilon}{dt} = D\exp\left(-\frac{U_{eff}}{RT}\right) \qquad (8.57)$$

where U_{eff} is the effective activation energy of the creep process and D a parameter depending on time and stress.

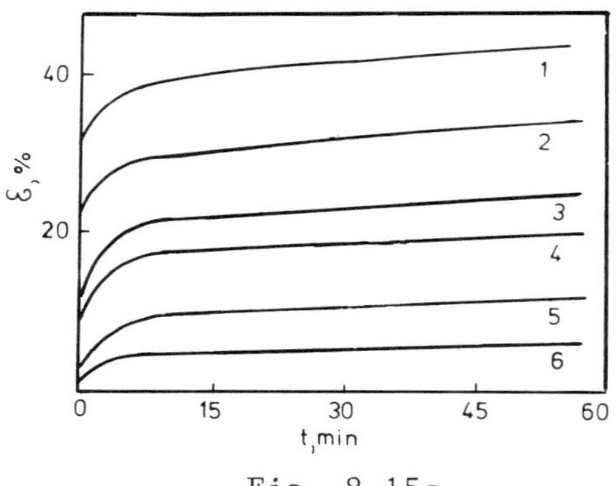
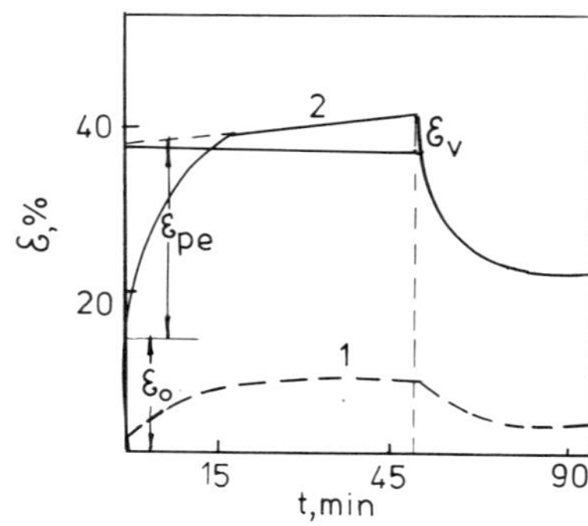

Fig. 8.15a   Fig. 8.15b

Fig. 8.15a. Creep curves of films of F-2 polyarylate in water (1-3)
            and in air (4-6) at 25°C at various loads: 1, 4) 90
            MPa; 2, 5) 80 MPa; 3, 6) 70 MPa.

Fig. 8.15b. Creep curves of films of F-2 polyarylate in air (1)
            and in water (2).

8.3.2. Influence of Physically Active Media on Creep

Under the action of mechanical loading the creep of polymer
articles in physically active media differs from that in air as a result
of the following processes: adsorption of the components of the aggressive medium, which leads to a decrease in the surface energy
at the polymer—medium interface; sorption of the components of the
aggressive medium, leading to swelling (a reversible process); and
irreversible change in the physical structure of the polymer.

Let us consider first the creep of polymers in single-component aggressive media.

Creep of Polymers in Liquid Media in the Absence of Diffusion
Restrictions. When the rate of diffusion of the liquid medium in
the polymer is higher than the rate of creep of the polymer in air,
the liquid medium rapidly fills up the accessible regions within
the matrix of the polymer, and there will be deformation of the polymer which is saturated with the medium. These conditions apply for
thin films of the majority of polymers in water.

Dependence of the formation of polymers on time. Various equations have been found for describing the curves of deformation as
a function of time, some being purely empirical, others being supported by theoretical grounds. For instance, Figs. 8.15a and b
show the creep curves of films of F-2 polyarylate (PA) [the product
of polycondensation of terephthalic acid and phenolphthalein] determined in air and in water [59]. These curves can be described by
the empirical equations

TABLE 8.6. Values of ε_∞, θ, m, and η in Eq.
(8.55), Which is Used to Describe
the Creep of F-2 Polyarylate in
Water and in Air at 20°C

σ, MPa	Water		Air
90	ε_∞, % θ, min m $\eta \cdot 10^{-12}$, Pa·sec	36.4 2.41 0.37 3.54	23.02 3.9 0.18 14.6
80	ε_∞, % θ, min m $\eta \cdot 10^{-12}$, Pa·sec	21.6 5.0 0.26 2.16	8.1 11.6 0.16 6.6
70	ε_∞, % θ, min m $\eta \cdot 10^{-12}$, Pa·sec	20.2 8.5 0.28 1.37	6.2 21 0.10 4.1

$$\varepsilon(t) = at^n \tag{8.58}$$

$$\varepsilon(t) = m_1 + m_2 \ln t \tag{8.59}$$

where a, n, m_1, and m_2 are empirical constants. The creep curves
of F-2 PA in water can be handled according to Eq. (8.55) by means
of a computer. The results are shown in Table 8.6.

Table 8.6 shows that in the creep of F-2 polyarylate in water,
irrespective of the stress value, the equilibrium deformation is
always considerably lower than that attained in air.

The creep curves of F-2 PA in water can also be handled according to Eq. (8.56) for any one retardation time. Figure 8.16
shows the handling of the initial section of the creep curve of
F-2 PA in water in accordance with the equation

$$\varepsilon(t) = \varepsilon_\infty[1 - \exp(-\tfrac{t}{\theta})] \tag{8.60}$$

which is a special case of the solution of Eq. (8.56). Describing
the creep curves in the initial section (delayed elastic deformation) by means of Eq. (8.60) (Fig. 8.16) seems the most successful
since this equation does not contain any empirical constants and
also has the least number of constants.

In addition, this approach leads to an explanation of the
mechanism of creep in liquid media.

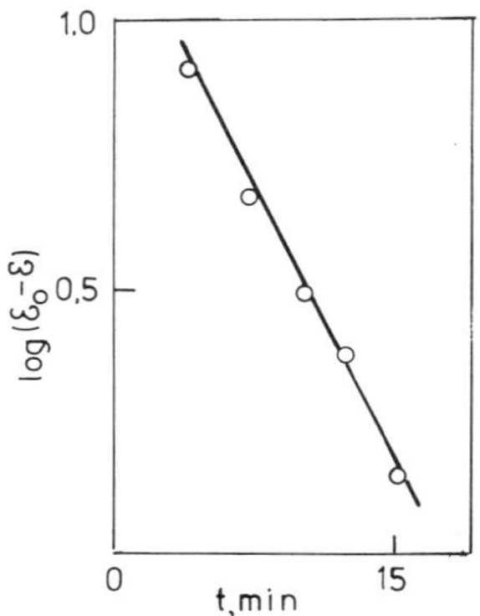

Fig. 8.16. Handling of the initial section of
           the creep curve of F-2 PA in water
           at 60°C and σ = 80 MPa using semi-
           logarithmic coordinates of Eq.
           (8.58) [59].

Equation (8.60) satisfactorily describes the creep of a variety of polymers in water [51, 59, 60, 61].

Table 8.7 shows the values of θ and ε_∞ for the creep of PETP
and PCA at various temperatures and stresses, as calculated according to Eq. (8.60).

For these polymers the retardation time changed only to a
slight extent with increase in temperature and stress. At the same
time these factors have a considerable influence on ε_∞. It is
known that hardly a single process of polymer creep in air can be
described by a simple exponential function. The possibility of describing the creep curves in the initial section by means of Eq.
(8.60) is evidence of the predominance in this instance of a single
mechanism of creep, which is connected with the influence of the
ambient medium.

<u>Dependence of the creep in polymers on the medium</u>. During the
process of creep in polymers there are scissions of some of the intermolecular bonds, and various conformational changes in the
macromolecules [62-65]; for these changes to be effected, a particular free volume is required. It may be suggested that ε_∞ will
be greater the weaker the intermolecular bonds and the greater the
free volume of the polymer—medium system [66].

Let us state the following relationship between ε_∞ and the
proportion of free volume in the polymer—medium system f:

TABLE 8.7.   Values of θ and ε_∞, as Calculated According to Eq. (8.60), for the Creep of PETP and PCA at Various Temperatures and Stresses

Polymer	Temperature, °C	$\sigma = 10^{-2}$, MPa	θ, min	ε_∞, %
PETP	20	1	2.2 ± 0.4	12
	40	2	1.8 ± 0.4	30
	40	1	2.2 ± 0.4	14
		2	1.6 ± 0.4	35
	50	1	2.0 ± 0.6	15
		2	2.0 ± 0.4	38
	60	1	1.4 ± 0.4	19
		2	1.6 ± 0.4	43
PCA	25	0.8	1.8 ± 0.2	31
		0.4	2.0 ± 0.2	
		0.15	2.6 ± 0.4	
	80	0.8	1.8 ± 0.2	
		0.4	2.2 ± 0.4	
		0.15	2.6 ± 0.4	

$$\varepsilon_\infty = A \exp(Bf) \qquad (8.61)$$

where A and B are parameters which are independent of the concentration of the medium in the polymer, and of the temperature.

The meaning of Eq. (8.61) is that the equilibrium deformation is not directly connected with the temperature and mechanical stress, but is determined, mainly, by the proportion of the free volume, the magnitude of which determines the possibility of conformational changes in the macromolecules, and which influences the molecular interaction [67].

The effect of the temperature, the mechanical stress, and the medium becomes evident through changes in the free volume, the free volume being a function of these parameters [68].

Taking as the standard state the equilibrium deformation of the polymer in air, we get

$$\varepsilon_\infty = \varepsilon_\infty^0 \exp[B(f - f^0)] \qquad (8.62)$$

where ε_∞^0 is the equilibrium deformation of the polymer in air at a given temperature and given stress.

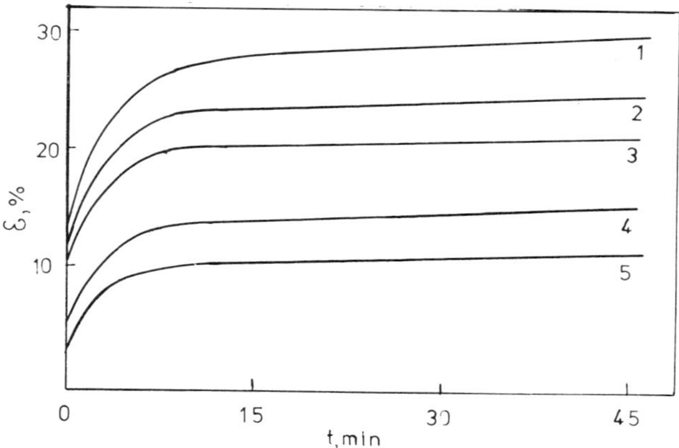

Fig. 8.17.   Creep curves of F-2 polyarylate at 20°C and
σ = 80 MPa in aqueous solutions of KNO_2 [66]:
1) H_2O; 2) 10% KNO_2; 3) 30% KNO_2; 4) 40%
KNO_2; 5) 60% KNO_2.

Using relationship (5.50) we get

$$\varepsilon_\infty = \varepsilon_\infty^0 \, \exp(B \cdot \beta \cdot \varphi_{solv}) \qquad (8.63)$$

The sorption of the solvents by the polymers increases with
increase in uniaxial stretching, and most frequently this takes
place according to a linear process [67-70]:

$$\varphi_{solv} = \varphi_{solv}^0 (1 + n\sigma) \qquad (8.64)$$

where φ_{solv} and φ_{solv}^0 are, respectively, the volume fractions of
the solvent in the polymer under stress and with no stress, and n
is a constant.

As already indicated (Sec. 5.4), there takes place for polymers which are at temperatures below T_g a double sorption, which
comprises the actual solubility of the low-molecular substances
and their sorption by the surface of the micropores. It may be expected that the low-molecular substances sorbed by the micropores
will hardly change the proportion of free volume in the polymer—
medium system and, consequently, will have far less influence on
the creep than will the low-molecular substances actually dissolved
in the polymer.

The general expression for the influence of the medium sorbed
by the polymer on ε_∞ takes the form

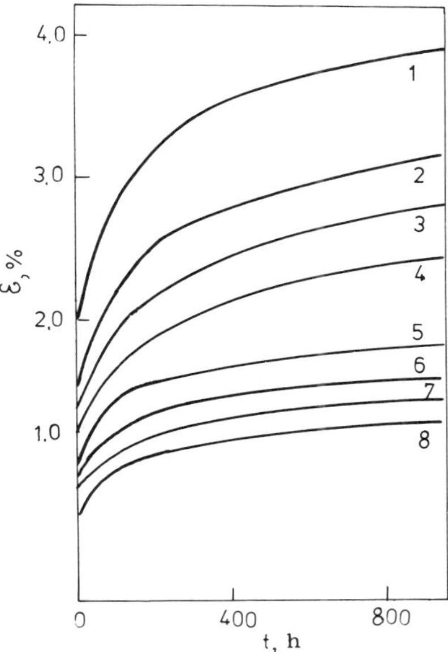

Fig. 8.18. Creep curves of PP in various media
at 60°C and σ = 5.2 MPa [61]: 1)
10% NaOH; 2) 2% OP-10; 3) 60%
CH_3COOH; 4) H_2O; 5) 30% H_2SO_4; 6)
60% H_2SO_4; 7) 80% H_2SO_4; 8) air.

$$\varepsilon_\infty = \varepsilon_\infty^0 \exp \{B\beta [\phi_{solv}^0 (1 + n\sigma) - \varphi_{pores}] \qquad (8.65)$$

where φ_{pores} is the volume fraction of the micropores in the polymer.

Influence of Various Factors on the Creep of Polymers. Let
us now consider the influence of various factors on the creep of
polymers in liquid media.

Surface activity of the medium. With the adsorption of the
components of the solution on the surface of the polymer there is
a change in the surface energy at the polymer—medium interface. It
is the micropores present in the polymer which are most subject to
adsorption action. These micropores may be regarded schematically
as drawn-out microcracks in a force field. The medium, as it penetrates into these microcracks, alters the surface energy at the
polymer—medium interface and creates a wedgelike pressure in the
openings of the microcracks, facilitating the deformation processes [69].

Aqueous electrolyte solutions are, as a rule, surface-inactive. Accordingly, with increase in the electrolyte concentra-

TABLE 8.8.   Values of $\gamma_{pol-sol}$, $C_{H_2O}^0$, and ε for the System PP–
             Aqueous H_2SO_4 at 20°C [51]

Concentration of H_2SO_4, mass %	$\gamma^*_{pol-sol}$, kJ/m^2	$c^0_{H_2O}$, mass %	ε, % †	Concentration of H_2SO_4, mass %	$\gamma^*_{pol-sol}$, kJ/m^2	$c^0_{H_2O}$, mass %	ε, % †
0	10	0.06	2.2	60	55	0.005	1.4
30	45	0.03	1.6	80	—	—	1.3

*Calculated by Young's equation, from the wetting angles [51].
†After 800 h at σ = 5.2 MPa.

tions, on the one hand there is an increase in the surface energy
at the polymer–medium interface while, on the other hand, the solubility of water in hydrophobic polymers decreases proportionally
to the change in activity of the water in solution (Chap. 5).
Thus, both these factors are bound to lead to a decrease in the
creep of polymers in electrolyte solutions as compared with water.

Figure 8.17 shows the creep curves of films of F-2 polyarylate at 20°C and σ = 80 MPa in water and in aqueous KNO_2 solutions
of various concentrations. A similar effect occurs in the creep
of polypropylene (PP) in aqueous solutions of sulfuric acid (Fig.
8.18 and Table 8.8).

It must be pointed out that comparison of the deformation
with $\gamma_{pol-solv}$ as calculated from the wetting angles is an arbitrary step since for PP $\gamma_{pol-solv}$ may change with time [70]. When
using substances which reduce the surface energy at the PP–solution
interface, the adsorption factor is determining. For instance, the
creep deformation of PP in 10% NaOH or the surfactant OP-10 is
greater than in water alone (Fig. 8.18).

It has been found experimentally that the sorption of water
by the polymer from these solutions will be practically the same
as from water alone. The base and OP-10 are likewise not sorbed
by PP to a noticeable extent. The action of solutions of NaOH and
OP-10 on PP under the conditions indicated gives rise to surface
cracks lying perpendicular to the direction of the force field,
whereas the action of the other media gives no such effect. This
suggests that in a force field in the presence of surfactants there
is an opening up of defects in the polymer, as a result of which
there are formed microcracks through which these substances penetrate into the polymer. A similar result is obtained in the study
of the creep of PETP in aqueous solutions of sodium dodecyl sulfate [71].

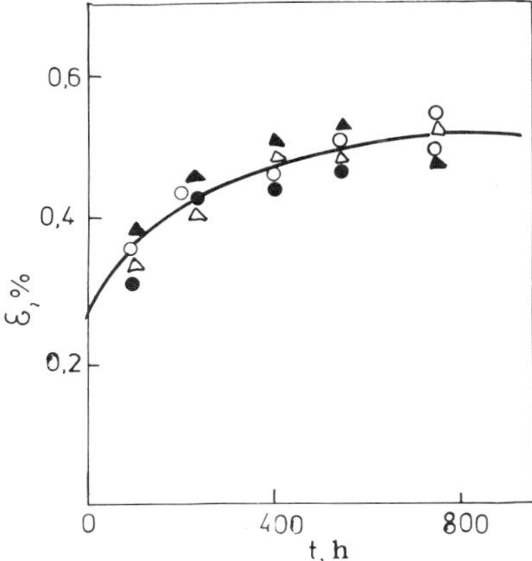

Fig. 8.19. Creep curves of PP in various media at 20°C and σ = 2.6 MPa [54]: $\circ$) air; $\bullet$) H_2O; $\triangle$) 60% H_2SO_4; $\blacktriangle$) 10% NaOH.

<u>Stress</u>. It is known that at certain stresses the creep curves of polymers in liquid media and in air coincide (Fig. 8.19). These data may be explained by the fact that PP sorbs little water. At low values of σ, φsolv $\leqq \varphi$pores, and according to Eq. (8.65), the equilibrium deformation is practically the same in the medium and in air. It may also be suggested that with stresses below the critical values there is no opening up of defects or penetration of the medium into them. It is interesting to note that similar critical stresses are characteristic of the creep of hydrophobic polymers in water: PP [54], PE [72], PETP [72], F-2 polyarylate [65], Diflon polycarbonate [73]; nevertheless, they are not present for polycaproamide, which dissolves a considerable amount of water [72].

<u>Nature of the Polymers</u>. If the polymer and solvent differ considerably in their solubility parameters, e.g., hydrophobic polymers and water, then two characteristic features of the creep of polymers may be pointed out.

1. Where the amount of sorbed low-molecular substance is commensurate with the total volume of the micropores, there are no significant differences in the creep of these polymers in air or in a particular liquid medium.

2. If the amount of sorbed low-molecular substance is greater than the volume of the micropores, the creep in the liquid medium will be greater than in air.

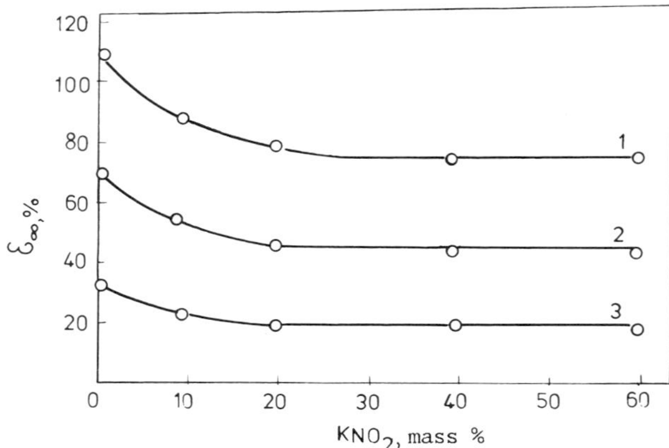

Fig. 8.20.   Influence of the concentration of KNO_2 solution on the equilibrium creep deformation of
Diflon polycarbonate at σ = 48 MPa and temperatures: 1) 90°C; 2) 60°C; 3) 20°C [73].

For instance, these postulates can be illustrated by Fig.
8.20, which shows the dependence of the equilibrium deformation of
Diflon polycarbonate in aqueous solutions of KNO_2.

Diflon polycarbonate is a hydrophobic polymer, in which KNO_2
is practically insoluble, while the amount of sorbed water as determined gravimetrically is proportional to the activity of the
water in the solution, i.e., it decreases with increase in the concentration of KNO_2.   In the polycarbonate the total volume of micropores, according to mercury porometry data, is $(3.0 \pm 0.5) \cdot 10^{-3}$
$cm^3 \cdot cm^{-3}$ [73].   When the concentration of water in the polymer is
less than this figure, the equilibrium deformation remains practically constant and does not differ greatly from ε_∞ in air.   On
the other hand, when the concentration of water in the polymer is
greater than this figure, the equilibrium deformation increases to
correspond with the increase in the proportion of free volume in
the polymer—medium system.

In the case of the creep of F-2 polyarylate at σ = 80 MPa,
there is, in all the KNO_2 solutions, a decrease in the equilibrium
creep with increase in KNO_2 concentration (Fig. 8.17).

The amount of water in the polymer was, as in the preceding
case, determined gravimetrically [59].   Under the conditions in
question the volume proportion of sorbed water was always greater
than that of the micropores in the polymer (0.022) [59], and, as
may be seen from Fig. 8.21, the experimental data are satisfactorily
described by Eq. (8.65).   The value of $B \cdot \beta$ is 40 $\pm$ 5, of n, (2.5
$\pm$ 0.5)$\cdot 10^{-3}$ MPa^{-1}.

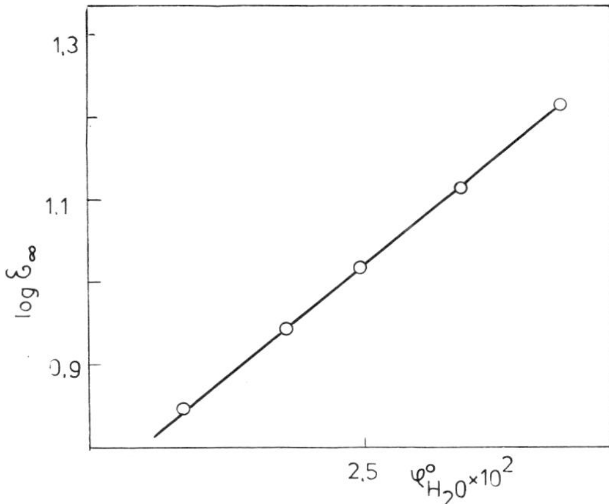

Fig. 8.21. Handling of experimental data according to Eq. (8.65). For explanation, see text.

Diflon polycarbonate and F-2 polyarylate are amorphous polymers, in which practically all the macromolecules may be subject
to deformation.

In partially crystalline polymers the fragments of macromolecules in the crystalline regions remain practically undeformed
[65, p. 138; 74], and thus the deformed fragments of macromolecules
will be those fragments of macromolecules in the amorphous regions
in which there is also concentrated the medium sorbed by the polymer.

Creep of Polymers in Liquid Media under Conditions of Diffusion Limitation. Where the rate of diffusion of the liquid medium
in the polymer is commensurate with or less than the rate of creep
of the polymer in air, there will be deformation with the distribution of the medium within the matrix of the polymer changing with
time, i.e., the creep in the initial stage will be limited by the
diffusion of the medium in the polymer.

On substituting in (8.60) the relationships (5.17) and (8.65),
we obtain an equation which describes the change with time in the
creep of a polymer film at constant temperature and stress under
diffusion limitation conditions:

$$\varepsilon(t) = \varepsilon_\infty^0 \exp\{B\beta[\varphi^\infty_{solv}(1 - \frac{8}{\pi^2}\exp(-\frac{\pi^2 D_{solv}t}{\ell^2}))(1 + n\sigma)$$
$$- \varphi_{pores}]\}[1 - \exp(-t/\theta)] \qquad (8.66)$$

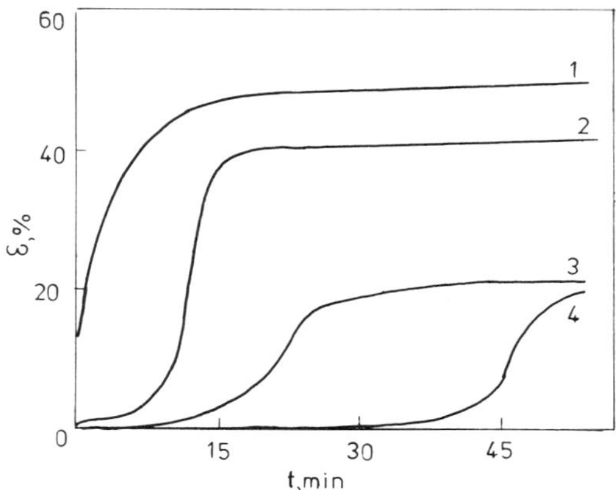

Fig. 8.22.   Creep curves of F-2 polyarylate at
40°C and σ = 20 MPa in the alcohols:
1) methanol; 2) ethanol; 3) n-propanol; 4) n-butanol [72].

Equation (8.66) contains two parameters, D_{solv} and θ, whose
values generally depend on the concentration of the medium in the
polymer. The concentration dependence of the diffusion coefficients
is discussed in Chap. 5, whereas the retardation time depends to
a lesser extent on the concentration of the medium in the polymer
[66].

Diffusion-limited creep occurs in the case of F-2 polyarylate
in alcohols (Fig. 8.22). Alcohols, unlike water, have a solubility parameter close to the solubility parameter of the polymer
[75, p. 21], and are sorbed in far larger amounts than water. In
addition, the alcohol molecules have larger linear dimensions than
the water molecules, and their diffusion coefficients are lower
than the diffusion coefficient of water in F-2 polyarylate.

The creep curves of films of F-2 polyarylate in ethyl, propyl,
and butyl alcohols are sigmoid in shape with the point of inflection approximately corresponding to the termination of the process
of sorption of these alcohols by the polymer under the conditions
in question. Methyl alcohol diffuses in F-2 polyarylate more
rapidly than the above-mentioned alcohols, and the creep curve is
of a shape typical of the creep of polymers where diffusion limitations are absent.

The value of the equilibrium deformation depends mainly on the
extent of sorption of the medium by the polymer, but we evidently
cannot exclude the influence on ε_∞ of the adsorption factor in the
case of alcohols with a large number of methylene groups, a factor
leading to a decrease in the surface energy at the polymer—medium
interface.

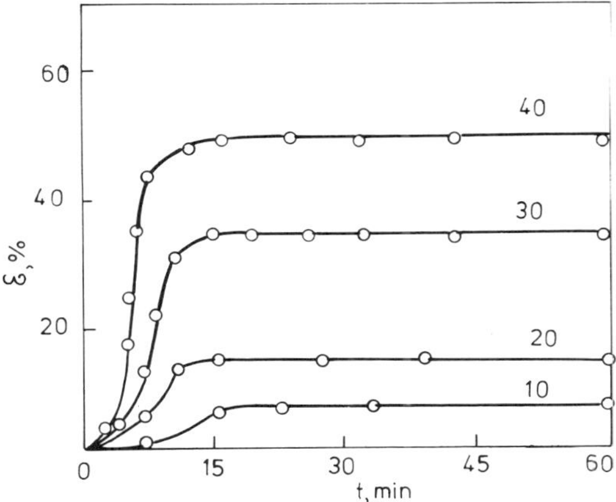

Fig. 8.23. Creep curves of F-2 polyarylate in
ethyl alcohol at 25°C at various
loads. The figures alongside the
curves give the load in MPa [60].

Let us now consider the influence of the various factors on
the creep of polymers under diffusion limitation conditions.

Stress. Stress applied to a polymer which is in a liquid
medium influences the diffusion (D, E_D) and sorption (S^0) parameters. In uniaxial stretching the values of D and S^0 increase with
increase in the stress [67, p. 80; 68]. Treloar [76] was the first
to state a theoretical basis for this in the case of highly swollen
polymers.

As Fig. 8.23 shows, with increase in load there is increase
in ε_∞, while the point of inflection of the sigmoid curve tends
toward shorter times. The points in Fig. 8.23 were calculated by
using a computer.

Temperature. With decrease in temperature there are increases
in the values of D and S^0. This also leads to an increase in the
equilibrium deformation and a shift of the point cf inflection of
the creep curves toward shorter times (Fig. 8.24).

Composition of the medium. The creep of polymers can be altered within wide limits by adding water to the alcohols. With
the addition of water there is a decrease both in D and in S^0, and
this leads to a considerable decrease in the creep of F-2 polyarylate (Fig. 8.25).

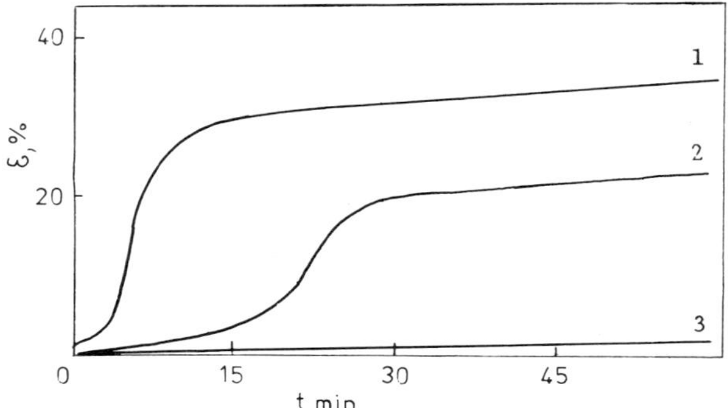

Fig. 8.24. Creep curves of F-2 polyarylate at σ = 20
            MPa in n-propanol at temperatures of: 1)
            60°C; 2) 40°C; 3) 20°C [72].

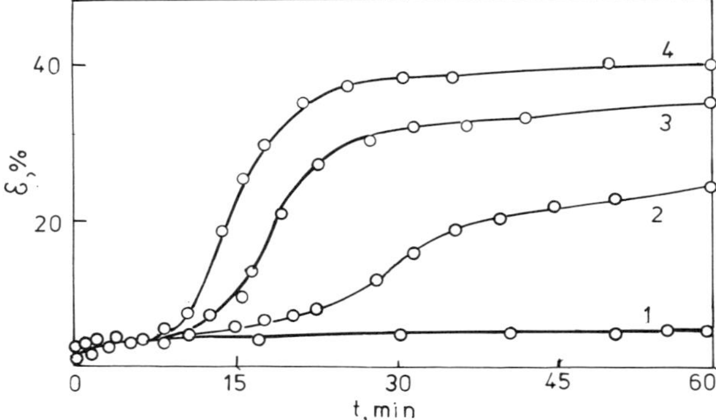

Fig. 8.25. Creep curves of films of F-2 polyarylate in
            mixtures of ethyl alcohol and water, of
            various concentrations at 25°C and σ = 40
            MPa: 1) 16 vol. %; 2) 33; 3) 50; 4) 66 vol.
            % [60].

8.3.3. Influence of Chemically Active Media on Creep

Sulfuric and acetic acids are, at low temperatures, physically
active media with respect to PP in a force field. Nevertheless,
as was shown in [51], at high temperatures there is perceptible
degradation of the polymer, and an increase in creep is brought
about, in the opinion of the authors, not only by the increase in
the temperature, but also by the chemical degradation.

In the testing of F-2 polyarylate in aqueous solutions of potassium hydroxide and sulfuric acid no great change in creep was observed, as compared with creep in an aqueous medium, but with increase in the concentration of these electrolytes there was a sharp reduction in the durability of this polyarylate.

8.3.4. Durability of Polymers under Conditions of Creep

With the simultaneous action of a force field and aggressive media, considerable deformation develops in a polymer article. Because of the reduction in cross section there is an increase in the stress, and failure of the polymers takes place under more complex conditions than described in Sec. 8.2.

In any particular case the shape retention of the polymer is an important factor.

The shape retention is the interval of time for which the inequality $\varepsilon(t) \leq \varepsilon_d$ holds, where ε_d is the permissible deformation of the polymer article.

REFERENCES

1. P. P. Kobeko, Amorphous Substances, Izd. Akad. Nauk SSSR, Moscow—Leningrad (1952).
2. I. Alfrey, Jr., Mechanical Behavior of High Polymers, Wiley, New York (1948).
3. G. M. Bartenev and Yu. S. Zuev, Strength and Failure of Viscoelastic Materials, Khimiya, Moscow (1964) [English translation: Pergamon Press, New York (1968)].
4. V. E. Gul' and V. N. Kuleznev, Structure and Mechanical Properties of Polymers, Khimiya, Moscow (1972).
5. V. A. Kargin and G. L. Slonimskii, Short Sketches on the Physical Chemistry of Polymers (2nd edn.), Khimiya (1967).
6. J. D. Ferry, Viscoelastic Properties of Polymers, Wiley, New York (1961) (2nd edn. publ. 1970).
7. A. V. Tobolsky, Structure and Properties of Polymers, Wiley, New York (1966).
8. V. E. Regel', A. I. Slutsker, and É. E. Tomashevskii, Kinetic Nature of the Strength of Solids, Nauka, Moscow (1974).
9. I. M. Ward, Mechanical Properties of Solid Polymers, Wiley, London (1971).
10. D. W. van Krevelen, Properties of Polymers; Correlation with Chemical Structure, Elsevier, Amsterdam (1972).
11. E. Orowan, Nature, 154, 341 (1944).
12. A. I. Gubanov and A. D. Chebychelov, Fiz. Tverd. Tela, 5, 91 (1963).
13. P. A. Rebinder, Z. Phys., 72, 191 (1931).

14. P. A. Rebinder, Izd. Akad. Nauk SSSR, Moscow (1958).
15. S. D. Gertsriken and I. Ya. Dekhtyar, Diffusion in Metals and Alloys in the Solid Phase, Fizmatgiz, Moscow (1960).
16. A. A. Tager, M. V. Tselipotkina, and D. A. Reshet'ko, Vysokomol. Soedin., A, 17, No. 11, 2566 (1977).
17. E. A. Sinevich, E. A. Ryzhkov, and N. F. Bakeev, Vysokomol. Soedin., B, 19, No. 9, 687 (1977).
18. S. A. Kazakevich and P. V. Kozlov, Fiz.-Khim. Mekh. Polim., 8, No. 5, 84 (1972).
19. K. Rogers, in: Physics and Chemistry of the Organic Solid State, D. Fox et al. (eds.), Wiley, New York (1963-1967).
20. K. Matsushige, E. Baer, and S. V. Radcliffe, J. Macromol. Sci., B, 11, No. 4, 565 (1975).
21. T. T. Daurova et al., Dokl. Akad. Nauk SSSR, 231, No. 4, 919 (1976).
22. T. E. Rudakova, Candidate's Dissertation, Inst. Khim. Fiz. Akad. Nauk SSSR (1974).
23. P. J. Flory, J. Am. Chem. Soc., 67, No.11, 2048 (1945).
24. P. I. Vincent, Polymer, 1, 425 (1960).
25. T. G. Plachenov, in: Adsorption and Porosity, M. M. Dubinin and V. V. Serpinskii (eds.), Nauka, Moscow (1976), p. 191.
26. P. Ludwik, Stahl. Eisen, 43, 1427 (1923).
27. A. A. Griffith, Philos. Trans. Roy. Soc. London, A, 221, 163 (1921).
28. H. A. Elliott, Proc. Phys. Soc., 59, 208 (1947).
29. F. C. Roesler, Proc. Phys. Soc., 69, 981 (1956).
30. J.P. Berry, J. Polym. Sci., 50, No. 153, 107 (1961).
31. N. L. Svensson, Proc.Phys. Soc., 77, 876 (1961).
32. A. N. Tynnyi, Strength and Failure of Polymers under the Action of Liquid Media, Naukova Dumka, Kiev (1975).
33. A. Tobolsky and H. Eyring, J. Chem. Phys., 11, 125 (1943).
34. F. Bueche, J. Appl. Phys., 28, 784 (1957).
35. J. P. Berry, in: Failure, Vol. 7, Part II [Russian translation], Mir, Moscow (1976), p. 7.
36. S. N. Zhurkov and B. N. Narzullaev, Zh. Tekh. Fiz., 23, 1677 (1953).
37. G. M. Bartenev and I. V. Razumovskaya, Fiz. Tverd. Tela, 6, 657 (1964).
38. G. B. Malenis et al., Vysokomol. Soedin., A, 19, No. 1, 86 (1977).
39. E. E.Tomashevskii, Fiz. Tverd. Tela, 12, 3202 (1970).
40. E. S. Pereverzev, Fiz.-Khim. Mekh. Mater., 8, No. 8, 57 (1972).
41. W. E. Statton, in: Newer Methods of Polymer Characterization, Bacon Ke (ed.), Wiley-Interscience, New York (1964).
42. S. P. Papkov and E. Z. Fainberg, Interaction of Cellulose and Cellulose Materials with Water, Khimiya, Moscow (1976).
43. R. P. Kambour, Nature, 195, 1299 (1962).
44. R. P. Kambour, Polymer, 5, 143 (1964).
45. G. M. Bartenev and I. V. Razumovskaya, Dokl. Akad. Nauk SSSR, 150, No. 4, 784 (1963).

46. L. L. Libatskii and S. E. Kovchik, Fiz.-Khim. Mekh. Mater.,
    3, No. 4, 458 (1967).
47. V. I. Manin et al., Plast. Massy, No. 1, 64 (1968); translated
    in Soviet Plast., No. 1, 66 (1968).
48. A. I. Tynnyi and A. I. Soshko, Fiz.-Khim. Mekh. Mater., 1, No.
    5, 522 (1965).
49. M. N. Bokshitskii, N. Ya. Klinov, and N. A. Bckshitskaya,
    Statistical Fatigue of Polyethylene, Mashinostroenie, Moscow
    (1967).
50. V. R. Regel', T. B. Boboev, and M. P. Vershinina, Fiz.-Khim.
    Mekh. Mater., 6, No. 5, 522 (1970).
51. V. A. Bershtein, L. M. Egorova, and V. V. Solcv'ev, Mekh.
    Polim., No. 5, 854 (1977).
52. J. Bailey, Glass Ind., 20, 21 (1939).
53. A. A. Askadskii, Deformation of Polymers, Khimiya, Moscow
    (1973).
54. A. A. Shevchenko, PhD Thesis, Inst. Neft. Gazov. Promst. (1972).
55. W. Holzmüller and K. Altenburg, Physics of Plastics, Akad.-
    Verlag, Berlin (1961).
56. V. Volterra, Linear Functions, Gauthier-Villars, Paris (1913).
57. A. A. Askadskii, V. G. Dashevskii, and Yu. S. Kochergin,
    Vysokomol. Soedin., B, 19, No. 2, 500 (1977).
58. I. I. Bugakov, Creep of Polymer Materials, Nauka, Moscow (1973).
59. N. I. Tikhonova et al., Vysokomol. Soedin., A, 20, No. 7, 1543
    (1978).
60. Yu. S. Zuev and A. Z. Borshevskaya, Dokl. Akad. Nauk SSSR, 144,
    879 (1962).
61. E. A. Skuratova, A. A. Shevchenko, and I. Ya. Klinov, in: Tr.
    Mosk. Inst. Khim. Mashinostr., Vol. 37 (1971), p. 108.
62. V. M. Vorobiov, I. V. Razumovskaya, and V. I. Vettegren,
    Polymer, 19, No. 11, 1267 (1978).
63. I. J. Hutchinson, I. M. Ward, and H. A. Willis, Polymer, 21,
    No. 1, 55 (1980).
64. W. Stach and K. Holland-Moritz, Contributed Papers from the
    14th European Congress on Molecular Spectroscopy, Frankfurt,
    Sept. 1979, Part 2, J. Mol. Struct., 60, 49 (1980).
65. V. A. Marikhin and L.P. Myasnikova, Supermolecular Structure
    of Polymers, Khimiya, Leningrad (1977).
66. T. E. Rudakova et al., Vysokomol. Soedin., A, 22, No. 2, 449
    (1980).
67. S. P. Papkov and E. S. Fainberg, Interaction of Cellulose and
    Cellulose Materials with Water, Khimiya, Moscow (1976).
68. L. V. Ivanova and G. E. Zaikov, Vysokomol. Soedin., A, 19,
    No. 3, 537 (1977).
69. V. I. Likhtman, P. A. Rebinder, and G. V. Karpenko, Influence
    of a Surfactant on the Deformation of Metals, Izd. Akad. Nauk
    SSSR, Moscow (1954).
70. A. L. Shterenzon et al., in: Trans. Ural Scientific-Research
    Institute of Chemistry, Sverdlovsk, Issue 30 (1973).

71. A. L.Porchkhidze et al., Vysokomol. Soedin., B, $\underline{22}$, No. 10
    (1980).
72. A. L. Porchkhidze, Candidate's Dissertation, Moscow Chemical
    Technology Institute, Moscow (1981).
73. A. G. Alekseev, Candidate's Dissertation, Moscow Chemical
    Technology Institute, Moscow (1981).
74. J. Militky and J. Jansa, Prepr. Short Commun., IUPAC Macro,
    Mainz, $\underline{3}$, s.1, s.a., 1401 (1979).
75. Yu. S. Zuev, Failure of Polymers under the Action of Aggres-
    sive Media, Khimiya, Moscow (1972).
76. L. R. G. Treloar, Proc. Roy. Soc., A, $\underline{200}$, No. 1061, 176 (1950);
    Trans. Faraday Soc., $\underline{46}$, No. 9, 783 (1950).

Chapter 9

INFLUENCE OF CHEMICAL DEGRADATION PROCESSES ON THE SERVICE PROPERTIES OF POLYMER ARTICLES

Chemical degradation in polymers is of twofold importance. On the one hand, chemical, physical, and mechanical properties of polymers may change to such a degree on contact with aggressive media that the polymer articles become unfit for service. On the other hand, chemical degragation is used advantageously for modification, rendering them useful for etching and other purposes, i.e., it is used advantageously for improving the particular service properties of polymer articles. In the former instance the problem of practical importance is reducing the rate of chemical degradation, i.e., increasing the chemical resistance of the polymers, while in the latter instance the opposite task arises, finding conditions for more rapid occurrence of these processes.

Let us now consider these problems.

9.1. INCREASING THE CHEMICAL RESISTANCE OF POLYMERS

The degradation of polymers in aggressive media includes a number of stages, the most important being adsorption and diffusion of the aggressive medium within the polymer and decomposition of chemically unstable groups under the action of the components of the aggressive medium. The last stage is, in the majority of cases, an ionic or molecular reaction. Generally the rate of this reaction, according to Eq. (6.1), depends on the rate constant, the concentration of chemically unstable groups, and the concentration of the components of an aggressive medium in the accessible regions of the polymer article.

The rate constant of the reaction can be altered by adding to the polymer salts (see Chap. 1), solvents, plasticizers, and chemicals which form complexes with the chemically unstable groups and thereby influence their reactivity.

However, this method of increasing the chemical resistance of polymers is not of practical significance since the incorporation of the above additives in small amounts (to retain the service properties of the polymer articles) has, to a considerable extent, no influence on the rate constants.

More radical methods of increasing the chemical resistance are those based on reducing the concentration of accessible groups and the concentration of the components of an aggressive medium, on the surface of the polymer if the degradation takes place under external diffusion-kinetic control (see 6.2.1), and within the polymer if the process is under internal diffusion-kinetic control.

Let us now consider these methods of increasing the chemical resistance of polymers in greater detail.

9.1.1. Increasing the Chemical Resistance of Polymers by Reducing the Concentration of Accessible Chemically Unstable Groups

The concentration of accessible chemically unstable groups may be reduced as follows.

1. By increasing the degree of crystallinity of the polymer. This method has been used to increase the resistance of PE to the action of hydrochloric acid [1], and that of PCA with respect to alcohols [2].

2. By the incorporation of fillers in the polymer. For instance, with the addition of rubbers to epoxy compounds there is a considerable increase in their resistance to alkalis [3].

3. By the incorporation into the macromolecules of substituents which offer steric hindrance to the accessibility of the components of the aggressive medium to the chemically unstable groups. For instance, in Sec. 7.5 we gave data showing the influence of various substituents in the ortho position to the carbonate group on the chemical resistance of polycarbonates in aqueous basic solutions [4]. The greatest effect of the substituents on the chemical resistance can be predicted quantitatively on the basis of the Taft Equation (7.38).

9.1.2. Increasing the Chemical Resistance
of Polymers by Reducing the Concentration
of Components of the Aggressive Medium
in the Polymer

If degradation takes place on the surface of the polymer article, the concentration of the components of the aggressive medium at the polymer-medium interface can be reduced as follows.

1. By applying, to the surface of the polymer, layers of other substances, which are resistant to the aggressive medium and which dissolve the components of the aggressive medium in very small amounts. To do this there are used most often low-MM PE, organosilicon liquids [5, 6], or paraffin waxes [7].

It must be noted that all these coatings are permeable to the components of the aggressive medium, i.e., they increase the chemical resistance of the polymers only up to a certain limit.

Metallic and enamel coatings, however, are impermeable to components of an aggressive medium. At the present time polymer articles are being covered with various chemically inert metals [8, 9].

This is a promising method of increasing the chemical resistance of polymers, but the application of the metallic coating to the polymer remains the principal problem here [10].

2. By the incorporation of inert fillers which are capable of diffusing to the surface of the polymer article, creating a high surface concentration at the interface. This method has been used extensively to improve the chemical resistance of rubbers [11, p. 194].

If degradation takes place within a polymer (under internal kinetic or diffusion-kinetic control) the concentration of the components of the aggressive medium can be reduced as follows.

1. By crosslinking the macromolecules by means of bifunctional compounds, with the formation of network structures. As a result of this operation there is a considerable reduction in the solubility of the components of the aggressive media in the polymer [12].

2. By increasing the degree of crystallinity and of orientation.

3. By altering the chemical structure of the polymer by, for instance, partial esterification and chlorination [13, p. 314].

4. By incorporating substances which combine with the components of the aggressive medium. This method is ordinarily used in

cases where aggressive media are formed in the thermal decomposition of the polymer. For instance, lead salts, fatty acid salts
of Groups I and II metals, or organotin compounds [14] can act as
active acceptors of the HCl which is evolved in the thermal degradation of PVC; amines, nitriles, and also a dicyanodiamine containing
cyano and amino groups [15] can act as acceptors of formaldehyde
and acetic acid formed in the decomposition of polyoxymethylene,
while mono- and dicarbodiimides [16] can act as acceptors of carboxyl groups formed in the hydrolysis of polycarbonates.

In this method of stabilization it is very important to add
an optimum amount of the acceptor which, on the one hand, effectively
retards catalytic decomposition of the polymer and, on the other hand,
does not greatly change the service properties of the polymer article.

A calculation of the optimum amount of acceptor was proposed
for the first time by Pudov [17]. In the thermal decomposition
of a polymer in the presence of an acceptor there are three
reactions: the noncatalytic formation of a catalyzing agent
(e.g., hydrogen chloride in the case of PVC, or acetic acid in the
case of POM) at rate w_0; autocatalytic decomposition of the polymer
with the formation of a catalyzing agent, taking place at rate
$\sim k c_{cat}$; and interaction of the acceptor with the catalyzing agent,
at rate $\sim k_1 c_{cat} c_{acc}$.

The general equation for the rate of formation of the catalyzing agent takes the form

$$w = w_0 + k c_{cat} - k_1 c_{cat} c_{acc} \tag{9.1}$$

It may be seen that the addition of an acceptor (i.e., a catalyst
acceptor) to the polymer is equivalent to reducing the rate of the
autocatalytic process.

With acceptor concentration

$$c_{acc} = \frac{k}{k_1} \tag{9.2}$$

Eq. (9.1) takes the form

$$w = w_0 \tag{9.3}$$

and the process changes from autocatalytic to steady-state conditions. The concentration of acceptor which corresponds to Eq.
(9.2) is called the critical concentration. With acceptor concentrations exceeding the critical amount the accumulation of catalyz-

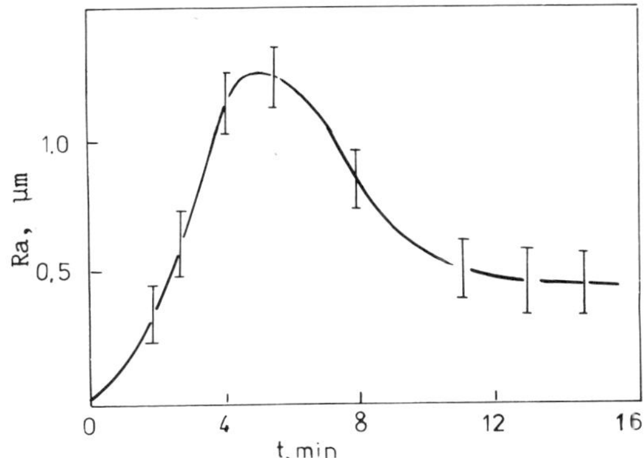

Fig. 9.1. Dependence of Ra on the time of etching of a PETP film (thickness 500 ±
20 μm and degree of crystallinity
40%) in 49% KOH at 108°C [19].

ing agent takes place at a constant rate and, moreover, the durability of the polymer increases sharply.

9.2. USE OF CHEMICAL DEGRADATION TO IMPROVE THE SERVICE PROPERTIES OF POLYMER ARTICLES

Chemical degradation processes are widely used in a number
of industrial chemical sectors for modifying, identifying, and preparing polymers for use. In this section we consider examples of the
use of aggressive media for chemical and physical modification of
polymer articles with the purpose of improving their particular
service properties.

In the majority of cases the surface properties of polymer articles
determine their service properties. Accordingly, much attention
is given to the surface modification of polymer articles which, besides
radiation grafting and chemical grafting, can be effected by means
of chemical degradation.

In the ideal case those reactions which lead to modification
of the surface of polymer articles ought to take place solely on those
macromolecules which are located directly on the surface, i.e.,
to a depth of around 10 Å. In fact, the reactions take place not
only on the surface but in a definite surface layer, the size of
which depends on the ratio of the rate of diffusion of the catalysts
and reagents and the rate of decomposition of the chemically unstable groups in the polymer. A theoretical discussion of this
question is given in Chap. 6.

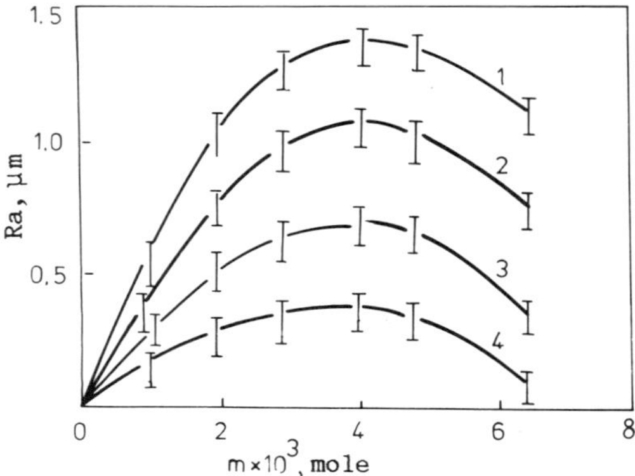

Fig. 9.2.   Dependence of Ra on the mass of layer
removed for films of PETP of various
degrees of crystallinity: 1) 0-5%;
2) 30%; 3) 40%; 4) 58% [19].

<u>9.2.1.   Optimum Conditions of Preparation
of Roughened Films of Polyethyleneterephthalate</u>

At the present time there is vigorous development of the technology of preparing films with a roughened surface for computers,
video and audio recording equipment, and synthetic paper. The
starting materials are polystyrene (PE), polyethyleneterephthalate
(PETP), and other polymers. One method of preparing synthetic
paper is special chemical treatment (etching) of polymer film.
Reference [18] describes the etching of PETP films by means of aqueous solutions of KOH and NaOH.

To find the optimum conditions for etching films of PETP, it
is necessary to know the mechanism of etching, the reasons for the
formation of the roughened surface, and the dependence of the degree of roughness on time, temperature, the thermodynamic parameters
of the medium, and the structure of the polymer.

As is shown in Fig. 9.1, during the course of etching a thick
film of PETP, the index of roughness Ra (the arithmetic root mean
deviation of the microirregularities from the baseline of the pro-
file) first increases, then passes through a maximum, and finally
decreases and remains practically unchanged up to complete disin-
tegration of the film. The maximum of roughness corresponds to
a thickness of an etched-away layer of 25-30 µm. The maximum of
Ra decreases with increase in the degree of crystallinity of the
original films (Fig. 9.2) and is practically independent of the
degree of orientation [19].

According to one common explanation, the formation of the roughened surface occurs with the preferential etching away of the amorphous phase (the exposed crystallites also form irregularities). However, the roughness which is obtained on amorphous films is greater than on crystalline ones. In addition, it has been found that PETP films, prepared under differing conditions but having similar degrees of crystallinity, gave different values for Ra when etched with basic solutions.

It may be suggested that the occurrence of roughness is brought about by the opening up of microdefects with etching, these being formed when the polymers are prepared from a melt [20].

According to present thinking, the surface layers of films have a more friable structure, lower density, and a greater number of defects. With increase in the degree of crystallinity there is a reduction in the total proportion of the amorphous phase, and a "healing" of the defects in the amorphous phase. All this leads to the formation of smoother surfaces when films of a high degree of crystallinity are being etched. The greatest number of defects is found at a depth of 25-30 μm which, in turn, causes maximum roughness when this particular layer is etched away.

The degree of roughness is decided solely by the thickness of the layer etched away, and depends neither on the temperature nor on the concentration of base. On thin films (5-20 μm) it is not possible to get a sufficiently rough surface since they evidently do not contain a sufficient number of microdefects.

By using Eq. (7.12), it is possible to obtain an expression for the change in Ra as a function of the time, temperature, and thermodynamic parameters of the medium for the section for which there is a rise in Ra:

$$\mathrm{Ra} = \frac{6.4 \cdot 10^6 \exp\left(-\dfrac{16\,500}{RT}\right) a_{H_2O}\, st}{1 + \dfrac{4 \cdot 10^2}{b_0}} \, \alpha \qquad (9.4)$$

where α is a coefficient of proportionality, depending on the nature of the polymer.

For producing a rough surface on PETP film the optimum processing variables are as follows: degree of crystallinity >10%, concentration of KOH 46-49%, temperature 105-110°C.

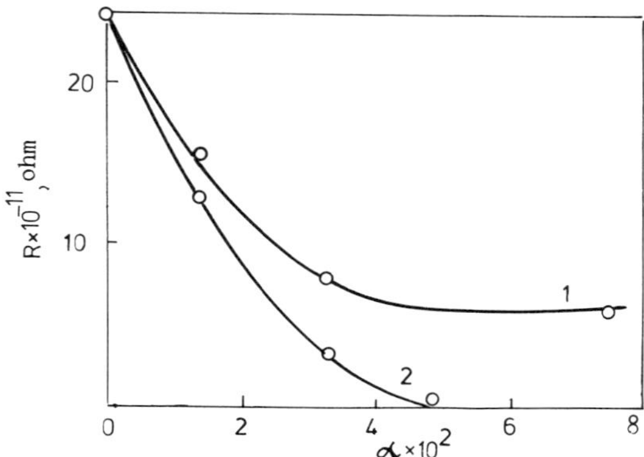

Fig. 9.3.   Dependence of the surface resistivity of a cellulose diacetate fabric
on the degree of dissociation of
acetate groups (α) under the action
of NaOH (1) and tetrabutylammonium
hydroxide (2) [22].

9.2.2. Modification of the Surface of Cellulose Diacetate Fibers by Means of Aqueous Basic Solutions

One shortcoming of synthetic, including polyester, fibers is
their high capability of electrification. The setting up of
charges of static electricity creates difficulties in production
and in service. For instance, charges on the surface of the polymer attract dust and other unwanted substances to the finished
article. The dirtying of clothing made from a fabric capable of
being electrostatically charged is several times greater than that
of cotton fabrics. There are also adverse consequences of the influence of static electricity on the human organism [21].

The incorporation of antistatic agents in polymers reduces
their static electricity but, as a rule, these same agents irritate
the skin and the mucous membranes of the eyes, and also leach out
of the polymer. It is possible to reduce the electrostatic charg-
ing if we replace threads in the surface layer containing hydrophobic ester groups by the hydrophilic hydroxyl groups of regenerated cellulose.

When cellulose diacetate fibers are treated with basic solutions a layer of regenerated cellulose is formed as a result of
the hydrolysis of the ester groups. At the same time there is no
hydrolysis of the glycoside linkages.

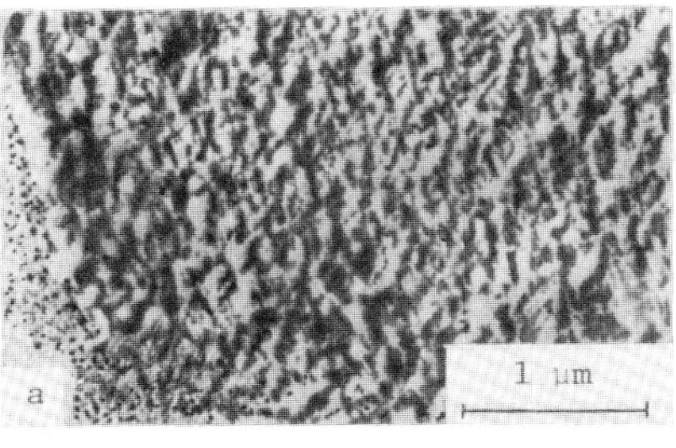

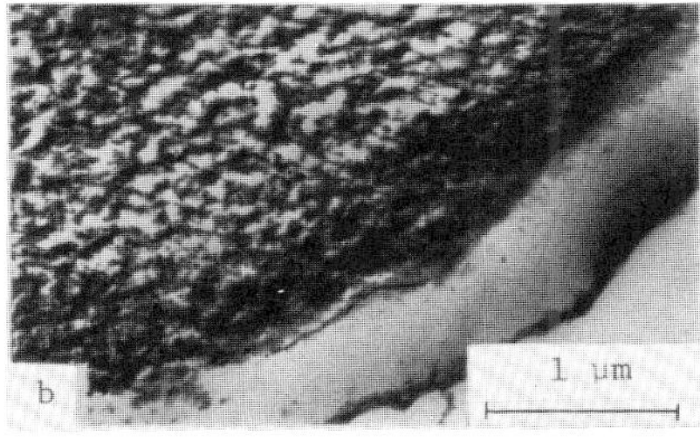

Fig. 9.4.   Photomicrographs of cross sections
of a cellulose diacetate fiber:
a) original fiber; b) after modification [22].

The concentration of decomposed ester groups in the cellulose
diacetate filament, and their distribution with respect to the diffusion coordinate, may be found from the system of equations

$$\frac{\partial c_n}{\partial t} = k\,(c_n^0 - c_n)\,c_{OH^-} \tag{9.5}$$

$$\frac{\partial c_{OH^-}}{\partial t} = D_{OH^-}\nabla^2 c_{OH^-} - k\,(c_n^0 - c_n)\,c_{OH^-} \tag{9.6}$$

There is no analytical solution of this system, but it can be
solved by using a computer.

The problem involves finding the conditions under which the
rate of hydrolysis would be much higher than the rate of diffusion,
i.e., the degradation would take place in the most narrow zone.

Such conditions can be achieved in two ways:

1. By using concentrated basic solutions (increasing the
rate of hydrolysis with D_{OH^-} remaining unchanged).

2. By using an alkali with a large bulky cation, e.g.,
$(C_4H_9)_3NOH$ (reducing D_{OH^-} while maintaining the rate of hydrolysis
practically unchanged).

Figure 9.3 shows the dependence of the surface resistivity of cellulose diacetate fabric on the degree of dissociation of the acetate groups under the action of various alkalis [22].

Figure 9.4 shows photomicrographs of cross sections of an elementary filament of cellulose diacetate before and after modification.

9.2.3. Changes in the Structure of Polycaproamide Films on Treatment with Aqueous Acidic Solutions

PK-4 PCA film contains two different crystalline structures: γ'-pseudohexagonal and α-monoclinic. The content of the α form can be increased by treating the polymer with polar solvents. But a practically complete polymorphic transition can be effected by using as polar media aqueous solutions of hydrochloric or sulfuric acid [23]. The transition is irreversible (the quantity of the α form remains unchanged after the acid is removed), and is due to the rupture of hydrogen bonds on the surface of the crystalline formations.

Since the resistance of PK-4 to thermooxidative and thermal degradation depends on the content of the α form [24, p. 93], it is possible to use the treatment of polyamide films with acids at low temperatures (so as to avoid chemical degradation) to regulate the resistance of this polymer to various physical and chemical factors.

9.2.4. Production of Fiters with a Given Size of Microchannels

In recent years nuclear filters have begun to be used for ultrafiltration, sterilization, and separation of chemical and biological media.

The production of these filters from Diflon polycarbonate and PETP film involves three stages [25]:

irradiation of the polymer film with heavy ions, which form local regions ("tracks") of degradable polymer;

exposure of the irradiated film in UV light to give more profound degradation of the oligomer and polymer molecules in the track zone;

treatment of the irradiated film with a basic solution, as a result of which there takes place degradation of the oligomer and polymer fragments in the track zone, and also of unaffected

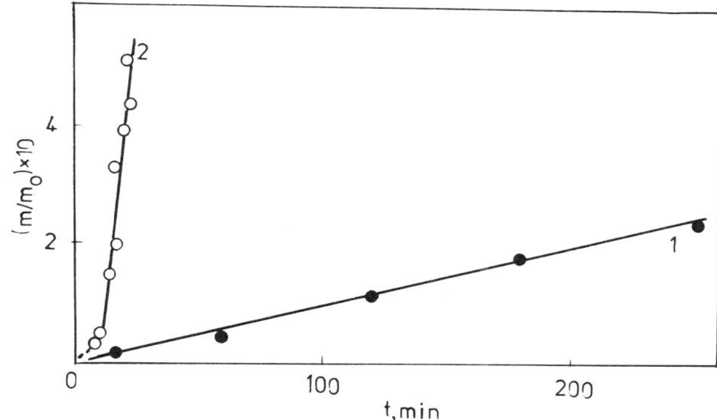

Fig. 9.5.   Dependence of the accumulation of decomposition products from original film (1) and
film previously irradiated with xenon ions
(2) [26].

polymer outside the track zone, and desorption of the decomposition
products into the surrounding medium.

The kinetic data show that the accumulation of decomposition
products from a film previously irradiated with xenon ions and UV
light takes place more rapidly than from nonirradiated film (Fig.
9.5). From this it may be concluded that the main loss of mass
of the film takes place as a result of degradation of the polymer
in the zone and from the sides of the tracks, and also by desorption
of soluble decomposition product into the surrounding medium [26].

Generally, an increase in the mass of the decomposition products takes place as a result of the occurrence of the following
degradation reactions:

on the surface of the polymer film, leading to a reduction
in its thickness, w_1;

in the track zone, w_2;

on the side of the tracks, leading to an increase in the diameter of the microchannels, w_3.

Let us now consider these reactions individually.

Degradation Reaction on the Surface of the Polymer Film. This
reaction is studied in detail in [27, 28]. It takes place in a
thin surface layer approximating a monolayer. The decomposition
of the ester group takes place according to a mechanism comprising
fast reversible attachment of a hydroxide ion to the carbonyl car-

bon with subsequent slow attack of the water molecule of the ionized form. When the reaction products dissolve and the ionization
of the ester group takes place only to a small extent, we have

$$w_1 = \frac{dn}{dt} = \frac{k_1}{K_{eq}} C_m b_0 a_{H_2O} S_s \qquad (9.7)$$

where k_1 is the rate constant; n the number of ester groups broken
down; C_m the concentration of ester groups on the film surface;
b_0 and a_{H_2O} the activities of the hydroxide ions and the water molecules on the film surface, respectively; K_{eq} the equilibrium constant for the formation of a complex between the ester group and
the hydroxide ion; and S_s the surface area of the film. The degree
of dissociation α may be expressed either as the number of ester
groups or else as the mass of the degradation products:

$$\alpha = n/n_\infty = m/m_0 \qquad (9.8)$$

where $n_\infty = (m_0 A/M)[1 - 1/(\bar{P}_n - 1)] \approx m_0 A/M$, m_0 the initial mass of
the polymer film, A Avogadro's number, M the mass of the molecular
unit, and $\bar{P}_n$ the number-average degree of polymerization.

On differentiating (9.8) and substituting the result in (9.7),
we get

$$m = \frac{k_1}{k_{eq}} \frac{M}{Az} C_m b_0 a_{H_2O} t S \qquad (9.9)$$

where z is a constant depending on the type of decomposition of
the PETP molecule and on the solubility of the degradation products
in the surrounding medium.

<u>Degradation Reaction in the Track Zone</u>. In the track zone
there are, as a result of nuclear and UV irradiation, various
degradation products which have so far hardly been detailed. It
may be supposed that further degradation of these products under
the action of the hydroxide ions will take place in a reaction zone
the dimensions of which depend on the ratio of the rates of diffusion of the hydroxide ions and of the degradation of the ester groups.

Special experiments have shown that the tracks are practically
impermeable to the salts KCl and NaCl. It may be expected that
the concentration of hydroxide ions in the zone of the tracks will
be extremely low, and that the latter likewise will not be permeable to the hydroxide ions.

Thus, degradation will in general take place from the end of
the tracks.

In [26-28] it is shown that the degradation of oligomers of PETP with $\bar{P}_n$ = 3-5 to monomers takes place more rapidly by several orders than degradation of the polymers. The rate of etching in the track zone differs by several orders from the rate of etching of the plastic [25]. However, it is not excluded that the diffusion of the degradation products out of the track zone will limit the rate of loss of mass of the polymer. Unfortunately, the mass of polymer in the track zone makes up only a very small part of the total mass of the polymer film (10^{-4} for N_0 = 10^8 cm^{-2}), and with degradation reactions taking place simultaneously on the surface of the film and on the side of the tracks it is practically impossible to investigate the kinetics of a particular reaction.

<u>Degradation Reaction on the Side of the Tracks</u>. The rate of this process can be determined from the equation

$$w_3 = \frac{\partial n}{\partial t} = \frac{k_3}{K_{eq}} C_m b_0 \alpha_{H_2O} S_{side} \tag{9.10}$$

We use n to express the mass of degradation products by analogy with Eq. (9.8):

$$\frac{dm}{dt} = \frac{k_3}{K_{eq}} \frac{M}{Az} C_m b_0 \alpha_{H_2O} S_{side} \tag{9.11}$$

On substituting in (9.11) $m_{tr} = \pi r^2 \ell \rho$ (m_{tr} = mass removal) and $S_{side} = 2\pi r \ell$, we get an equation for the rate of change in radius of a track:

$$\frac{dr}{dt} = \frac{k_3}{K_{eq}} \frac{M}{Az} C_m b_0 \alpha_{H_2O} \frac{1}{\rho} \tag{9.12}$$

On integration of Eq. (9.12) we get an expression for the change in radius of a track as a function of the time and thermodynamic parameters of the medium:

$$r = r_0 + \frac{k_3}{K_{eq}} \cdot \frac{M}{Az} C_m b_0 \alpha_{H_2O} \frac{1}{\rho}, \tag{9.13}$$

where r_0 is the initial radius of the track.

From (9.13) we can obtain an equation for the change in mass of the products which are formed by degradation on the side of the tracks:

$$m^{1/2} = \frac{k_3}{K_{eq}} \frac{C_m b_0 \alpha_{H_2O} Nt (N_0 S_S)^{1/2} (\pi \ell)^{1/2}}{Az \rho^{1/2}} \tag{9.14}$$

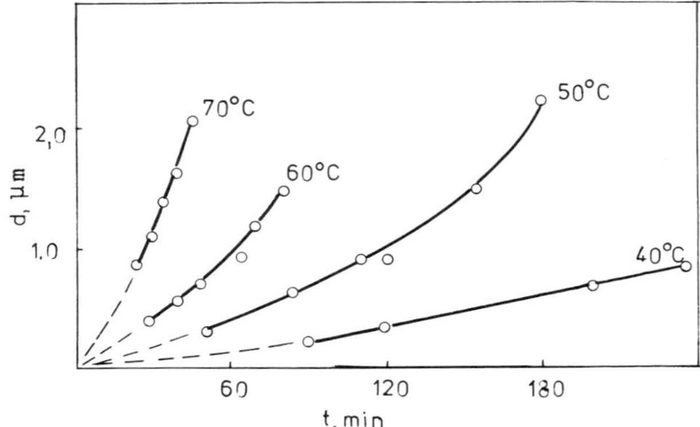

Fig. 9.6. Dependence of the change in diameter of the
           pores as a function of time in 3 N NaOH at
           40, 50, 60, and 70°C.

where $N_0 S_s$ is the total number of tracks per specimen.

From Eqs. (9.9) and (9.14) we get a general expression for
the increase in mass of the degradation products as a function of
the time and the thermodynamic parameters of the medium:

$$m = \frac{k_1}{K_{eq}}\frac{M}{Az}C_m b_0 a_{H_2O} t S_s + \left[\frac{k_3}{K_{eq}}\cdot\frac{C_m b_0 a_{H_2O} M t}{Az}\right]^2 \frac{N_0 S_s \pi \ell}{\rho} \quad (9.15)$$

Figure 9.6 shows typical experimental data for the change in
diameter of the pores as a function of time in a 3 N aqueous solution of NaOH at different temperatures. The relationships are of
linear character and are evidently governed by a fusion of the
openings with long processing times, as may be seen in the photomicrographs obtained using a Philips scanning microscope (Fig.
(9.7). The effective rate constants $k_{eff} = (k_3/K_{eq})(M/Az)(C_m b_0 a_{H_2O})(1/\rho)$
calculated from the initial sections of the curves; their values
are given in Table 9.1. The data are satisfactorily expressed by
the Arrhenius equation. The activation energy is 17 ± 1 kcal/mole
and is independent of the concentration of the base. When we col-
late $k_3 eff$ at 25°C with a_{H_2O} and b_0, we determine the value

$$(k_3 eff/K_{eq})MC_m/Az\rho = (2.0 \pm 0.3)\cdot 10^{-4}$$

TABLE 9.1.   Rate Constants for the Increase in Pore Diameter of a
             PETP Film Irradiated with Xenon Ions, in Treatment in
             Aqueous NaOH Solutions of Various Concentrations and
             Temperatures [29]

T, °C	NaOH conc., mass %	$k_3$eff, μm/min	T, °C	NaOH conc., mass %	$k_3$eff, μm/min
25	3.8	$3.5 \cdot 10^{-5}$	25	19.7	$2 \cdot 10^{-3}$
40		$1.58 \cdot 10^{-4}$	40		$8 \cdot 10^{-3}$
50		$3.9 \cdot 10^{-4}$	50		$2.2 \cdot 10^{-3}$
60		$1 \cdot 10^{-3}$	60		$2.6 \cdot 10^{-2}$
70		$2 \cdot 10^{-3}$	70		$1 \cdot 10^{-1}$
25	10.7	$6.9 \cdot 10^{-4}$	25	25.1	$4 \cdot 10^{-3}$
40		$3.6 \cdot 10^{-3}$	40		$1.8 \cdot 10^{-2}$
50		$7.2 \cdot 10^{-3}$	50		$5.2 \cdot 10^{-2}$
60		$1.5 \cdot 10^{-2}$	60		$1 \cdot 10^{-1}$
70		$3.5 \cdot 10^{-2}$	70		

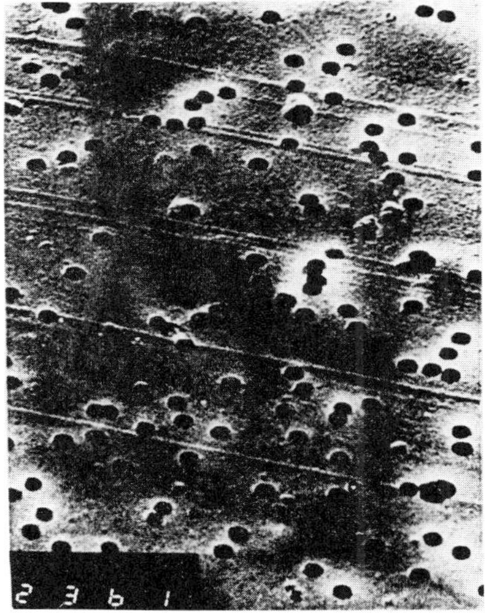

Fig. 9.7.   Photomicrograph of a nuclear filter
            obtained by means of a Philips scan-
            ning microscope. λ = 3000 μm, 15 kV,
            slope 30°, scale 3 mm = 1 μm, d = 0.57-
            0.58 μm, Xe 56.

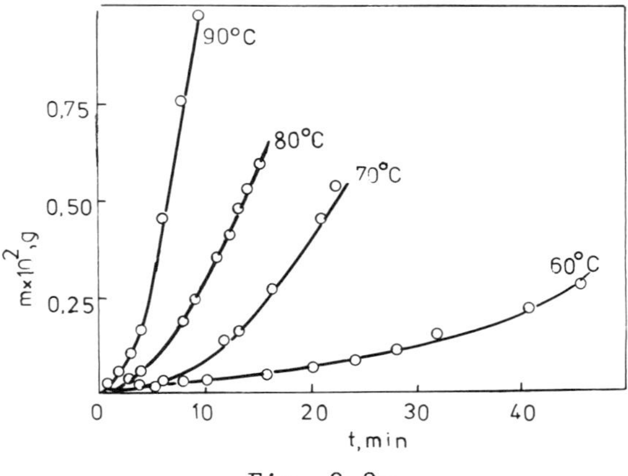

Fig. 9.8a

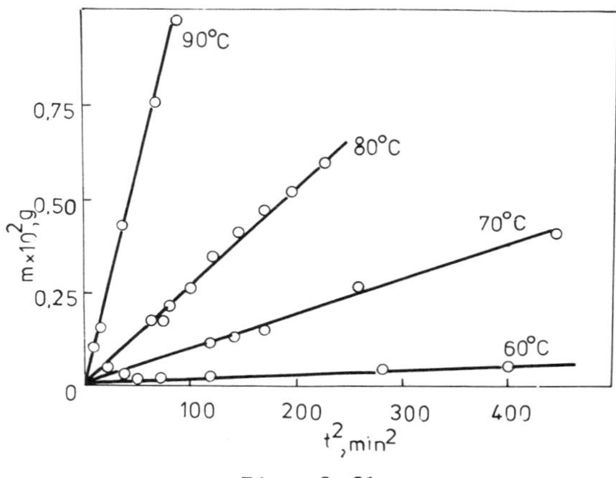

Fig. 9.8b

Fig. 9.8a.   Kinetic curves of accumulation of a
salt of terephthalic acid in 5% KOH
at 60, 70, 80, and 90°C vs. t.

Fig. 9.8b.   Kinetic curves of accumulation of a
salt of terephthalic acid in 5% KOH
at 60, 70, 80, and 90°C vs. t².

Thus, the initial change as a function of the time, temperature, and thermodynamic parameters of the medium may be represented
by the relationship

$$r = r_0 + 3.5 \cdot 10^8 \exp\left(-\frac{17,000}{RT}\right) a_{H_2O} b_0 t \qquad (9.16)$$

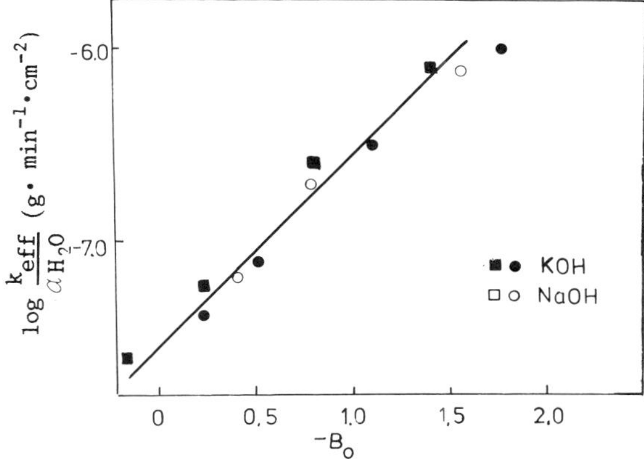

Fig. 9.9. Dependence of $\log (k_{eff}/a_{H_2O})$ on the alkalinity function of aqueous solutions of KOH and NaOH for the degradation of a PETP film.

TABLE 9.2. Rate Constants of Accumulation for the Products of Degradation of a PETP Film Irradiated with Xenon Ions, in Treatment in Aqueous KOH Solutions of Various Concentrations and Temperatures [30, 31]

T, °C	KOH conc., mass %	$k_3$eff, g/cm²·min	T, °C	KOH conc., mass %	$k_3$eff, g/cm²·min
25	5	$3.3 \cdot 10^{-8}$	25	20	$5 \cdot 10^{-7}$
50		$1.59 \cdot 10^{-7}$	30		$8.3 \cdot 10^{-7}$
70		$1.32 \cdot 10^{-6}$	40		$2 \cdot 10^{-6}$
90		$5.0 \cdot 10^{-6}$	50		$5 \cdot 10^{-6}$
25	10	$1.26 \cdot 10^{-7}$	25	30	$9.1 \cdot 10^{-7}$
50		$1.29 \cdot 10^{-6}$	30		$1.4 \cdot 10^{-6}$
70		$5.0 \cdot 10^{-6}$	40		$3.16 \cdot 10^{-6}$
90		$2.5 \cdot 10^{-5}$	50		$8.9 \cdot 10^{-6}$

Figures 9.8a, b show typical kinetic curves of the increase of the degradation product (a salt of terephthalic acid) in a 5% aqueous solution of KOH at various temperatures. After a particular initial period of time the experimental data can be handled satisfactorily using the coordinates of Fig. 9.9; this indicates the

dominant role of the degradation on the side of the tracks. The values of

$$k_3 eff = \frac{k_3}{K_{eq}} = \frac{MC_m}{Az} b_0 a_{H_2O} \left(\frac{N_0 S_S \pi \ell}{\rho}\right)^{1/2}$$

are given in Table 9.2.

The activation energy is practically independent both of the concentration of, and the nature of, the base, at 16.5 ± 0.5 kcal $\times$ $(mole)^{-1}$. The value of $(k_3/K_{eq})(MC_m/Az)$ was calculated graphically (Fig. 9.9) and is $(1.6 \pm 0.4)\cdot 10^{-8}$ g/cm^2·min.

It is interesting to note that both the activation energy and the above-mentioned constant, which is independent of the catalyst concentration, practically coincide with the corresponding parameters for the degradation on the surface of the film.

Thus, the change in mass of the polymer as a function of the time, temperature, and thermodynamic parameters of the medium can be calculated from the following equation:

$$m_n = m_0 - 1.6\cdot 10^4 \exp\left(-\frac{16,500}{RT}\right) b_0 a_{H_2O} S_{st}$$

$$- \left(1.6\cdot 10^4 \exp\left(-\frac{16,500}{RT}\right) b_0 a_{H_2O} t\right)\frac{N_0 S_S \pi \ell}{\rho} \qquad (9.17)$$

REFERENCES

1. D. F. Kagan, Investigation of the Properties of, and Design Calculations, of Polyethylene Tubes, Stroiizdat, Moscow (1964).
2. C. D. Weiske, Kunststoffe, 54, NO. 10, 626 (1964).
3. A. F. Nikolaev et al., Plast. Massy, No. 8, 5 (1977).
4. A. A. Khokhlov et al., Vysokomol. Soedin., B, 20, No. 3, 231 (1978).
5. J. Galperin and W. Arnheim, J. Appl. Polym. Sci., 11, No. 7, 1259 (1967).
6. A. L. Iordanskii, Candidate's Dissertation, Inst. Khim. Fiz., Akad. Nauk SSSR, Moscow (1974).
7. I. Ya. Klinov, in: Tr. Mosk. Inst. Khim. Mashinostr., Vol. 22 (1960), pp. 159, 179.
8. M. I. Shalkauskas et al., Chemical Metallization of Plastics, Khimiya, Leningrad (1972).
9. N. V. Katts, Metallization of Fabrics, Khimiya, Moscow (1972).
10. A. A. Berlin and V. E. Basin, Fundamentals of Polymer Adhesion, Khimiya, Moscow (1974).

11. Yu. S. Zuev, Failure of Polymers under the Action of Aggressive Media, Khimiya, Moscow (1972).
12. A. L. Iordanskii, Candidate's Dissertation, Inst. Khim. Fiz., Akad. Nauk SSSR, Moscow (1974).
13. S. P. Papkov, Physicochemical Fundamentals of Processing Polymer Solutions, Khimiya, Moscow (1971).
14. K. S. Minsker and G. T. Fedoseeva, Degradation and Stabilization of Polymers, Khimiya, Moscow (1979).
15. N. S. Enikolopov and S. A. Vol'fson, Chemistry and Technology of Polyformaldehyde, Khimiya, Moscow (1968).
16. W. Neumann et al., 4th Rubber Technol. Conf., London (1962), p. 738.
17. V. S. Pudov, Plast. Massy, No. 2, 18 (1976).
18. British Patent No. 845850.
19. T. E. Rudakova et al., Vysokomol. Soedin., A, $\underline{17}$, No. 8, 1791 (1975).
20. M. Shen and M. Bever, J. Mater. Sci., $\underline{7}$, 742 (1972).
21. B. I. Sazhin, Electrical Properties of Polymers, Khimiya, Moscow (1970).
22. V. V. Pashkevichus et al., in: Nauchno-Issled. Tr. Litov. Nauchno-Issled. Inst. Tekst. Promst., $\underline{3}$, 165 (1974).
23. A. N. Machyulis et al., in: Tr. Promst., Akad. Nauk Litov, $\underline{1(64)}$, 151 (1971).
24. A. N. Machyulis and E. E. Tornau, Diffusion Stabilization of Polymers, Miitis, Vil'nyus (1974).
25. T. E. Rudakova, S. S. Kuleva, and L. I. Samoilova, Vysokomol. Soedin., A, $\underline{22}$, No. 2, 443 (1980).
26. Y. Komaki, Nucl. Tracks (formerly Nucl. Track Detection), $\underline{3}$, Nos. 1-2, 33 (1979).
27. G. V. McKinley, Radiat. Effects, $\underline{37}$, Nos. 3-4, 199 (1978).
28. A. P. Akishina, V. S. Barashenkov, L. I. Samoilova, S. P. Tret'yakova, and N. B. Khitrova, Determination of Pore Diameters of Nuclear Filters by the "Bubble" Method, Joint Inst. for Nuclear Research (Nuclear Reaction Laboratory), Dubna (1974).
29. T. E. Rudakova, Yu. V. Moiseev, A. E. Chalykh, and G. E. Zaikov, Vysokomol. Soedin., A, $\underline{14}$, No. 2, 449 (1972).
30. G. N. Akar'ev, V. S. Barashenkov, L. I. Samoilova, S. P. Tret'yakova, and V. A. Shchegolev, Preparation of Nuclear Filters, Joint Inst. for Nuclear Research (Nuclear Reaction Laboratory), Dubna, B 1-14-8214 (1974).
31. T. E. Rudakova, S. S. Kuleva, and L. I. Samoilova, Vysokomol. Soedin., A, $\underline{22}$, No. 2, 443 (1980).

Chapter 10

PREDICTION
OF THE SERVICE PROPERTIES
OF POLYMER ARTICLES

If in service or storage a polymer article comes into contact with aggressive media, under the action of these media and also under the action of thermal and mechanical energy, the following fundamental processes may occur:

chemical degradation of the polymer, leading to a change in the MM;

sorption of components of the aggressive medium by the polymer;

solution of the polymer;

desorption of various additives (stabilizers, plasticizers, pigments, and fillers) from the polymer;

change in the physical structure of the polymer.

The occurrence of these processes usually leads to a change (generally for the worse) in the major service properties: mechanical, sorption, diffusion, and changes in the mass and outward appearance of the polymer.

It is of practical importance to be able to predict the service properties of a polymer article functioning in an aggressive medium, i.e., to forecast the durability of the polymer under the stated conditions.

If the polymer article, in contact with an aggressive medium under the stated conditions, suffers a change in its service properties within a period of short duration, it is no great problem to fore-

cast its durability. However, as a rule, polymers have quite a
high chemical resistance under service conditions, and any change
in their properties is so slow that it is exceptionally difficult
to determine the durability experimentally under any given conditions.

In such a case the testing of the article is usually carried
out under more rigorous conditions (high temperature, high catalyst
concentration, and high mechanical stresses) and the durability
of the article under service conditions is then calculated from
that point.

This is definitely the correct way and usually the only one
possible. However, it is now necessary to know what parameters
linked with the service properties of the article should be determined
and under what conditions these tests should be carried out.

To answer the first question it is necessary to clarify which
of the above processes play a significant role in the changes in
the service properties of polymers.

10.1. IDENTIFICATION OF THE CHEMICAL AND PHYSICAL PROCESSES LEADING TO CHANGES IN THE SERVICE PROPERTIES OF POLYMER ARTICLES

Let us now consider how one can identify the chemical and
physical processes which take place on contact of polymers with
aggressive media and which lead to changes in the service properties of the articles made from them.

10.1.1. Chemical Degradation

Chemical degradation causes the most extensive failure of polymer articles. A characteristic sign of chemical degradation is a change
(generally a reduction) in the MM of the polymer. But one cannot
confirm that a process, which is accompanied by a change in the
MM of the polymer, actually is chemical degradation. One such process can be desorption of low-molecular fractions of the polymer
into the aggressive medium which is in contact with it.

Chemical degradation, as indicated in Chap. 6, can take place
in three regions: external diffusion-kinetic, internal diffusion-
kinetic, and internal kinetic (different expressions for these concepts are indicated earlier in this translation).

Let us now consider how the MM and mass of the polymer article
change with the occurrence of chemical degradation in these regions.

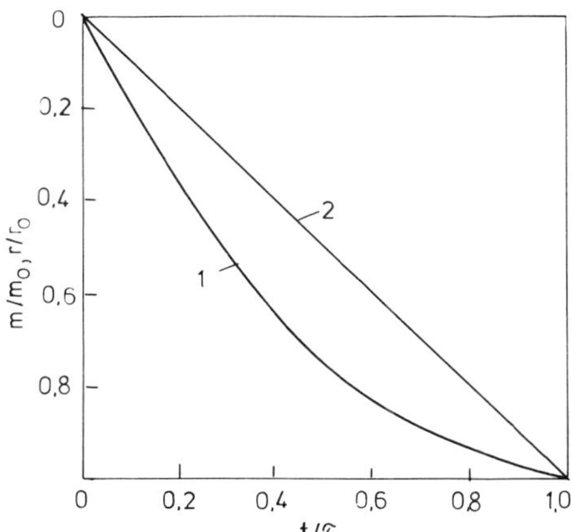

Fig. 10.1. Dependence of the relative mass (1)
and relative radius (2) of a poly-
mer filament on the duration of the
degradation process in the external
diffusion-kinetic region (where τ
is the time of conclusion of the degradation process).

External diffusion-kinetic region. The process takes place
in a thin reaction zone of constant dimensions, which proceeds into the depth of the specimen at a constant rate. Determining the
dimensions of the reaction zone is difficult experimentally since
the magnitude is usually much less than a micrometer. Most often
multiply frustrated total internal reflection is used.

When degradation takes place in the external diffusion-kinetic
region the two following cases are possible:

1. The degradation products are soluble in the aggressive
medium. The course of the process may be followed either from the
increase in the amount of degradation products in the surrounding
medium, or else from the loss in mass of the polymer article, if there is
no significant swelling of the article in one of the components
of the aggressive medium. If the dimensions of the reaction zone
are much less that a micrometer, the MM of the polymer throughout
the whole volume of the film or elementary filament remains unchanged during the experiment.

The mass of these articles will change during the course of
degradation in accordance with Eq. (6.29) for film or (6.35) for
a filament. Figure 10.1 shows the change in mass and in the dimensions of the articles on dimensionless coordinates.

It must be borne in mind that the process of solution of the polymers can be described by analogous equations and, accordingly, additional experiments are necessary to identify the products accumulating in the surrounding medium.

2. The degradation products are practically insoluble in the aggressive medium. The course of degradation can be followed by periodically extracting the polymer article from the aggressive medium and removing the degradation products either mechanically or chemically (by dissolving them in a solvent in which the polymer itself does not dissolve). It is possible also to recorded the movement of the reaction zone by one of the methods described in Chap. 5. The degradation products accumulating on the surface of the article may prevent diffusion of the aggressive medium toward the reaction zone, which will lead to a more complex pattern of change in mass and dimensions of the articles with time, as compared with Eqs. (6.29) and (6.35). The MM of the film or elementary filament will be reduced in this case, but two characteristic peaks occur on the gel chromatogram, that of the original polymer and that of the degradation product.

Internal diffusion-kinetic and internal kinetic region. The degradation takes place in a reaction zone whose dimensions increase and, in the limiting case (internal kinetic region), become equal to those of the article itself. With degradation taking place in these regions we must differentiate two cases:

1. The reactivity of the chemically unstable bonds remains practically identical. This applies in amorphous polymers. In the reaction zone, the dimensions of which are generally commensurate with those of the polymer article, there is a change in the MM. Accordingly, if in the course of the experiment there is a change in the MM in the film, filament, or surface section cut from the polymer slab, it may be supposed that the degradation occurs in the internal diffusion-kinetic or internal kinetic region.

For kinetic calculations we need to know the time of decomposition of the molecules in the solid polymer. If the decomposition is random, then the number of bonds broken can be determined either by physical methods directly within the matrix of the polymer (as a rule the increase in the number of terminal groups is determined) or else from the change in the number-average MM according to an equation analogous to (4.2).

Figure 10.2 shows the change in the degree of dissociation of chemically unstable groups in a polymer film as a function of the reduced time, in accordance with Eq. (6.18). A similar relationship between α and t/τ is obtained, for instance, in the case of a filament. In this type of decomposition the polymer article, as a rule, loses mechanical strength (it disintegrates into separate

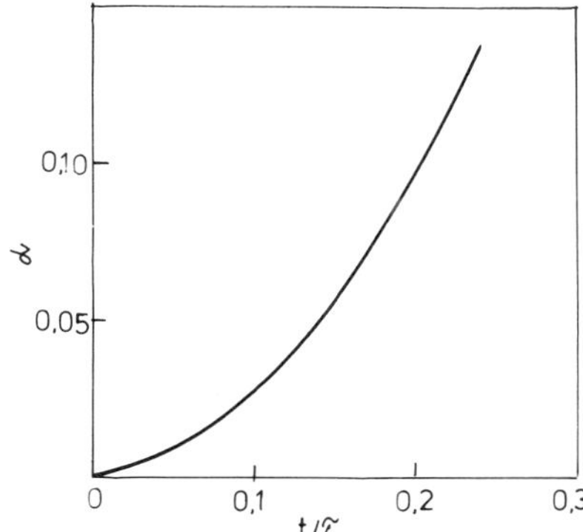

Fig. 10.2. Dependence of the degree of dissociation of chemically unstable groups in
a polymer film on the duration of the
degradation process in the internal
diffusion-kinetic region.

lumps) at degrees of dissociation 0.10-0.15. At such degrees of
dissociation the mass of the polymer article decreases only very
slightly if, for instance, the resulting low-molecular products dissolve
in the aggressive medium. If, however, these products are not soluble in the aggressive medium, or else are desorbed very slowly,
the mass of the article remains practically unchanged. If depolymerization of the molecules takes place by a terminal group, the MM
of the polymer changes less sharply, whereas the mass of the article decreases perceptibly (see Chap. 4).

2. The reactivities of the chemically unstable bonds differ significantly. This is typical of partially crystalline polymers.
The number and nature of the chemically unstable bonds with differing reactivities can be determined by careful analysis of the kinetic results and of the change in the MMD of the polymer during
the course of degradation (see Chap. 4).

Generally, the kinetics must be "stepwise," but since, in practice, the investigation of chemical degradation is carried out only
to quite low degrees of dissociation, the resulting kinetic curves
generally relate to chemically unstable bonds of a particular reactivity.

Thus there are enough signs by which it is possible not only
to establish the occurrence of chemical degradation but also to
determine within which region a given process takes place.

10.1.2. Sorption Processes

For the contact of polymer articles with aggressive media there takes place sorption of the components of the aggressive media by the polymers. The magnitude of sorption is determined by the properties of the polymers and of the aggressive medium (see Chap. 5).

The main indication of the occurrence of sorption is the increase in the mass of the polymer articles. We can specify the following features of polymers as sorbents.

1. Relatively low porosity of the polymer materials. Usually the overall volume of the micropores does not exceed 0.1 cm^3/g. Sorption can occur according to differing mechanisms, depending on the character of the interaction between the polymer and aggressive medium. If micropores are practically absent from the polymer, or are closed, and also if the polymer has a low affinity to the components of the aggressive medium, there is practically zero sorption (e.g., water in elastomers). Polymers which have open micropores can sorb aggressive media in which they do not swell; in this case the limiting values of sorption are determined by the total volume of accessible micropores. If the polymer swells in the aggressive medium, the limiting value of sorption can be estimated from Eq. (5.32); in this case there is redistribution of the micropores.

2. As a consequence of the small volume of the micropores in polymer materials produced by the usual methods, the processes of adsorption and capillary condensation play a lesser role than absorption processes. Distinguishing between the processes of adsorption and absorption of aggressive media by polymers is a difficult experimental task. There are only a few instances of this distinction in the literature [1].

3. The dependence of the sorption on the degree of crystallinity of the polymer in the specimen in question. The sorption of aggressive media, as of other low-molecular substances, occurs in amorphous regions and at imperfections in the crystallites of the polymer. Generally there is a linear relationship between the amount of sorption and the degree of crystallinity (Fig. 10.3), but the sorption is determined by the accessibility, and its relationship with the structural parameters of the polymers may be more complex [4].

4. There is a possibility of change in the structure of the polymer during the course of sorption. If with sorption of the components of the aggressive medium the polymer passes from a glassy to a highly elastic state, the process may be complicated by crystallization, which will lead to an extremal dependence of the amount of sorption on the time. A similar effect occurs with

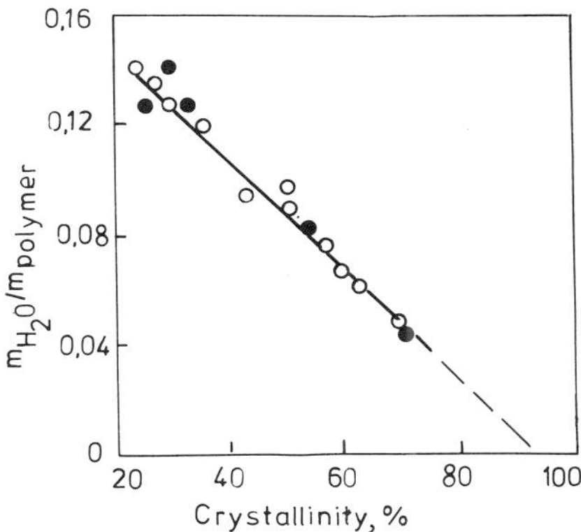

Fig. 10.3. Dependence of the sorption of water
            (at 60% relative air humidity) on
            the degree of crystallinity of cell-
            ulose: ●) [2]; ○) [3].

contact of polychlorotrifluoroethylene with water [5]. Sometimes
during the course of sorption the degree of crystallinity of the
polymer decreases, which leads to an increase in the amount of sorption. Such an effect occurs, for instance, with systems of aliphatic polyamides and aqueous solutions of phenol [6, p. 17].

With the sorption of components of aggressive media by polymers there may be slighter changes in structure, for instance,
transition from one crystalline modification to another (e.g., the
γ-α transition in a polycaproamide under the action of aqueous
acidic solutions [7]). It must be noted that a change in the supermolecular structure of polymers under the action of aggressive media
may have a significant influence on the service properties of the
articles made from them.

10.1.3. Solution of Polymers

Polymers may dissolve in aggressive media when in contact with
them. For instance, in concentrated solutions of sulfuric acid
many hetero-chain polymers will dissolve.

The process of solution begins with the formation of a swollen
surface layer. The relationship between the time of swelling
(τ_{sw}) and the thickness of the swollen surface layer (δ) is described as follows [8]:

$$\tau_{sw} = \delta^2/6\overline{D} \tag{10.1}$$

where $\overline{D}$ is the diffusion coefficient of the solvent in the polymer.

After this the process takes place under steady-state conditions, and the change in mass of the polymer can be found from the equation

$$m = m_0 - \beta t \tag{10.2}$$

where $\beta = (s\bar{D}\rho/\delta)\Delta v_{solv}$ is the solution coefficient; Δv_{solv} the concentration differential of the solvent, in parts by volume, between the liquid and the surface layer of the polymer; s the surface of the polymer article; and ρ the density of the polymer.

The magnitude of δ depends on the character of flow of the solvent, which is determined by the Reynolds number (N_{Re}) and the MM of the polymer.

For polymers with "normal" MMs,

$$\delta \simeq 0.35 \cdot 10^{-2} M_n^{1/2} (1 + 0.35 \cdot 10^{-3} N_{Re}) \tag{10.3}$$

For instance, the process of solution of phenol-formaldehyde oligomers in basic solutions with pH 12-13 [9] is expressed by Eqs. (10.2) and (10.3).

Thus it is extremely difficult to distinguish between the process of solution and that of degradation taking place in the external diffusion-kinetic region, and special experiments are necessary to determine the MM of substances in a solvent making contact with a polymer.

10.1.4. Desorption of Various Additives from Polymers

When polymer articles are in service, various additives (stabilizers, pigments, plasticizers, or fillers) may be desorbed from them. Moreover, the rate of desorption depends on the temperature, the dimensions of the molecules, and the character of interaction of these additives with the components of the aggressive media. For instance, phenolic stabilizers may leach out actively from polymer articles and, accordingly, contact with aggressive media will lead to an increase in the thermooxidative degradation.

In [10] examples are given of the solution of fillers and stabilizers in clusters of hydrochloric acid in polyethylene, and of their subsequent desorption from the polymer.

10.2. QUANTITATIVE EXPRESSION FOR THE CHANGE IN SERVICE PROPERTIES OF POLYMER ARTICLES IN CONTACT WITH AGGRESSIVE MEDIA

After determining the main physical and chemical processes taking place in a polymer article in contact with aggressive media it is necessary to clarify which parameters should be measured to describe the change in service properties with time.

The service properties of polymer articles may be divided conventionally into two groups:

1) Structure-sensitive properties, where a very slight change in the structure of the polymer material has a large and disproportionate influence on the properties. One example of such a structure-sensitive property is the strength, which depends on the defects (e.g., microcracks) in the polymer article.

2) Properties which average out through the volume, where change in the structure of the polymer material leads, as a rule, to a proportionate change in the service properties. Such properties are the diffusion, dielectric, and certain mechanical properties, etc.

It is exceptionally difficult to forecast the changes in the structure-sensitive service properties of polymer articles when these are used in aggressive media. Preliminary tests are required for this.

It is, however, possible to predict service properties averaged out over the volume. Table 10.1 gives equations which make it possible to predict the major service properties of polymers (film or filament) when physical and chemical processes take place in the polymers. These equations were obtained for the simplest cases:

chemical degradation in the internal kinetic region taking place at random;

the sorption of the components of the aggressive medium being of a specific character;

the polymer materials having no structure differentials, apart from the degree of crystallinity.

It must be noted that these equations include rate constants (where chemical processes are taking place) or coefficients which have the meaning of rate constants (where physical processes are taking place). Thus, when we have determined these constants from independent experiments, it is possible to predict the major service properties of the polymer articles.

TABLE 10.1. Changes in Various Service Properties of Polymers with the Occurrence of Various Chemical and Physical Processes

Main processes	Service properties		
	Mass of the polymer article	Tensile stress	Mass of substance passing through the film
Chemical degradation in the external diffusion-kinetic region	$m = m_0\left(1 - \dfrac{k_{\mathrm{eff(surf)}}t}{l_0\rho}\right)$ film $m = m_0\left(1 - \dfrac{k_{\mathrm{eff(surf)}}t}{r_0\rho}\right)^2$ filament	$\sigma = \sigma_0\left(1 - \dfrac{k_{\mathrm{eff(surf)}}t}{l_0\rho}\right)$ film $\sigma = \sigma_0\left(1 - \dfrac{k_{\mathrm{eff(surf)}}t}{r_0\rho}\right)^2$ filament	$Q = \dfrac{P_0 st}{l_0 - \dfrac{k_{\mathrm{eff(surf)}}t}{\rho}}$
Chemical degradation in the internal kinetic region	$m = m_0\left\{1 - 2\left[1 - \left(1 - \dfrac{1}{\bar{P}_{n_0}}\right) \times\right.\right.$ $\left.\left. \times \exp - (k_{\mathrm{eff}}t)\right]^2 + \dfrac{2}{\bar{P}_{n_0}^2}\right\}$	$\sigma = \sigma_0\,(1 - B k_{\mathrm{эф}}t)$ where B is a constant	$Q = \dfrac{Q_0(1 + k_{\mathrm{eff}}t)^2\,st}{l_0}$
Sorption of the components of the aggressive medium	$m = m_0 + M_\infty\left(1 - \dfrac{8}{\pi^2}\sum_{n=0}^{\infty}\dfrac{\exp - \alpha t}{(2n+1)^2}\right)$ where $\alpha = \dfrac{\pi^2 D(2n+1)^2}{l_0^2}$ film	$\sigma = \sigma_0\left[1 - \gamma\left(1 - \dfrac{8}{\pi^2}\sum_{n=0}^{\infty}\dfrac{\exp - \alpha t}{(2n+1)^2}\right)\right]$ film	

	$m = m_0 + M_\infty \left(1 - \dfrac{4}{r_0^2} \displaystyle\sum_{m=1}^{\infty} \dfrac{1}{\alpha_m^2}\, \exp \alpha_m^2\, Dt\right)$ where α_m are the roots of the equation $I_0(r\alpha_m) = 0$	$\sigma = \sigma_0 \left[1 - \gamma \left(1 - \dfrac{4}{r^2} \displaystyle\sum_{m=1}^{\infty} \dfrac{1}{\alpha_m^2}\, \exp - \alpha_m^2 Dt\right)\right]$ where γ is a coefficient proportional to M_∞	
Change in the degree of crystallinity		$\sigma = \sigma(\text{amorph}) + \delta(1 - \varphi)$ where δ is a coefficient proportional to $\sigma_{\text{cryst}} - \sigma_{\text{amorph}}$	$Q = \dfrac{P_{(0)}\,(\text{amorph})\,\varphi^2 st}{l_0}$ where φ is the volume fraction of amorphous regions
Solution of the polymer	$m = m_0(1 - \beta t)$ film $m = m_0(1 - \beta t)^2$ filament where β is the coefficient of solution	$\sigma = \sigma_0(1 - \beta t)$ film $\sigma = \sigma_0(1 - \beta t)^2$ filament	$Q = \dfrac{P_0 st}{l_0(1 - \beta t)}$

Let us now consider the changes in the major service properties indicated in Table 10.1.

Change in Mass. The equations describing the change in mass
of polymer films and filaments accompanying various types of chemical degradation and sorption of components of aggressive medium
are given, respectively, in Chaps. 6 and 5. The change in mass
of polymer articles during the process of solution is discussed in
Sec. 10.1.3.

Change in Mechanical Properties. At the present time, quantitative expressions can be obtained only for the change in tensile
strength in chemical degradation, unless there is a considerable
change in the contour of the surface,e.g., the formation of microcracks (Chap. 8).

Corresponding expressions may be obtained for describing the
influence of solution upon the change in TS (tensile stress).

If the sorption of the components of the aggressive medium
leads to a weakening of the molecular interaction, the TS decreases
in accordance with the equation given in Table 10.1 [11].

Change in Diffusion Permeability. The mass of substance passing through the film increases when chemical degradation takes
place. Moreover, if this process takes place in the external diffusion-kinetic region, the mass of substance which has diffused
through increases further on account of a reduction in the thickness of the polymer film:

$$Q = \frac{P_0 s t}{l_0 - \dfrac{k_{\text{eff(surf)}} t}{\rho}} \qquad (10.4)$$

where P_0 and l_0 are, respectively, the initial permeability and
the thickness of the polymer film, and s the surface of contact
with the diffusing substance.

When chemical degradation takes place in the internal diffusion-kinetic region, the diffusion coefficient and the distribution
coefficient usually increase linearly with increase in the degree
of dissociation of the chemically unstable bonds in the polymer
[12]:

$$D = D_0 (1 + k_{\text{eff}} t) \qquad (10.5)$$

$$K_{\text{distr}} = K_{(\text{distr})_0} (1 + k_{\text{eff}} t) \qquad (10.6)$$

where D_0 and $K_{(\text{distr})_0}$ are, respectively, the diffusion and distribution coefficients in the original polymer film.

Since the permeability coefficient is the product of D and $K(distr)_0$, we obtain the following relationship:

$$P = P_0 (1 + k_{eff}t)^2 \qquad\qquad (10.7)$$

where P_0 is the permeability of the original polymer film.

In a similar way we obtain a relationship linking Q with the volume fraction of the amorphous regions in the polymer.

10.3. TESTING OF POLYMER ARTICLES UNDER MORE RIGOROUS CONDITIONS

Under service conditions polymer articles usually exhibit sufficiently high chemical resistance and, accordingly, it is expedient to carry out the tests under more rigorous conditions and then calculate the changes in service properties and in durability under service conditions.

By durability we mean the time during which a polymer material, under defined service conditions, changes its properties to a level laid down in the technical specification.

Let us now consider how raising the temperature and increasing the concentration of the components of the aggressive medium influence the chemical and physical processes which lead to changes in the service properties.

10.3.1. Influence of Temperature

The temperature has a big influence on the physical and chemical processes which lead to changes in the service properties of polymer articles.

The influence of temperature on the rate constants and the equilibrium constants are satisfactorily described by the Arrhenius and van't Hoff equations but, nevertheless, these equations are rightfully used within a definite temperature range.

Degradation Processes. The rate constants of the decomposition of chemically unstable bonds are satisfactorily expressed in the temperature range below 100°C by the Arrhenius equation although for some reactions there is a slight change in the activation energy with temperature. The following factors influence the activation energy of decomposition of chemically unstable bonds in a polymer.

TABLE 10.2.   Relationships for the Effective Activation Energies
              Corresponding to Mechanisms A-1, A-2, and B-2 under
              Conditions of Low and High Ionization of the Reagents

Mechanism	Kinetic equation	Relationship
A-1	$k_{eff} = \dfrac{k_{act}}{k_{BH^+}} h_x$	$E_{eff} = E_{act} - \Delta H_{BH^+} + \Delta H_{h_x}$
	$k_{eff} = k_{act}$	$E_{eff} = E_{act}$
A-2	$k_{eff} = k_{act} c_{HS}^+$	$E_{eff} = E_{act} + \Delta H_{HS}^+$
	$k_{eff} = k_{act} c_{HS}^+ k_{BH^+}$	$E_{eff} = E_{act} + \Delta H_{HS}^+ - \Delta H_{BH^+}$
B-2	$k_{eff} = \dfrac{k_{act}}{k_{BOH^-}} b_0 a_{H_2O}$	$E_{eff} = E_{act} + \Delta H_{B_0} + \Delta H_{a_{H_2O}} - \Delta H_{BOH^-}$
	$k_{eff} = k_{act} a_{H_2O}$	$E_{eff} = E_{act} + \Delta H_{a_{H_2O}}$

Note. E_{act} is the actual activation energy, ΔH_{BH^+}, ΔH_{BOH^-} are, respectively, the enthalpies of attachment of a proton and hydroxide ion to the reagent, and ΔH_{HS^+}, $\Delta H h_x$, ΔH_{B_0}, and $\Delta H H_2O$ are the enthalpies of change in the concentrations and activities of aqueous acidic and basic solutions with temperature.

<u>The dependence of the dielectric permittivity on temperature</u>

$$\varepsilon = \varepsilon_0 \exp(-LT)$$

where L is a constant, characteristic of the medium in question.

This factor arises with degradation processes whose activation energy depends on permittivity (see Chap. 7).

<u>The complex character of the dependence of k_{eff} on the kinetic and thermodynamic parameters.</u> Table 10.2 shows the kinetic equations corresponding to mechanisms A-1, A-2, and B-2, and the relationships for the effective activation energies under conditions of low and high ionization of the reagents.

It may be seen that the effective activation energies of decomposition of chemically unstable bonds differ according to the degree of ionization of the reagents.

<u>Diffusion Processes</u>.   In the temperature range below 100°C, there is usually an exponential dependence of the diffusion coefficient on temperature.

Deviations from this exponential dependence may be brought about by a change in the physical state of the polymer in the temperature range in question, for instance a transition of the polymer from the glassy to the viscous flow state with increase in temperature.

Processes of Sorption and Desorption. The temperature dependence of the distribution coefficient of the diffusant between the polymer and the external phase is expressed as follows:

$$K_{dist} = K_{distr}^{0} \exp - \left(\frac{\Delta H}{RT} \right) \qquad (10.8)$$

where K_{distr}^{0} is a constant, not depending on the temperature within a quite small temperature range, and ΔH the enthalpy of solution of the low-molecular substance in the polymer.

In the general case ΔH is the sum of the heats of condensation (ΔH_c) and of mixing (ΔH_{mix}):

$$\Delta H = -\Delta H_c + \Delta H_{mix}$$

The enthalpies of solution may be either negative or positive, depending on the nature of the aggressive medium and also on the nature and physical state of the polymer. Their pattern of behavior is as follows:

the enthaopies of solution of readily condensing aggressive media (NH_3, SO_2) from the gaseous phase are, as a rule, negative ($\Delta H_c > \Delta H_{mix}$) [13, p. 247 of Russian translation];

the enthalpies of solution from aggressive media in polymers which are in the viscous flow state, from the liquid phase, are usually positive (the principal part being played by ΔH_{mix});

for polymers which are in the glassy state, the enthalpy of solution may differ in sign, depending on the relationship between the heats of scission of the bonds between the molecules of the polymer and of the formation of the bonds between the polymer and the aggressive medium.

Solution of Polymers. With increase of temperature there is in general an increase in the thickness of the swollen surface layer of a polymer dissolving in an aggressive medium [8]:

$$\delta = \delta_0 \exp \left(- \frac{\Delta H_{solv}}{RT} \right) \qquad (10.9)$$

The enthalpy of solution of the polymers does not exceed 12.6-16.8 kJ/mole.

10.3.2. Influence of the Concentration
of the Aggressive Medium

Aggressive media may be single-component (water, NH_3, SO_3, HCl) or multicomponent (aqueous solutions of acids, bases, or neutral salts) in nature. If a single-component medium is in the gaseous state, its concentration in a polymer article can be increased by increasing its vapor pressure. If a multicomponent medium is being used, then, irrespective of its state of aggregation, increasing the concentration of one of its components will lead to a reduction of the other in the external phase and in the polymer article. This needs to be taken into account since the different components of the aggressive medium may play a determining role in the various processes, leading to a change in the service properties of the articles.

Let us now consider how increase in the concentration of the components of the aggressive medium influence the course of the various processes.

<u>Chemical Degradation</u>. If degradation takes place in the internal diffusion-kinetic and kinetic regions, then with increase in the catalyst concentration (acid or base) in the aggressive medium the rate of the process, measured by the increase in the amount of monomers (the depolymerization reaction) or in the number of terminal groups (chain decomposition reactions), will rise. Nevertheless, at the same time there may be other factors which should be taken into account in calculating the durability of the polymer articles with concentrations of aggressive medium under service conditions.

An anomalous increase in the rate of degradation may be brought about by an increase in the accessibility of chemically unstable bonds as a consequence of a change in the physical state of the polymers (e.g., sorption of aggressive media reducing the glass transition temperature), and also by an increase in the rate of the process taking place in the external diffusion-kinetic region (for reactions taking place by mechanism A-1 with increase in the concentration of acid, $h_X \gg c^0_{cat}$).

An anomalous reduction in the rate of degradation may occur as a consequence of a reduction in the accessibility of chemically unstable bonds as a result of increase in the degree of crystallinity of the polymer (e.g., sorption of aggressive media brings about crystallization of the polymer), and also as a result of exceeding the isotherm of sorption of catalyst in the polymer (for polymers whose terminal groups interact with the catalyst).

If the degradation takes place in the external diffusion-kinetic region, the rate of the process will change (increasing n most cases) with increase in the concentration of catalyst, in accordance with the kinetic equations given in Chap. 6.

Nevertheless, there is also the possibility of the simultaneous occurrence of degradation in the internal diffusion-kinetic or internal kinetic region, leading to an increase in the overall rate.

Sorption Processes. The behavior pattern of the sorption of electrolytes by polymers was set out in Chap. 6. Electrolyte solutions with low vapor pressure are sorbed by hydrophobic polymers in exceptionally small amounts. With increase in the concentration of the electrolytes in the external phase the sorption of solvents in the polymer decreases considerably. The limiting extent of sorption depends on the thermodynamic activity of the solvents in the aggressive medium which is in contact with the polymer.

Processes of Solution of Polymers. Many polymers dissolve in aggressive media; the dissolving capacity of the medium may depend to a great extent on the content of electrolyte in it. For instance, polycaproamide dissolves in 35-60% H_2SO_4, this being brought about by protonation of the amide groups in the polymer (in 40% H_2SO_4 practically all the amide groups are in the protonated form). In more concentrated solutions of sulfuric acid there is marked formation of ion pairs between the protonated form and the bisulfate ion, and this leads to considerable reduction in the solubility of the polymer. In 85% H_2SO_4 the concentration of HSO_4^- decreases, which brings about considerable solution of the polycaproamide.

Thus, in contact with aggressive media the service properties of polymer articles may change (generally for the worse) as a consequence of chemical degradation of the polymer, of sorption of components of aggressive medium by the polymer articles, of changes in the physical structure of the polymer, of solution of the polymer, and of desorption of various additives from the polymer article.

If the polymer article has sufficiently high chemical resistance under service conditions, its durability can be calculated from data obtained in testing the article under more rigorous conditions (high temperature and high catalyst concentration) but in such cases it is necessary that there should be no change in the physical structure of the polymer and no occurrence, to any perceptible extent, of those processes which do not take place under actual service conditions.

REFERENCES

1. A. A. Tager, M. V. Tselipotkina, and D. A. Reshet'ko, Vysokomol. Soedin., A, <u>17</u>, No. 11, 2566 (1975).

2. R. Jeffries, J. Appl. Polym. Sci., 8, No. 10, 1213 (1964).
3. J. Mann and H. J. Marrinan, Trans. Faraday Soc., 52, No. 3, 492 (1956).
4. S. V. Lin, Fibre Sci. Technol., 5, No. 4, 303 (1972).
5. S. Okuda, Plast. Age, 13, No. 5, 49 (1967).
6. S. P. Papkov and E. Z. Fainberg, Interaction of Cellulose and Cellulose Materials with Water, Khimiya, Moscow (1976).
7. A. N. Machyulis et al., in: Tr. Akad. Nauk Litov. SSR, B, $1(64)$, 151 (1971).
8. K. Ueberreiter, in: Diffusion in Polymers, J. Crank and G. S. Park (eds.), Academic Press, London (1968), p. 219.
9. Yu. I. Kol'tsov, N. V. Bershov, and D. D. Mozzhukin, Vysokomol. Soedin., B, 20, No. 9, 644 (1978).
10. A.M. Grantsbergs et al., in: Diffusion Phenomena in Polymers, Riga (1977), p. 265.
11. L. V. Andreev, R. M. Khasanov, and N. A. Sharova, in: Diffusion Phenomena in Polymers, Riga (1977), p. 275.
12. M. I. Artsis et al., Vysokomol. Soedin., A, 17, No. 1, 128 (1975).
13. K. C. Rodgers, in: Physics and Chemistry of the Organic Solid State, Wiley, New York (1963-1967).

CONCLUSION

In 1968, Professor D. J. Cram, of the University of California
at Los Angeles, wrote in the Foreword to a monograph by Michael Szwarc
entitled "Carbanions, Living Polymers and Electron Transfer Processes," "As in all endeavours, chemistry has its scholars, investigators, craftsmen, prospectors, speculators, adventurers, en-
trepreneurs, statesmen, and chroniclers whose combined functions
give movement and substance to the science. In the early stages
of development of a branch of chemistry occasionally one person
integrates some of these activities. But as the field expands,
the hard economics of time drive investigators into specialization, both of <u>subject matter</u> and <u>function</u>" (emphasis added by
the present authors). This statement justifies our position
with respect to the chemistry of polymers — specialization in
the field of polymer degradation and stabilization. Polymers
now constitute an extensive field of activity. In fact, on the
problem of the degradation and stabilization of polymers there
are already thousands of scientists working in many scientific
centers all over the world. In particular, we list the following centers and researchers:

<u>West Germany</u>. The South German Polymer Center (Würzburg, Prof.
W. Woebken), the German Plastics Institute (Darmstadt, Prof. D.
Braun), University of Duisburg (Prof. H. J. Moseler), Hoechst Re-
search Center (Frankfurt a.M., Prof. J. Voigt), the Dechema Insti-
tute (Frankfurt a.M., Prof. C. M. von Meysenbug), and the Original
Hanau-Hereus Research Center (Hanau, Dr. D. Kockott).

<u>Austria</u>. The Austrian Polymer Institute (Prof. H. Tschamler,
Dr. K. Binder), Fa. Chemie-Linz Research Center (Linz, Dr. H. P.
Frank).

363

Switzerland. Ciba-Geigy Research Center (Basel, Prof. H. Batzer and Prof. H. Müller), the Federal Polytechnic (Lausanne, Prof. H. H. Kausch), the Polytechnic Institute (Zurich, Prof. P. Pino).

Czechoslovakia. The Corrosion Institute (Prague, Dr. B. Dolezel), the Macromolecular Chemistry Institute (Prague, Prof. J. Kalal and Prof. J. Postisie), the Polymer Institute (Bratislava, Prof. A. Romanov).

East Germany. Leipzig University (Prof. K. Thinius), Leuna-Werke Research Center (Merseburg, Prof. M. Fedtke), Institute of Macromolecular Chemistry (Teltov, Prof. H. Zimmerman).

The Netherlands. AKZO Research Laboratories (Arnhem, Dr. C. R. H. I. de Jonge).

Hungary. The Polymer Institute (Budapest, Prof. D. Hardy), Central Chemical Institute (Budapest, Prof. F. Tüdos).

Bulgaria. Organic Chemistry Institute (Sofia, Prof. D. M. Shopov), University of Sofia (Prof. N. A. Natov), and Polymer Institute (Sofia, Dr. K. Rakovski).

Poland. The Polymer Institute (Zabrze, Prof. Z. J. Jedlinskii).

France. Rhone-Poulenc Research Center (Lyons, Dr. J. Guidott).

Italy. University of Naples (Prof. J. Nicolais), Montedisson Research Center (Ferrara, Prof. J. Cresby).

Canada. National Research Center (Ottawa, Dr. D. J. Carlsson), University of Toronto (Prof. J. Gillett).

The United States. University of Michigan (Ann Arbor, Prof. C. Overberger), Polytechnic Institute of New York (Brooklyn, Prof. H. Mark, Prof. H. Morawetz), Stanford Research Institute (San Francisco, Prof. F. Mayo), University of California (Prof. S. Hammond), University of Syracuse (Prof. M. Szwarc), Ciba-Geigy Research Center (New Jersey, Dr. P. Klemchuk).

Great Britain. University of Glasgow (Prof. N. Grassie), University of Aston in Birmingham (Prof. G. Scott), University of Sussex (Brighton, Prof. J. Jenkins), University of London (Prof. B. C. Callis), University of Aberdeen (Dr. M. Cameron), University of St. Andrews (Scotland, Prof. J. R. MacCallum).

Japan. University of Tokyo (Prof. I. Kamiya and Prof. E. Niky), University of Kyoto (Prof. T. Kagiya), Nippon-Soda Research Center (Dr. Hiraoka), Gunma University (Kiryu, Prof. Z. Osawa).

In the Soviet Union, work on the degradation and stabilization of polymers was being spearheaded by Academician N. M. Émanuél'.*

*N. M. Émanuél' died December 8, 1984 (Prof. G. E. Zaikov is his successor).

At the Institute of Chemical Physics work is being carried out on the thermal, thermooxidative, and photooxidative degradation of polymers, on the ozone aging of polymers, on the hydrolysis of hetero-chain polymers, on the degradation of polymers in biological media (the living organism), on the combustion of polymers, on the stabilization of polymers, on the synthesis of stabilizers, and so on. There are more than 50 labs (about 2000 scientists) where degradation and stabilization of polymers are being studied.

Of course, we have not listed here even all of the best-known centers dealing with this problem. The list merely demonstrates the dissemination of this work over various countries.

Today there are practically no firms and institutes concerned with the synthesis and production of polymers that are not also studying their stability and seeking ways of prolonging the useful life of polymer articles.

Here there are two principal tasks:

1. Prolonging the service life (and storage!) of polymer articles. Indeed, prolonging life, shall we say by a factor of two, is equivalent to doubling the production of these articles (or considerably raising the quality). With present-day scales of production even a slight prolongation of the life of polymers promises great advantages to the economy, particularly with the rising cost of petroleum, the starting material for polymer production.

2. A strictly quantitative prediction of the life of polymer articles and the development of methods of rapid and reliable estimation of the time over which, under service or storage conditions, the polymers undergo no significant change in those properties which determine their serviceability.

This second task is almost as important as the first. Indeed, if we err on the safe side and estimate the service life on the low side, this will be equivalent to throwing good and serviceable articles on the scrap heap. If, however, we reassure laymen that the article is going to function for a very long time, and give estimates which are on the optimistic side, the article may fail in service, which may lead to a variety of accidents (including disastrous ones).

Unfortunately, there are as yet no inhibitors devised for the ionic reactions which govern the hydrolytic failure of polymers in aggressive media. Such inhibitors exist only for free radical chain reactions; it is by this mechanism, as a rule, that thermal, thermooxidative, and photooxidative degradation processes take place. Accordingly, in this monograph we set out to solve this

second task, the prediction of the safe service life of hetero-
chain polymers operating in aggressive media. The method we used
is that of chemical kinetics: On the basis of study of the chemical
kinetics of the hydrolysis of low-molecular analogs of the polymers,
of establishing the special features of the polymeric state of the
substance, of study of the kinetic behavior pattern and transfer
mechanism of a multicomponent aggressive medium in the polymers,
we seek to devise equations which describe the degradation of the
polymers and the articles made from them.

Thus, the subject matter of our study (we may recall the words
of Prof. Cram quoted at the beginning of these concluding remarks)
is hetero-chain polymers and their hydrolytic decomposition, while
the function is the chemical kinetics and the mechanism of the process. The purpose of the investigation is the prediction of the
properties. It is, in fact, with respect to the function that
this monograph, in our view, differs from others in the field.

The physicist and Nobel Prize winner Richard Feynman said recently, "It was brought home to us in an unusual way that we live
in an age when it is possible to make discoveries... . In our day
we feel joy, enormous joy, that we can foresee how Nature will behave under new conditions as yet unseen by the human eye." Our
tasks were, without doubt, more modest. But we are happy that we
can foresee how polymers will behave in aggressive media. It seems
to us that the scientific foundations of prognosis (foresight, if
one prefers) are a very interesting and poetic field to which it
is worthwhile to devote not merely a monograph but a lifetime.